LOGIC COLLOQUIUM '84

STUDIES IN LOGIC

AND

THE FOUNDATIONS OF MATHEMATICS

VOLUME 120

NORTH-HOLLAND
AMSTERDAM • NEW YORK • OXFORD • TOKYO

LOGIC COLLOQUIUM '84

Proceedings of the Colloquium held in Manchester, U.K.
July 1984

Editors

J. B. PARIS
A. J. WILKIE
G. M. WILMERS

Department of Mathematics
University of Manchester
Manchester, U.K.

1986

NORTH-HOLLAND
AMSTERDAM • NEW YORK • OXFORD • TOKYO

ISBN: 0 444 87999 4

Published by:

Elsevier Science Publishers B.V.
P.O. Box 1991
1000 BZ Amsterdam
The Netherlands

Sole distributors for the U.S.A. and Canada:

Elsevier Science Publishing Company, Inc.
52 Vanderbilt Avenue
New York, N.Y. 10017
U.S.A.

Library of Congress Cataloging-in-Publication Data

Logic Colloquium (1984 : Manchester, Greater Manchester)
Logic Colloquium '84.

(Studies in logic and the foundations of mathematics ; v. 120)
Bibliography: p.
1. Logic, Symbolic and mathematical--Congresses.
I. Paris, J. B. II. Wilkie, A. J. (Alec J.)
III. Wilmers, G. M. IV. Series.
QA9.A1L63 1984 511.3 86-4462
ISBN 0-444-87999-4

PRINTED IN THE NETHERLANDS

Dedicated to

Alfred Tarski, 1901–1983

PREFACE

Logic Colloquium '84, the European Summer Meeting of the Association for Symbolic Logic, was held at the University of Manchester from 15th July to 24th July, 1984. The main themes of the conference were the model theory of arithmetic, and the semantics of natural languages. The present volume constitutes the proceedings of this conference.

Invited lectures at the conference were given by:

Z. Adamowicz (Warsaw), **J. Barwise** (Stanford), **P. Clote** (Boston), **J. Denef** (Leuven), **C. Di Prisco** (Caracas), **J.E. Fenstad** (Oslo), **L. Harrington** (Berkeley), **H. Kamp** (Stanford), **L. Kirby** (New York), **J. Knight** (Notre Dame), **A. Macintyre** (Yale), **B. Poizat** (Paris), **P. Pudlák** (Prague), **J. Saffe** (Freiburg), **P. Schmitt** (Heidelberg), **S. Simpson** (Munich), **R. Solovay** (Berkeley), **S. Thomas** (Freiburg), **C. Toffalori** (Florence), **L. Van den Dries** (Stanford), **A. Wilkie** (Manchester), **H. Wolter** (Berlin), **C. Wood** (Connecticut), **H. Woodin** (Cal. Tech.), **A. Woods** (Kuala Lumpur).

In addition to the invited lectures there were many contributed papers. Abstracts of most of these may be found in the report of the conference in the Journal of Symbolic Logic.

Most, but not all, of the papers corresponding to the invited lectures are in this volume. In addition there are some papers in the volume which do not correspond to contributions made at the conference.

The organizing committee of Logic Colloquium '84 consisted of P.H.G. Aczel, J.B. Paris, A.J. Wilkie, G.M. Wilmers, and C.E.M. Yates. The conference was supported financially by the Bertrand Russell Memorial Logic Conference Fund, the British Academy, the British Council, the British Logic Colloquium, the Logic, Methodology and Philosophy of Science Division of the International Union for the History and Philosophy of Science, the London Mathematical Society, the Royal Society, and the University of Manchester.

On behalf of the organizing committee we wish to thank the above-mentioned institutions for their support. We also owe a debt of gratitude to all those people who by their generous help and sound advice contributed to the success of the conference. In particular we would like to thank the following secretarial staff of the mathematics department at Manchester University: Mrs. P. McMunn, Rosemarie Horton, Kath Smith, Beryl Sweeney and Stephanie Worrall.

Jeff Paris
Alex Wilkie
George Wilmers

CONTENTS

LOGIC COLLOQUIUM '84
J.B. Paris, A.J. Wilkie, and G.M. Wilmers (Editors)

SOME RESULTS ON OPEN AND DIOPHANTINE INDUCTION

Zofia Adamowicz

Institute of Mathematics
of the Polish Academy of Sciences
Warsaw, Poland

The paper contains some results on extending a $\mathbb{Z}$-ring by adding to it a zero of a given polynomial. The main of them is Lemma 1. These results are applied to build a model for a special fragment of diophantine induction having a bounded set of primes. The technics of the paper is based on a purely number-theoretic result which is the sublemma of Lemma 1.

Fix an ω_1-saturated model M of Peano arithmetic, such that $M \equiv \mathbb{N}$. Let M^* be the fraction field of M, $\hat{M}$ the real closure of M.

Definition 1.
Let $x, y \in \hat{M}$. We say that y is much bigger than x and write $x \ll y$ if for every $n \in \mathbb{N}$, $x^n < y$.

Definition 2.
Let $p(t, x_1 \ldots x_n)$ be a polynomial. We call p totally unbounded if there is a $k \leq n$ and algebraic functions

$$\theta_1(t, x_1 \ldots x_k) \ldots \theta_{n-k}(t, x_1 \ldots x_k)$$

such that $p(t, x_1 \ldots x_k, \theta_1(t, x_1 \ldots x_k) \ldots \theta_{n-k}(t, x_1 \ldots x_k) \equiv 0$ and for some $t, x_1 \ldots x_k \in M$ satisfying

$$\mathbb{N} < t \ll x_1 \ll x_2 \ldots \ll x_k$$

we have

$$\theta_i(t, x_1 \ldots x_k) \in \hat{M}.$$

Here by the term "algebraic function" we mean a given branch of an algebraic function.

Remark 1.
In the last definition the assumption "for some $t, x_1 \ldots x_k \in M$ tisfying $\mathbb{N} < t \ll x_1 \ll \ldots \ll x_k$" can be equivalently replaced by "for all $t, x_1 \ldots x_k \in M$ satisfying $\mathbb{N} < t \ll x_1 \ll \ldots \ll x_k$".

This follows from the fact that the theory of real closed fields (of $\hat{M}$) admits the elimination of quantifiers and that all $t, x_1 \ldots x \in M$

such that $\mathbb{N} < t \ll x_1 \ll \dots \ll x_k$ satisfy the same open formulas in $\hat{M}$.

Definition 3.
Here we define a certain class ϕ of formulas. A typical example of a formula in ϕ, as we shall see later, is the diophantine formula describing (via the Matiasiewicz theorem) the existence of consecutive solutions of a Pell equation $x^2 - (a^2 - 1)Y^2 = 1$. There is a diophantine formula $\varphi_0^a(t)$ of the form

$$(\exists x_1 \dots x_n)(p_0(t,a,x_1,x_2 \dots x_n) = 0 \ \& \ x_1 \geq t)$$

which is (in $I\Sigma_1$) equivalent to "there is a $\langle x,y \rangle$ such that y is the second coordinate of the t'th solution of the equation $X^2 - (a^2 - 1)Y^2 = 1$".

The study of ϕ has been inspired by the study of φ_0.

Let $\varphi(t) \in \phi$ if there are polynomials p,q such that φ is of the form

$$(\exists x_1 \dots x_n)(p(t,x_1 \dots x_n) = 0 \ \& \ q(t,x_1 \dots x_n) \geq 0)$$

and the following is satisfied:

1) p is totally unbounded.
2) Assume that $\Theta_i(t,x_1 \dots x_k)$ parametrize p as in Def. 2. Then for every $t \in M - \mathbb{N}$ there are $x_1 \dots x_n \in M$ such that

$$t \ll x_1 \ll \dots \ll x_k, \quad x_{k+i} = \Theta_i(,x_1 \dots x_k)$$

and

$$q(t,x_1 \dots x_n) \geq 0.$$

3) For every $t \in M^*$ there are arbitrarily large $x_1 \dots x_n \in M^*$ such that

$$t \ll x_1 \ll \dots \ll x_k, \quad x_{k+1} = \Theta_i(t,x_1 \dots x_k)$$

and

$$q(t,x_1 \dots x_n) \geq 0.$$

4) For every $t \in M^*$ there are $x_1 \dots x_n \in M^*$ such that $x_1 \dots x_n$ are values of $\mathbb{Z}$-polynomials at t,

$$x_{k+i} = \Theta_i(t,x_1 \dots x_k).$$

5) $\Theta_i(t,x_1 \dots x_k)$ is of the form

$$\sum_{r=1}^{l_i} g_{r,i}(t)\eta_{r,i}(x_1 \dots x_k) \quad \text{where}$$

$g_{r,i}$ are rational functions of t and $\eta_{r,i}(x_1 \dots x_k)$ are algebraic functions of $x_1 \dots x_k$.

Example.
Consider the formula $\varphi_0^a(t)$, for a fixed $a \in \mathbb{Z}$. The polynomial p_0

is defined as follows:

$$p_0(a,t,x_1,x_2 \dots x_{10}) = 0 \quad \text{iff}$$
$$x_4^2 - (a^2 - 1)x_1^2 = 1$$
$$x_2 = 2x_4^2x_1^2x_8$$
$$x_5^2 - (a^2 - 1)x_2^2 = 1$$
$$x_6 = a + (x_5^2 - a)x_5^2$$
$$x_7^2 - (x_6^2 - 1)x_3^2 = 1$$
$$x_3 - x_1 = x_9x_5^2$$
$$x_3 - t = 2x_1x_{10}.$$

This definition has been taken from [2], Chapter 6, where we have changed the notation $\langle y,y_1,y_2,x,x_1,A,x_2\rangle$ to $\langle x_1,x_2,x_3,x_4,x_5,x_6,x_7\rangle$.

Take $k = 3$. Then t,x_1,x_2,x_3 become independent parameters and the parametrizing functions have the form

$$x_4 = \theta_1(t,x_1,x_2,x_3) = \theta_1(x_1) = \sqrt{1+(a^2-1)x_1^2}$$
$$x_5 = \theta_2(t,x_1,x_2,x_3) = \theta_2(x_2) = \sqrt{1+(a^2-1)x_2^2}$$
$$x_6 = \theta_3(t,x_1,x_2,x_3) = \theta_3(x_2) = a + (\theta_2^2(x_2)-a)\theta_2^2(x_2)$$
$$x_7 = \theta_4(t,x_1,x_2,x_3) = \theta_4(x_2,x_3) = \sqrt{1+(\theta_3^2(x_2)-1)x_3^2}$$
$$x_8 = \theta_5(t,x_1,x_2,x_3) = \theta_5(x_1,x_2) = \frac{x_2}{2\theta_1^2(x_1)x_1^2}$$
$$x_9 = \theta_6(t,x_1,x_2,x_3) = \theta_6(x_1,x_2,x_3) = \frac{x_3-x_1}{\theta_2^2(x_2)}$$
$$x_{10} = \theta_7(t,x_1,x_2,x_3) = \theta_7(t,x_1,x_3 = \frac{x_3-t}{2x_1} = \frac{x_3}{2x_1} - \frac{1}{2x_1}t$$

Observe that (1)-(5) of Definition 3 are satisfied for φ_0.

(1) If $\mathbb{N} < t \ll x_1 \ll x_2 \ll x_3$ then $\theta_i(t,x_1,x_2,x_3) \in M$.

(2) We take x_1 to be the second coordinate of the t'the solution of the equation $X^2-(a^2-1)Y^2=1$, x_2 to be a sufficiently big solution of this equation divisible by $2\theta_1^2(x_1)x_1^2$, and x_3 to be the second coordinate of the t'th solution of the equation $X^2-(\theta_3^2(x_2)-1)Y^2=1$. Then $t \ll x_1 \ll x_2 \ll x_3$, $\theta_i(t,x_1,x_2,x_3) \in M$ and $x_1 \geq t$.

(3) Let $t \ll x_1 \ll x_2 \ll x_3$ be sufficiently large elements of M such

that x_1 is the second coordinate of a certain solution of $X^2-(a^2-1)Y^2=1$, x_2 is the second coordinate of another sufficiently large solution of this equation, and x_3 is the second coordinate of any sufficiently large solution of $X^2-(\Theta_3^2(x_2)-1)Y^2=1$. Then $x_4=\Theta_1(x_1)$, $x_5=\Theta_2(x_2)$, $x_6=\Theta_3(x_2)$ and $x_7=\Theta_4(x_2,x_3)$ are in M and $x_8=\Theta_5(x_1x_2)$, $x_9=\Theta_6(x_1,x_2,x_3)$, $x_{10}=\Theta_7(t,x_1,x_3)$ are in M*. Moreover $x_1 \geq t$.

(4) Let $x_1=1$, $x_2=0$, $x_3=t$, $x_4=a$, $x_5=1$, $x_6=1$, $x_7=1$, $x_8=0$, $x_9=t-1$, $x_{10}=0$. Then $x_{3+i}=\Theta_i(t,x_1,x_2,x_3)$ and $x_1 \dots x_{10}$ are as required.

(5) Only Θ_7 dependends on t and it is of the required form.

We have the following theorem:

Theorem 1.
There is a model N of open induction in which the set of primes is bounded, every number > 1 has a prime divisor, $N^* \equiv Q$ and $(\forall t)\varphi(t)$ holds in N for every $\varphi \in \phi$. Especially, N satisfies induction for $\varphi \in \phi$. Hence follows that $N \models (\forall t)\varphi_0^a(t)$ for every $a \in \mathbf{Z}$. Moreover $N \not\models (\forall a)(\forall t)\varphi_0^a(t)$.

Before proving the theorem let us explain its content.

Work in $I\Sigma_1$. In this theory we can define the following relations:

$R_0(a,t,y) \Leftrightarrow y$ is the second coordinate of the t'th solution of the equation $X^2-(a^2-1)Y^2=1$

$$R_1(a,t,y) \Leftrightarrow y = a^t$$

$$R_3(a,t,y) \Leftrightarrow y = \binom{t}{a}$$

$$R_4(t,y) \Leftrightarrow y = t!.$$

For every relation there are polynomials p_i and q_i such that

$$R_i(a,t,y) \Leftrightarrow (\exists x_1 \dots x_n)(p_i(a,t,y,x_1 \dots x_n) = 0 \ \& \ q_i(a,t,y,x_1 \dots x_n) \geq 0)$$

If we define

$$S_i(a,t,y) \Leftrightarrow (\exists x_1 \dots x_n)(p_i(a,t,y,x_1 \dots x_n) = 0 \ \& \ q_i(a,t,y,x_1 \dots x_n) \geq 0)$$

then $I\Sigma_1 \vdash R_i \equiv S_i$.

We say that S_i pretends R_i. It is R_i in $I\Sigma_1$ but not necessarily in a weaker theory.

From Theorem 1 it follows, as we shall see later, that the totality of S_0,S_1,S_2,S_3 in the sense that $(\forall a,t)(\exists y)(S_i(a,t,y))$, together with open induction and the existence of prime divisors does not prove the infinity of the set of primes, unlike the totality of R_0, R_1, R_2, R_3.

In fact, we will prove a stronger theorem.

Theorem 2.
For every countable **Z**-ring A such that $A \subseteq M^*$ there is a countable **Z**-ring B such that $A \subseteq B \subseteq M^*$ and B satisfies open induction + the true theory of rationals + every number larger than 1 has a prime divisor + set of primes is bounded + $(\forall t)\varphi(t)$ + + $(\forall a)(\forall t)\varphi_0^a(t)$ for $\varphi \in \phi$. Moreover, there is a ring homomorphism from B onto A.

For the notion of a **Z**-ring see [3].

The proof of the theorem is based on the following lemmas:

Lemma 1.
Let $A \subseteq M^*$ be a countable **Z**-ring. Let $t \in A$, $\varphi \in \phi$, $a \in A$. Then there is a countable **Z**-ring B such that $A \subseteq B \subseteq M^*$, A is a homomorphic image of B and $B \models (\forall t)(\varphi(t) \& \varphi_0^a(t))$.

Lemma 2.
Let $A \subseteq M^*$ be a countable **Z**-ring, and let $x_1 \ldots x_n \in M^*$. Then there is a countable **Z**-ring B such that $A \subseteq B \subseteq M^*$ A is a homomorphic image of B and $x_1 \ldots x_n \in B^*$.

Lemma 3.
Let J be a non-standard initial segemnt of M. Let $A \subseteq M^*$ be a countable **Z**-ring such that $A \not\subseteq J$, $A \cap J \subseteq M$. Let $x_0 \in A - J$. Then there is a countable **Z**-ring B such that $A \subseteq B \subseteq M^*$, $B \cap J \subseteq M$, x_0 has a prime factor u in B such that $u \in J$ and A is a homomorphic image of B.

Lemma 4 (Wilkie)
Let $A \subseteq M^*$ be a countable **Z**-ring, and let $\alpha \in M$ be real algebraic over A. Then there is a countable **Z**-ring B such that $A \subseteq B \subseteq M^*$, A is a homomorphic image of B and there is a $y \in B$ such that $y \leq \alpha < y + 1$.

The proof of the Theorem consists in an appropriate iterated use of the lemmas. We start from a **Z**-ring $A_0 \subseteq M^*$. Lemma 1 serves to add solutions of $p(t,x_1 \ldots x_n) = 0$ & $q(t,x_1 \ldots x_n) \geq 0$, Lemma 2 to close fraction fields of the obtained **Z**-ring under Skolem function in M^*, Lemma 3 to destroy primes bigger than J and adding prime factors to them (note that since we add prime factors in J, and the parts of the **Z**-ring that lie in J consist of integers, the prime factors remain prime during the whole construction) and Lemma 4 serves to close the resulting **Z**-ring under integer parts of algebraic reals. The required model M_0 is the union of all **Z**-rings obtained in the construction. If we apply the lemmas (construct the appropriate extensions $A_0 \subseteq A_1 \subseteq \ldots$), then we easily ensure that M_0 satisfies open induction (via Shepherdson's theorem), that all its primes lie in J, that $M_0 \models \forall t\varphi(t)$ for $\varphi \in \phi$ and that A_0

is a homomorphic image of M_o (we summarize the homomorphisms occurring in the construction). To show that M_o satisfies the true theory of rationals we show that $M_o^* < M^*$.

Proof of Lemma 1.
Assume that $\varphi \in \phi$ and

$$\varphi(t) \Leftrightarrow (\exists x_1, \ldots, x_n)(p(t,x_1, \ldots, x_n) = 0 \ \& \ q(t,x_1, \ldots, x_n) \geq 0).$$

Let $\theta_1, \ldots, \theta_{n-k}$ parametrize p as in Definition 2. If $f(x_1, \ldots, x_n)$ is a polynomial with coeficients in Q then we consider the following algebraic function

$$\theta_f(T,X,\ldots,X_k)=f(X_1,\ldots,X_k,\theta_1(T,X_1,\ldots,X_k),\ldots,\theta_{n-k}(T,X_1,\ldots X_k)$$

We need the following sublemma which is formulated and proved in the usual analysis over $\mathbb{R}$.

Sublemma.
For every $d \in \mathbb{N}$ there is a constant $c \in \mathbb{N}$ such that for every polynomial $f \in Q[X_1, \ldots, X_n]$ of degree less than d and for all $\langle t,x_1, \ldots, x_k\rangle \in \mathbb{R}^{k+1}$ satisfying $t \in Q$, $c < t$, $(t_1t_2)^c < x_1$ for some $t_1t_2 \in \mathbb{N}$ such that $t = \frac{t_1}{t_2}$, (parameters of $f)^c < x_1$, $x_i^c < x_{i+1}$ we have either $\theta_f(t,X_1, \ldots, X_k)$ does not depend on $X_1 \ldots X_k$ or $|\theta_f(t,x_1, \ldots, x_k)| > 1$.

Proof.
Let $s_1, \ldots, s_n < d$. Consider the following function:

$$X_1^{s_1} \ldots X_k^{s_k}\, \theta_1(T,X_1, \ldots, X_k)^{s_{k+1}} \ldots \theta_{n-k}(T,X_1, \ldots, X_k)^{s_n}.$$

By (5) of Definition 3 this function is of the form

$$\sum_{r=1}^{l} g_{rs_1\ldots s_n}(T)\,\zeta_{s_1\ldots s_n r}(X_1,\ldots,X_k)$$

where $g_{rs_1\ldots s_n}$ are rational functions and $\zeta_{s_1\ldots s_n r}$ are algebraic functions:

We can expand $\zeta_{s_1\ldots s_n r}(X_1, \ldots, X_k)$ in a descending fraction power series

$$\sum_{r=m}^{-\infty} \eta_r(X_1, \ldots, X_{j-1}, X_{j+1}, \ldots, X_k)\cdot X_j^{\frac{r}{s}}$$

for every $j = 1, \ldots, k$, where η_r are algebraic functions, $s \in \mathbb{N} - \{0\}$, and there is a constant c_1 such that for all $\langle X_1, \ldots, X_k\rangle$ satisfying $x_i^{c_1} < x_{i+1}$, $c_1 < x_1$,

$$\zeta_{s_1 \ldots s_n, r}(x_1, \ldots, x_k) = \sum_{r=m}^{-\infty} \eta_r(x_1 \ldots x_{j-1}, x_{j+1} \ldots x_k) x_j^{\frac{r}{s}}$$

Indeed we first expand $\zeta_{s_1 \ldots s_n, r}$ as a function of x_k "in infinity". We obtain

$$\sum_{r=m}^{-\infty} \eta_{r,k}(X_1, \ldots, X_{k-1}) X_k^{\frac{r}{s}}$$

Then we expand every $\eta_{r,k}$ as a function of X_{k-1} "in infinity".

We continue.
For an appropriate constant c_1, for all $\langle x_1, \ldots, x_k \rangle$ satisfying $c_1 < x_1$, $x_i^{c_1} < x_{i+1}$, every series we obtain is absolutely convergent and represents $\zeta_{rs_1 \ldots s_n}$, $\eta_{r,k}$, ... respectively. Changing suitably the order of the summation and groupping the terms we obtain the required expansions. We have

$$f(X_1, \ldots, X_k, \Theta_1(T, X_1 \ldots X_k) \ldots \Theta_{n-k}(T, X_1, \ldots X_k)) =$$

$$= \sum_{s_1 \ldots s_n < d} a_{s_1 \ldots s_n} \sum_{r=1}^{1} g_{rs_1 \ldots s_n}(T) \zeta_{s_1 \ldots s_n r}(X_1 \ldots X_k).$$

Substituting the expansions for $\zeta_{s_1 \ldots s_n r}$ we obtain the following expansions for Θ_f for $j = 1, \ldots k$:

(¤) $\Theta_f(T, X_1, \ldots, X_k) =$

$$= \sum_{s_1 \ldots s_n < d} a_{s_1 \ldots s_n} g_{rs_1 \ldots s_n}(T) \sum_{r'=m}^{-\infty} \eta_{r'rjs_1 \ldots s_n}(X_1 \ldots X_{j-1} X_{j+1} \ldots X_k) X_j^{\frac{r'}{s}}$$

The equality holds for all $\langle x_1 \ldots x_k \rangle \in \mathbb{R}^k$ satisfying $c_1 < x_1$, $x_i^{c_1} < x_{i+1}$. Groupping suitably the terms we also obtain expansions for $j = 1 \ldots k$:

$$\Theta_f(T, X_1, \ldots, X_k) = \sum_{r=m}^{-\infty} \xi_r^{fj}(T, X_1 \ldots X_{j-1}, X_{j+1} \ldots X_k) X_j^{\frac{r}{s}}$$

for certain algebraic function ξ_r^{fj}.

Let r' be fixed. Then the coefficient of the exponent $\frac{r'}{s}$ of X_j is

$$\sum_{s_1 \ldots s_n < d} a_{s_1 \ldots s_n} \sum_{r=1}^{1} g_{rs_1 \ldots s_n}(T) \eta_{r'rjs_1 \ldots s_n}(X_1 \ldots X_{j-1}, X_{j+1} \ldots X_k)$$

Claim.
There is a constant $c_2 \in \mathbb{N}$ such that for every $t > c_2$ and every

polynomial $f \in [X_1, \ldots, X_n]$ either $\Theta_f(t, X_1, \ldots, X_k)$ does not depend on $X_1, \ldots, X_k$ or there is a $j = j(f,t)$ such that the expansion $(\square)$ of $\Theta_f(T, X_1, \ldots, X_k)$ for $j = j(f,t)$, $T = t$ contains a non-zero term with a positive exponent of $X_{j(f,t)}$ and $\Theta_f(t, X_1, \ldots, X_k)$ does not depend on $X_{j(t,t)+1}, \ldots, X_k$.

Proof of the claim.
Consider the family of algebraic functions

$$\{\eta_{r'rjs_1 \ldots s_n} : j = 1 \ldots k,\ r = 1 \ldots l,\ s_1 \ldots s_n < d,$$
$$r' = 1, \ldots, m\} \cup \{1\}.$$

We choose a maximal subset of this family which is linearly independent over $\mathbf{Q}$ as a set of functions of $X_1, \ldots, X_k$.

Let it be $\{\eta_1, \ldots, \eta_p\}$.

Then for every $j, r, s_1 \ldots s_n, r' \geq 1$ there are rational $b^1_{r'rjs_1 \ldots s_n}, \ldots, b^p_{r'rjs_1 \ldots s_n}$ such that

$$\eta_{r'rjs_1 \ldots s_n}(X_1 \ldots X_{j-1} X_{j+1} \ldots X_k) \equiv$$
$$\equiv b^1_{r'rjs_1 \ldots s_n} \eta_1(X_1, \ldots, X_k) + \ldots + b^p_{r'rjs_1 \ldots s_n} \eta_p(X_1, \ldots, X_k)$$

Let c_2 be so big, that any polynomial equation involving T and all the rational numbers which occured so far and which do not depend on T, $a_{s_1 \ldots s_n}$, $X_1 \ldots X_k$ (i.e. involving the parameters of $g_{rs_1 \ldots s_n}(T)$, $b^i_{r'rjs_1 \ldots s_n}$ for $j = 1 \ldots k$, $i = 1 \ldots p$, $r = 1 \ldots l$, $s_1 \ldots s_n < d$, $r' = 1 \ldots m$) with powers at most d and with coeficients at most d being satisfied for a $t > c_2$ is satisfied identivally in T.

Let $t > c_2$ and f be given.
Assume that $\Theta_f(t, X_1 \ldots X_k)$ depends on some of $X_1 \ldots X_k$. Let $j(f,t)$ be the largest j such that $\Theta_f(t, X_1 \ldots X_k)$ depends on X_j. Denote later $j(f,t)$ by j_o. We show that there is a non-zero term with a positive exponent of X_{j_o} in the expansion $(\square)$ of $\Theta_f(t, X_1 \ldots X_k)$ for $j = j_o$.

Suppose, otherwise, that all terms with positive exponents of X_{j_o} vanish for all $X_1, \ldots, X_{j_o - 1}, X_{j_o + 1} \ldots X_k$. Also all terms with positive exponents of $X_{j_o+1}, \ldots, X_k$ vanish, by the choice of j_o in suitable expansions $(\square)$ for $j = j_o + 1, \ldots, j = k$ respectively.

On the other hand there is a negative exponent $\frac{r'}{s}$ of X_{j_o} such that its coefficient does not vanish.

It follows that the rational numbers $a_{s_1\dots s_n}$ for $s_1\dots s_n < d$ satisfy the following system of linear equations

$$\bigwedge_{j\ge j_0}\ \bigwedge_{r'=1}^{m}\ \sum_{s_1\dots s_n<d} a_{s_1\dots s_n}\Big(\sum_{r=1}^{l} g_{rs_1\dots s_n}(t)\ b^1_{r'rjs_1\dots s_n}\,\eta_1(X_1\dots X_k)+\dots+b^p_{r'rjs_1\dots s_n}\,\eta_p(X_1\dots X_k)\Big) = 0$$

and satisfy the non-identity

$$\sum_{s_1\dots s_n<d} a_{s_1\dots s_n}\sum_{r=1}^{l} g_{rs_1\dots s_n}(t)\,\eta_{r'rj_0s_1\dots s_n}(X_1\dots X_{j_0-1}X_{j_0+1}\dots X_k))\ne 0.$$

Hence $a_{s_1\dots s_n}$ satisfy

$$(*)\qquad \bigwedge_{i=1}^{p}\ \bigwedge_{j\ge j_0}\ \bigwedge_{r'=1}^{m}\ \sum_{s_1\dots s_n<d} a_{s_1\dots s_n}\sum_{r=1}^{l} g_{rs_1\dots s_n}(t)\,b^i_{r'rjs_1\dots s_n}=0$$

and

$$(**)\qquad \sum_{s_1\dots s_n<d} a_{s_1\dots s_n}\Big(\sum_{r=1}^{l} g_{rs_1\dots s_n}(t)\,\eta_{r'rj_0s_1\dots s_n}(X_1\dots X_{j_0-1}X_{j_0+1}-X_k)\Big)\ne 0.$$

Replacing the functions

$$\eta_{r'rj_0s_1\dots s_n}(X_1\dots X_{j_0-1},X_{j_0+1}\dots X_k) \text{ by}$$

certain **Q**-linear combinations of $\eta_1 \dots \eta_p$ and some of other linearly independent $\eta_{p+1}(X_1,\dots,X_k)$, ... $\eta_q(X_1, \dots ,X_k)$ we can replace (**) by

$$\sum_{i=1}^{q}\ \sum_{s_1\dots s_n<d} a_{s_1\dots s_n}\Big(\sum_{r=1} g_{rs_1\dots s_n}(t)\,b^i_{r'rj_0s_1\dots s_n}\,\eta_i(X_1\dots X_k)\Big) \ne 0$$

where $b^i_{r'rj_0s_1\dots s_n}$ are the rational coeficients of the combinations representing $\eta_{r'rj_0s_1\dots s_n}$ in the basis $\{\eta_1, \dots ,\eta_q\}$. We conclude (***);

$$\sum_{s_1\dots s_n<d} a_{s_1\dots s_n}\Big(\sum_{r=1}^{l} g_{rs_1\dots s_n}(t)\,b^{i_0}_{r'rj_0s_1\dots s_n}\Big) \ne 0$$

for a fixed $i_0 \in \{1, \dots ,q\}$.

We infer that the system of equations (*) has a solution $a_{s_1\dots s_n}$ for $s_1\dots s_n<d$ which satisfies the unequation (***).

Now by the Kronnecker Capelli theorem the existence of a solution of (*) satisfying (***) can be expressed by a system (or possibly an alternative of systems) (Δ) of polynomial equations and non-identities involving $g_{rs_1\dots s_n}(t)$, $b^i_{r'rjs_1\dots s_n}$ $b^{i_0}_{r'rj_0s_1\dots s_n}$ for $r = 1\dots l$, $j = k$, $s_1\dots s_n <d$, $i = 1\dots p$ with powers at most d

and with coeficients at most d.

Note that $b^{i_o}_{r'rj_0 s_1 \ldots s_n}$ do not occur in equations.

Since $t > c_2$ we infer that (Δ) is satisfied for every t. Now we have to define a new constant $c_3 \in \mathbb{N}$. Let $c_3 > c_1$ be such that

1) if $(t_1 t_2)^{c_3} < x_1$, for some $t_1 t_2 \in \mathbb{N}$ such that $t = \frac{t_1}{t_2}$ $a_{s_1 \ldots s_n} = \frac{a'_{s_1 \ldots s_n}}{a''_{s_1 \ldots s_n}}$ and $(a'_{s_1 \ldots s_n})^{c_3} < x_1$, $(a''_{s_1 \ldots s_n})^{c_3} < x_1$ for $s_1 \ldots s_n < d$, $x_i^{c_3} < x_{i+1}$ then no polynomial-equation with parameters less than d between the functions $\eta_i(X_1, \ldots, X_k)$ and the numbers $b^1_{r'rs_1 \ldots s_n}$, $b^i_{r'rj_0 s_1 \ldots s_n}$, $a_{s_1 \ldots s_n}$, $g_{rs, \ldots, s_n}(t)$ for $r = 1, \ldots, l$, $s_1, \ldots, s_n < d$, $j = 1, \ldots, k$, $i = 1, \ldots, q$ holds unless it holds identically in $X_1, \ldots, X_k$.

2) if $\xi_r^{f_j}(t, X_1, \ldots, X_{j-1}, X_{j+1}, \ldots, X_k)$ does not depend on $X_{j+1}, \ldots, X_k$ then $|\xi_r^{f_j}(t, x_1, \ldots, x_{j-1})| < x_{j-1}^{c_3}$ for every $r \leq m$. Moreover for every $j = 1, \ldots, k$

$$\sum_{r=m}^{-\infty} \xi_r^{f_j}(t, x_1, \ldots, x_{j-1}) x_j^{\frac{r}{s}} | > 1$$

provided it contains a non-zero term with a positive exponent of x_j.

3) If t, $a_{s_1 \ldots s_n}$, $x_1, \ldots, x_k$ are as in (1) and $a \in \mathbb{N}$, $a^{c_3} < x_1$ then

$$a\xi_0^{f_j}(t, x_1, \ldots, x_{j-1}) - [a\xi_0^{f_j}(t, x_1, \ldots, x_{j-1})] > \frac{1}{x_j^{\frac{1}{3s}}}$$

provided that $\xi_0^{f_j}(t, X_1, \ldots X_{j-1}, X_{j+1}, \ldots, X_k)$ does not depend on $X_{j+1}, \ldots, X_k$ and that $a\xi_0^{f_j}(t, x_1, \ldots, x_{j-1})$ is not integer.

4) For every $t > c_2$ if there is a solution $a_{s_1 \ldots s_n} = \frac{a'_{s_1 \ldots s_n}}{a''_{s_1 \ldots s_n}}$ of $(*) \wedge (***)$ then there is one with $a'_{s_1 \ldots s_n}$, $a''_{s_1 \ldots s_n} < t^{c_3}$.

Now we choose $\bar{t} \in \mathbb{N}$ so that

i) $\bar{t}$ satisfies all the non identities occuring in the system (Δ)

ii) there are $\bar{x}_1, \ldots, \bar{x}_n \in \mathbf{Z}$ such that $\bar{t}^{c_3^2} < \bar{x}_1$, $\bar{x}_i^{c_3} < \bar{x}_{i+1}$,

$\bar{x}_{k+i} = \theta_i(\bar{t},\bar{x}_1, \ldots ,\bar{x}_k)$ and $g(\bar{t},\bar{x}_1, \ldots ,\bar{x}_k) \geq 0$. Such a $\bar{t}$ exists by (2) of Definition 3 and "underspill". Then (Δ) is satisfied for $t = \bar{t}$ because it's equations are satisfied for every t. It follows that the system (*) has a solution $\bar{a}_{s_1 \ldots s_n}$, $s_1 \ldots s_n < d$ which satisfies (***).

We can take $\bar{a}_{s_1 \ldots s_n}$ rational because the coeficients of (*),(***) are rational (they are of the form

$$\sum_{l=1}^{r} g_{rs_1 \ldots s_n}(\bar{t}) b^{i}_{r'rjs_1 \ldots s_n},$$

$$\sum_{l=1}^{r} g_{rs_1 \ldots s_n}(\bar{t}) b^{i_o}_{r'rj_o s_1 \ldots s_n}).$$

By the choice of c_3 we can assume $\bar{a}'_{s_1 \ldots s_n}$, $\bar{a}''_{s_1 \ldots s_n} < \bar{t}^{c_3}$ where

$$\bar{a}_{s_1 \ldots s_n} = \frac{\bar{a}'_{s_1 \ldots s_n}}{\bar{a}''_{s_1 \ldots s_n}}.$$

Take $\bar{x}_1, \ldots ,\bar{x}_n$ as in (ii). We have

$$\bar{t}^{c_3} < x_1, (\bar{a}_{s_1 \ldots s_n})^{c_3} < \bar{x}_1, (\bar{a}''_{s_1 \ldots s_n})^{c_3} < \bar{x}_1, \bar{x}_i^{c_3} < \bar{x}_{i+1}.$$

Then form (1) and (***) we can infer (**) for $\bar{a}_{s_1 \ldots s_n}$, $\bar{t},\bar{x}_1,\ldots,\bar{x}_k$. We infer that $j = j_o, j_o + 1, \ldots ,k$

$$\sum_{s_1 \ldots s_n < d} \bar{a}_{s_1 \ldots s_n} \sum_{r=1}^{1} g_{rs_1 \ldots s_n}(\bar{t}) \sum_{r'=m}^{-\infty} \eta_{r'rjs_1 \ldots s_n}(\bar{x}_1 \ldots \bar{x}_{j-1}\bar{x}_{j+1} \ldots \bar{x}_k)\bar{x}_j^{\frac{r'}{s}}$$

contains no non-zero term with a positive exponent of $\bar{x}_{j_o}$, $\bar{x}_{j_o+1}, \ldots ,\bar{x}_k$ respectively and for $j = j_o$ it contains the non-zero term corresponding to the exponent $\frac{r'}{s}$ of $\bar{x}_{j_o}$.

But the above series is equal to $\theta_{\bar{f}}(\bar{t},\bar{x}_1, \ldots ,\bar{x}_k)$ where $\bar{f}$ is the folynomial determined by $\bar{a}_{s_1 \ldots s_n}$ for $s_1 \ldots s_n < d$.

Let j_1 be the biggest j such that $\theta_{\bar{f}}(\bar{t},X_1, \ldots ,X_k)$ depends on X_j. It follows that $j_1 \geq j_o$ and that the expansion

$$\sum_{s_1 \ldots s_n < d} \bar{a}_{s_1 \ldots s_n} \sum_{r=1}^{1} g_{rs_1 \ldots s_n}(\bar{t}) \eta_{r'rj_1 s_1 \ldots s_n}(\bar{x}_1 \ldots \bar{x}_{j_1-1} x_{j_1+1} \ldots \bar{x}_k)\bar{x}_{j_1}^{\frac{r'}{s}}$$

contains no non-zero term with a positive exponent of x_{j_1}. On the the other hand we have

$$\Theta_f(\bar t,\bar x_1\ldots\bar x_k) = \sum_{r=m}^{-\infty} \xi_r^{fj_1}(\bar t,\bar x_1\ldots\bar x_{j_1-1}\bar x_{j_1-1}\bar x_{j_1+1}\ldots\bar x_k)\bar x_{j_1}^{\frac{r}{s}} =$$

$$= \sum_{r=0}^{-\infty} \xi_r^{fj_1}(\bar t,\bar x_1\ldots\bar x_{j_1-1})\bar x_{j_1}^{\frac{r}{s}}.$$

Indeed, by the choice of j_1, ξ_r^{f,j_1} cannot depend on $\bar x_{j_1+1},\ldots,\bar x_k$ and $\xi_r^{fj_1}$ vanish for positive r.

But

$$\Theta_{\bar f}(\bar t,\bar x_1\ldots\bar x_k)=\bar f(\bar x_1,\ldots,\bar x_k,\Theta_1(\bar t,\bar x_1\ldots\bar x_k),\ldots,\Theta_{n-k}(\bar t,\bar x_1\ldots\bar x_k))$$

which is the value of the polynomial $\bar f$ with the coeficients $\bar a_{s_1\ldots s_n}$ at integer arguments. Thus $\Theta_{\bar f}(\bar t,\bar x_1\ldots\bar x_k)$ can be presented as a rational with the denominator $\prod_{s_1\ldots s_n<d}(\bar a_{s''_1\ldots s_n})$. Denote the last product by a. Then $a\,\Theta_{\bar f}\ \bar t,\bar x_1\ldots\bar x_k \in \mathbf{Z}$. We have

$$a\,\Theta_{\bar f}(\bar t,\bar x_1\ldots\bar x_k) = a\xi_0^{\bar f j_1}(\bar t,\bar x_1\ldots\bar x_{j_1-1})$$

$$a \sum_{r=-1}^{-\infty} \xi_r^{\bar f j_1}\ \bar t,\bar x_1\ldots\bar x_{j_1}\ldots\bar x_{j_1-1})\,\bar x_{j_1}^{\frac{r}{s}}$$

All values $|\xi_r^{f,r}(\bar t,\bar x_1,\ \ldots\ ,\bar x_{j-1})|$ are bounded by $\bar x_{j-1}^{c_3}$.

Since

$$\bar x_j^{-\frac{r}{s}} > (\bar x_{j-1})^{c_3}\bar x_j^{-\frac{r}{2s}} \qquad \text{for } r \geq 1$$

we have

$$0 \neq a\Big|\sum_{r=-1}^{-\infty}\xi_r^{\bar f j_1}(\bar t,\bar x_1,\ \ldots\ ,\bar x_{j_1-1})\bar x_{j_1}^{\frac{r}{s}}\Big| \leq$$

$$\leq a\sum_{r=-1}^{-\infty}|\xi_r^{\bar f j_1}(\bar t,\bar x_1,\ \ldots\ ,\bar x_{j_1-1})|\,\bar x_{j_1}^{\frac{r}{s}} <$$

$$< a\sum_{r=-1}^{-\infty}\bar x_{j_1-1}^{c_3}\frac{1}{\bar x_{j_1-1}^{c_3}\bar x_{j_1}^{-\frac{r}{s}}}\quad\sum_{r=-1}^{\infty}\frac{1}{\bar x_{j_1}^{\frac{m}{2s}}} =$$

$$= \frac{a}{\bar x_{j_1}^{\frac{1}{2s}-1}} < \frac{a}{\bar x_{j_1}^{\frac{1}{3s}}} < 1.$$

If $a\xi_0^{\bar f j_1}(\bar t,\bar x_1,\ \ldots\ ,\bar x_{j_1-1}) \in \mathbf{Z}$, this gives a contradiction. If

$a\xi_o^{\bar{f}j_1}(\bar{t},\bar{x}_1\ldots\bar{x}_{j_1-1}) \notin \mathbf{Z}$ then

$$(a\xi_o^{\bar{f}j_1}(\bar{t},\bar{x}_1\ldots\bar{x}_{j_1-1}) - [a\xi_o^{fj_1}(\bar{t},\bar{x}_1\ldots\bar{x}_{j_1-1})] > \frac{1}{\bar{x}_j^{\frac{1}{3s}}}$$

by 3 of the choice of c_3.

Again we obtain a contradiction with the fact that $a\,\theta_{\bar{f}}(\bar{t},\bar{x}_1\ldots\bar{x}_k) \in \mathbf{Z}$. Thus we have proved the claim.

Now let us prove the sublemma.

Let c be such that $c > \max c_1, c_2, c_3$. Then for every expansion $(\square)$ of $\theta_f(t,X_1, \ldots ,X_k)$ for $j = 1, \ldots ,k$, f of degree$<d$, $t \in \mathbf{Q}$ for every coeficient of the form:

$$\sum_{s_1\ldots s_n<d} a_{s_1\ldots s_n} \sum_{r=1}^{1} g_{rs_1\ldots s_n}(t)\eta_{r'rjs_1\ldots s_n}$$

$$(X_1,\ldots,X_{j-1},X_{j+1},X_k) \qquad \text{for } r > 0$$

if the coeficient does not vanish identically in $X_1, \ldots ,X_{j-1}, X_{j+1}, \ldots ,X_k$ then it does not vanish for $x_1, \ldots ,x_k$ provided

$$(t_1 t_2)^c < x_1 \quad \text{for some } t_1 t_2 \in N \text{ such that } t = \frac{t_1}{t_2}$$

$$(a'_{s_1\ldots s_n}\, a''_{s_1\ldots s_n})^c < x_1 \quad \text{for } s_1 \ldots s_n<d,\ x_i^c < x_{i+1}\,.$$

Let $f,t,x_1, \ldots ,x_k$ be given such as required in the sublemma. Consider $\theta_f(t,X_1, \ldots ,X_k)$. If it does not depend on $X_1,\ldots,X_k$ then we are done. If $\theta_f(t,X_1, \ldots ,X_k)$ depends on some of $X_1, \ldots ,X_k$ then take $j(f,t)$ provided by the claim. Then the expansion $(\square)$ of $\theta_f(t,X_1, \ldots ,X_k)$ for $X_i = x_i$ contains a non-zero term with a positive exponent of $x_{j(f,t)}$ and $\theta_f(t,X_1, \ldots ,X_k)$ does not depend on $X_{j(f,t)+1)} \ldots X_k$. Hence

$$\theta_f\; t,x_1,\ldots,x_k = \sum_{r=m}^{-\infty} \xi_r^{fj(f,t)}(t,x_1, \ldots ,x_{j(f,t)-1})x_{j(f,t)}^r$$

and a positive power of $x_{j(f,t)}$ does not vanish. It follows that $|\theta_f(t,x_1\ldots x_k)| > 1$ by the fact that $c > c_3$.

Proof of the lemma.
Let A, $t \in A$ be given. If $t \in \mathbf{N}$ then

$$A \models (\exists x_1, \ldots ,x_n)(p(t,x_1, \ldots ,x_n) = 0 \;\&$$
$$q(t,x_1, \ldots ,x_n) \geq 0)$$

since $\mathbb{N} \subseteq A$ and $\mathbb{N} \models (\forall t)\varphi(t)$. So assume that $t > \mathbb{N}$.

Let $x_1, \ldots x_k \in M^*$ be such that

$$(\forall a)_A \qquad (a \ll x_1 \ll x_2 \ldots \ll x_k),$$

$$\theta_i(t, x_1, \ldots, x_k) \in M^* \quad \text{and}$$

$$q(t, x_1 \ldots x_k, \theta_1(t, x_1 \ldots x_k) \ldots \theta_{n-k}(t, x_1 \ldots x_k)) \geq 0.$$

Such $x_1 \ldots x_k$ exist by 3 of Definition 3. Let $d \in \mathbb{N}$. Note that we have $c < t$, $(\text{parameters } f)^c < x_1$, $x_1^c < x_2, \ldots, x_{k-1}^c < x_k$ where c is as required in the sublemma.

Now we want to show that we can add $x_1 \ldots x_k$, $\theta_i(t, x_1 \ldots x_k)$ to A without spoiling the discrete ordering and so that

$$A[x_1, \ldots, x_k, \theta_1\ t, x_1, \ldots x_k \ \ldots\ \theta_{n-k}\ t, x_1 \ldots x_k]$$

is embaddable in a **Z**-ring B so that A is a homomorphic image of B.

Let $x_1, \ldots x_n$ be a solution of p for t, given by **Z**-polynomials at t. Let $r_m^1, \ldots, r_m^n$ be such that

$$r_m^i \in \mathbb{N} \cup \{0\}, \quad r_m^i < m \quad \text{and} \quad x_i \equiv r_m^i \bmod m \quad \text{in} \quad A.$$

Then $x_{k+i} = \theta_i(t, x_1, \ldots x_k) \equiv r_m^{k+i} \bmod m$ in A.

Since A is a **Z**;ring, there are such $r_m^1, \ldots, r_m^n$.

Consider

$$B = \bigcup_{m \in \mathbb{N}} A\left[\frac{x_1 - r_m^1}{m}, \ldots, \frac{x_k - r_m^k}{m}, \frac{\theta_1(t, x_1 \ldots x_k) - r_m^{k+1}}{m}, \ldots, \frac{\theta_{n-k}(t, x_1 \ldots x_k) - r_m^n}{m}\right].$$

To show that B is a **Z**-ring we have to show that B is discretely ordered. So let g be a polynomial over **Z**, $a_1 \ldots a_l \in A$ and let $d-1$ be the degree of g. Consider

$$g\left(a_1 \ldots a_l, \frac{x_1 - r_{m_1}^1}{m_1}, \ldots, \frac{x_k - r_{m_1}^k}{m_1}, \frac{\theta_1(\ldots) - r_{m_1}^{k+1}}{m_1}, \ldots \frac{\theta_{n-k}(\ldots) - r_{m_1}^n}{m_1}, \right.$$
$$\left. \ldots \frac{x_1 - r_{m_1}^1}{m_1} \ldots \frac{x_k - r_{m_1}^k}{m_1}, \frac{\theta_1(\ldots) - r_{m_1}^{k+1}}{m_1}, \ldots, \frac{\theta_{n-k}(\ldots) - r_{m_1}^n}{m_1}\right).$$

Then $g(\ldots) = f(x_1 \ldots x_k, \theta_1(t, x_1 \ldots x_k) \ldots \theta_{n-k}(t, x_1 \ldots x_k))$ for an A^* polynomial f. We can use the sublemma since $M \equiv \mathbb{N}$, $M \equiv \mathbb{R}$. By the sublemma, two cases are possible:

1) f and g depend only on t, not on $X_1 \ldots X_k$

2) $|f(x_1 \ldots x_k, \theta_1(t,x_1 \ldots x_k), \ldots, \theta_{n-k}(t,x_1 \ldots x_k)| > 1$

Consider (1).
Then

$$g(a_1 \ldots a_l, \frac{x_1 - r^1_{m_1}}{m_1} \ldots \frac{\theta_{n-k}(t,x_1 \ldots x_k) - r^n_{m_1}}{m_1}) =$$

$$= g(a_1 \ldots a_l, \frac{x_1 - r^1_{m_1}}{m_1} \ldots \frac{\theta_{n-k}(t,x_1 \ldots x_k) - r^n_{m_1}}{m_1}) \in A$$

Consider (2).
Then $|g(a_1 \ldots a_l \ldots)| > 1$. It follows that B is disretely ordered. It remains to show that A is a homomorphic image of B.

Define

$$h(g(a_1 \ldots a_l, \frac{x_1 - r^1_{m_1}}{m_1} \ldots \frac{\theta_{n-k}(t,x_1 \ldots x_k) - r^n_{m_1}}{m_1}) =$$

$$= g(a_1 \ldots a_l, \frac{x_1 - r^1_{m_1}}{m_1} \ldots \frac{\theta_{n-k}(t,x_1 \ldots x_k) - r^n_{m_1}}{m_1}).$$

Clearly h is a homomorphism.

From the above proof it follows that for every $t \in A$ and $a \in \mathbf{Z}$ we can find a $\mathbf{Z}$-ring B extending A in which $\varphi^a_o(t)$ holds. Consider now the case of an arbitrary $a \in A$. To prove the existence of the appropriate B in this case we have to use the particular form of polynomial p_0. It is difficult to impose any more general requirement on a polynomial p that would imply the existence of B for φ^a_o for $a \notin \mathbf{Z}$. Certainly from the case of an arbitrary a, treated below, follows the case of an $a \in \mathbf{Z}$. Thus to prove the existence of B satisfying $\varphi^a_o(t)$ for any a we do not need the general considerations concerning the class ϕ. We have given them because they seem interesting in their own right.

Let $a,t \in A$ be given. First we will reenumerate the varibles in p_0, more convenient for our new purpose.

We have

$$(\exists x_1 \ldots x_{10})\ p_0(a,t,x_1 \ldots x_{10}) = 0 \Leftrightarrow (\exists x_1 \ldots x_9)($$

1 $x_4^2 - (a^2-1)\ x_1^2 = 1$ &

2 $x_2 = 2x_4^2 x_1 x_8$ &

3 $x_5^2 - (a^2-1) x_2^2 = 1$ &

4 $x_6^2 = a + (x_5^2 - a) x_5^2$ &

5 $x_7^2 - (x_6^2-1) x_3^2 = 1$ &

6) $x_3-x_1=x_9x_5^2$
7) $x_3 \equiv t \mod 2x_1)$

We can replace (7) by

7') $x_9 + x_1 \equiv t \mod 2x_1$

so that (1) ... (6),(7) $\Leftrightarrow$ (1) ... (6),(7) in open induction. Indeed form (6), $x_3=x_9x_5^2+x_1$. From (3), $x_5^2=1+(a^2-1)x_2^2$ and so by (2) $x_5^2 - 1 \equiv 0 \mod 2x_1$. Hence $x_9x_5^2+x_1 \equiv x_9+x_1 \mod 2x_1$ and so $x_9+x_1 \equiv t \mod 2x_1$ from (7). To show the converse implication the argument is similar. Thus

$$(\exists x_1 \ldots x_{10})\ p_0(a,t,x_1\ldots x_{10}) = 0 \Leftrightarrow (\exists x_1 \ldots x_9)((1)\&\ldots(7)) \Leftrightarrow$$
$$(\exists x_1 \ldots x_{10})((1\& \ \ldots(6)\&\ x_9+x_1-t=2x_1x_{10}).$$

Let now

$$y_1=x_1,\ y_2=x_8,\ y_3=x_9+x_1-t \ \text{(whence } x_9=y_3-y_1+t),\ y_4=x_4,\ y_5=x_2,$$
$$y_6=x_5,\ y_7=x_6,\ y_8=x_3,\ y_9=x_7,\ y_{10}=x_{10}.$$

We obtain

$$(\exists x_1 \ldots x_{10})\ p_0(a,t,x_1\ldots x_{10}) = 0 \Leftrightarrow (\exists y_1 \ldots y_{10})($$

1) $y_4^2 - (a^2-1)y_1^2 = 1$ &

2) $y_5 = 2y_4^2y_1^2y_2$ &

3) $y_6^2 - (a^2-1)y_5^2 = 1$ &

4) $y_7 = a+(y_6^2-a)y_6^2$ &

5) $y_9 - (y_8^2-1)y_7^2 = 1$ &

6) $y_8 - y_1 = (y_3-y_1+t)y_6^2$ &

7) $y_3 = 2y_1y_{10})$

We treat t,y_1,y_2,y_3 as independent parameters and define

$$y_4 = \bar{\theta}_1(a,y_1) = \sqrt{1+(a^2-1)y_1^2}$$
$$y_5 = \bar{\theta}_2(a,y_1,y_2) = 2\ \bar{\theta}_1^2\ (a,y_1)\ y_1^2y_2$$
$$y_6 = \bar{\theta}_3(a,y_1,y_2) = \sqrt{1+(a^2-1)\ \theta_2^2(a,y_1,y_2)}$$
$$y_7 = \bar{\theta}_4(a,y_1,y_2) = a+(\bar{\theta}_3^2(a,y_1,y_2)-a)\ \bar{\theta}_3^2\ (a,y_1,y_2)$$
$$y_8 = \bar{\theta}_5(a,t,y_1,y_2,y_3) = y_1+(y_3+y_1-t)\ \bar{\theta}_3^2(a,y_1,y_2)$$
$$y_9 = \bar{\theta}_6(a,t,y_1,y_2,y_3) = \sqrt{1+(\bar{\theta}_4^2(a,y_1,y_2)-1)\,\theta_5^2(a,t,y_1,y_2,y_3)}$$
$$y_{10}= \bar{\theta}_7(y_1,y_3) = \frac{y_3}{2y_1}$$

It is not difficult to see that (1),(2),(3),(4) of Def 3 are satisfied.

Let now $y_1, y_2, y_3 \in M^*$ be such that $(\forall a)_A (a \ll y_1 \ll y_2 \ll y_3)$ and $\bar{\theta}_i(a,t,y_1,y_2,y_3) \in M^*$. We shall extend A in stages.

First we shall show that $A[y_1, \bar{\theta}_1(a,y_1)]$ is included in a countable $\mathbf{Z}$-ring $A_1 \subseteq M^*$. Then we shall show that $A_1[y_2, \bar{\theta}_3(a,y_1,y_2)]$ is included in a countable $\mathbf{Z}$-ring $A_2 \subseteq M^*$. It follows that $\bar{\theta}_2(a,y_1,y_2)$, $\bar{\theta}_4(a,y_1,y_2) \in A_2$. Finally we shall show now that $A_2[y_3, \bar{\theta}_6(a,t,y_1,y_2,y_3), \bar{\theta}_7(y_1,y_3)]$ is included in a countable $\mathbf{Z}$-ring $B \subseteq M^*$. It follows that $\bar{\theta}_5(a,t,y_1,y_2,y_3) \in B$.

Consider $A[y_1, \bar{\theta}_1(a,y_1)]$. Let us show that it is discretely ordered. Let f be an A-polynomial and consider $f(y_1, \bar{\theta}_1(a,y_1))$. Then there is an A-polynomial g such that $f(y_1, \bar{\theta}_1(a,y_1)) = g(y_1, \bar{\theta}_1(a,y))$ and $\bar{\theta}_1$ occures in g with the exponent 0 or 1. Indeed, every occurence of $\bar{\theta}_1$ with an even exponent can be replaced by an A-polynomial of y_1 since it is equal to a polynomial of a, y_1. Thus

$$g(y_1, \bar{\theta}_1(a,y_1)) = g_0(y_1) + \bar{\theta}_1(a,y_1)\, g_1(y_1)$$

where g_0, g_1 are A-polynomials. The expansion of $\bar{\theta}_1(a,y_1)$ "in infinity" w.r.t. y_1 has the leading term $\sqrt{a^2-1}\, y_1$. Assume first that $\deg g_o = \text{def } g_1 + 1$. Hence $\deg g_0 \geq 1$. Then the leading coeficient of the expansion of $g(y_1, \bar{\theta}_1(a,y_1))$ "in infinity" w.r.t. y_1 is of the form $a_0 + a_1\sqrt{a^2-1}$ where a_0, a_1 are the leading coeficients of g_0, g_1 respectively. Since $a_0 \neq 0$ ($\deg g_0 \geq 1$) and $a_0, a_1 \in M^*$, $\sqrt{a^2-1} \notin M^*$ we have $a_0 + a_1\sqrt{a^2-1} \neq 0$. Hence $g(y_1, \bar{\theta}(a,y_1))$ is greater that every element of A.

If $\deg g_0 > \deg g_1 + 1$ then the leading term of the expansion of $g(y_1, \bar{\theta}_1(a,y_1))$ is the leading term of g_0 and again $g(y_1, \bar{\theta}_1(a,y_1))$ is greater than every element of A.

If $\deg g_0 < \deg g_1 + 1$ then either $g_1 \equiv 0$ and then $g(y_1, \bar{\theta}_1) = g_0(y_1) \in A$, or $a_1 \neq 0$ and the leading term of the expansion of $g(y_1, \bar{\theta}_1(a,y_1))$ is $a_1\sqrt{a^2-1}\, y_1^m$, $m \geq 1$. Again $g(y_1, \bar{\theta}_1(a,y_1))$ is greater than every element of A.

We define $A_1 = \bigcup_m A\left[\dfrac{y_1 - r_m^1}{m}, \dfrac{\bar{\theta}_1(a,y_1) - r_m^2}{m}\right]$ where r_m^i are chosen as before and correspond to y_1, $\bar{\theta}_1(a,y_1)$ where y_1 is taken from (4) of Def. 3. Similarly as before we show that A_1 is a $\mathbf{Z}$-ring.

Consider $A_1[y_2, \bar{\theta}_3(a,y_1,y_2)]$. We have $\bar{\theta}_3(a,y_1,y) =$

$= \sqrt{1+(a^2-1)\bar{\theta}_2^2(a,y_1,y_2)} = \sqrt{1+(a^2-1)\, 4\bar{\theta}_1^4(a,y_1)\, y_1^4 y_2^2}$. The expansion of of $\bar{\theta}_3$ "in infinity" w.r.t. y_2 has the leading term $\sqrt{a^2-1}\;\; 2\bar{\theta}_1^2\, y_1^2\, y_2$. Using the fact that $\bar{\theta}_1 \in A_1$ and reasonning as before with A_1 in place of A we prove the existence of the **Z**-ring A_2.

Finally consider $A_2[y_3, \bar{\theta}_6(a,t,y_1,y_2,y_3), \bar{\theta}_7(y_1,y_3)]$. We have $\bar{\theta}_7(y_1,y_3) = \frac{y_3}{2y_1}$ and $\bar{\theta}_6(a,t,y_1,y_2,y_3) =$

$= \sqrt{1+(\bar{\theta}_4^2(a,y_1,y_2)^{-1})\bar{\theta}_5^2(a,y_1,y_2,y_3)} = \sqrt{1+(b^2-1)(y_1+(y_3+y_1-t)c)^2}$ for certain $b,c \in A_2$, namely $b = \bar{\theta}_4$, $c = \bar{\theta}_3(a,y_1,y_2)$. The leading term of the expansion of $\bar{\theta}_6$ "in infinity" w.r.t. y_3 is $\sqrt{b^2-1}\;\; y_3$.

Let g be an A-polynomial and consider $g(y_3, \frac{y_3}{2y_1}, \bar{\theta}_6(a,y_1,y_2,y_3))$.

As before we can assume that $\bar{\theta}_6$ occurs in g with exponent at most 1. Thus

$$g(y_3,\frac{y_3}{2y_1}, \bar{\theta}_6(a,y_1,y_2,y_3)) = g_0(y_3,\frac{y_3}{2y_1}) + \bar{\theta}_6\, g_1(y_3,\frac{y_3}{2y_1})$$

where g_0, g_1 are A-polynomials. If $g_1 \not\equiv 0$ we argue as before and we show that $g(y_3, \frac{y_3}{2y_1}, \bar{\theta}_6(a,y_1,y_2,y_3))$ is greater than every element of A. Assume $g_1 \equiv 0$. Either $\deg g_0(y_3,\frac{y_3}{2y_1})$ w.r.t. y_3 is positive and then $g(y_3,\frac{y_3}{2y_1},\bar{\theta}_6)$ is greater than every element of A, or $g_0(y_3,\frac{y_3}{2y_1})$ does not depend on y_3. But then the denominator vanishes as-well in $g_0(y_3,\frac{y_3}{2y_1})$ and so $g_0(y_3,\frac{y_3}{2y_1}) \in A_2$. We define

$$B = \bigcup_m A_2\left[\frac{y_3 - r_m^3}{m}, \frac{\bar{\theta}_6 - r_m^6}{m}, \frac{\frac{y_3}{2y_1} - r_m^7}{m}\right]$$ and we complete the proof as before.

q.e.d.

The proof of Lemma 2 can be found in [1] in §4, only it is not shown there that A is a homomorphic image of the resulting B. To show this define $h(U) = 0$ and extend h canonically onto B. Then h is the required homomorphism.

Proof of Lemma 3.

This is an adaptation of Lemma 3.1 of [1]. Let $u \in M - \mathbb{N}$ be chosen sufficiently small, $u \in J$, such that $u \equiv 1 \bmod n$ for $n \in \mathbb{N}$, u is transcendental over A, and u is a prime. We construct B almost as in Lemma 3.1 of [1].

Let $X_0 = \{x_0, x_0\, x_1 \ldots x_0 x_k, \ldots x_{i_1} \ldots x_{i_k} \ldots\}$ be a basis of A over $\mathbb{Q}$ built as in the proof of Lemma 3.1 in [1].

$B = \{x \in M^*:\;\; x$ if of the form

$$X = \frac{1}{n}\left(\frac{f_0(u)x_0+f_1(u)x_0x_1+\ldots+f_k(u)x_0x_k}{m} + f_{i_1}(u)x_{i_1}+\ldots+f_{i_k}(u)x_{i_k}\right.$$

where $f_0, \ldots, f_k, f_{i_1}\ldots f_{i_k} \in \mathbf{Z}[U]$,

$$n \mid f_0(u)x_0+f_1(u)x_0x_1+\ldots+f_k(u)x_0x_k+f_{i_1}(u)x_{i_1}+\ldots f_{i_k}(u)x_{i_k} \text{ in } A\}.$$

As in [1] we show that for sufficiently small non-standard u, B is a $\mathbf{Z}$-ring extending A in which $x_0 = u\frac{x_0}{u}$. Moreover if $x_0 \notin J$ then $B \cap J \subseteq M$. It remains to show a homomorphism h from B onto A. Let $h(u) = 1$ and let h be extended canonically onto B. Then h is as required.

q.e.d.

The proof of lemma 4 is an easy modification of Wilkie [3]. Finally observe the following:
From the totality of S_0 follows in open induction the totality of S_1, S_2, S_3. Indeed, following [2] Chapter 6 we have

$$\begin{aligned} & S_1(a,t,y) \Leftrightarrow \\ & \Leftrightarrow (\exists x_1,x_2,x_3,x_4)(S_0(x_1,t+1,x_2) \quad \& \\ & \& \ S_0(ax_1,t+1,x_3) \quad \& \quad S_0(a,t+1,x_4) \quad \& \\ & \& \ a > 1 \quad \& \quad x_1 > 4t(x_4+1) \quad \& \\ & \& \ 0 \le \frac{x_3}{x_2} - (y-1) \le 1). \end{aligned}$$

Having a,t we take x_4 by the totality of S_0, x_1 sufficiently big and x_2,x_3 by the totality of S_0. Then we take y so that $y-1$ is $[\frac{x_3}{x_2}]$. The existence of such a y is provable in open induction. We have

$$\begin{aligned} & S_2(a,t,y) \Leftrightarrow (\exists x_1,x_2,x_3,x_4)(\\ & (S_1(t,a,X_2) \quad \& \quad S_1(x_3+1,t,x_4) \quad \& \\ & \& \quad y \equiv x_1 \bmod x_3 \quad \& \quad x_3 > x_2 \quad \& \quad x_3 > \quad \& \\ & \& \quad t \ge a \quad \& \quad x_1 = [\frac{x_4}{x_2}]). \end{aligned}$$

We take x_2 by the totality of S_1, x_3 sufficiently large, x_4 by the totality of S_1, x_1 to be $[\frac{x_4}{x_2}]$ and then y to be the remainder of the divisibility of x_1 by x_3, which exists by open induction. We have

$$\begin{aligned} & S_3(t,y) \Leftrightarrow (\exists x_1,x_2,x_3,x_4)(S_1(2t,t+1,x_2) \quad \& \\ & \& \quad S_1(x_1,t,x_3) \quad \& \quad S_2(t,x_1,x_4) \quad \& \\ & \& \quad x_1 > x_2 \quad \& \quad y = [\frac{x_3}{x_4}]. \end{aligned}$$

It follows that S_3 is total.

REFERENCES

[1] Adamowicz, Z., Open induction and the true theory of rationals, submited to the Journal of Symbolic Logic.

[2] Manin, Yu.I., A course in Mathematical Logic, New Jork 1977, Springer-Verlag 8°, Graduate texts in Mathematics, 52.

[3] Wilkie, A.J., Some results and problems on weak systems of arithmetic, Logic Colloquium 77, 1977, North Holland.

LOGIC COLLOQUIUM '84
J.B. Paris, A.J. Wilkie, and G.M. Wilmers (Editors)

Situations, Sets and the Axiom of Foundation

Jon Barwise

Abstract: In this paper the rudiments of a theory of structured situations, construed as comprehensible parts of reality, is outlined. Some relations between situations and sets are discussed. It is argued that situations are not necessarily wellfounded under the constituent-of relation. It is then suggested that this gives an alternative conception of set (which we dub "hyperset") under which hypersets are not necessarily wellfounded under the membership relation $\in$. Connections with the axiom AFA of anti-foundation from Aczel [1] are briefly discussed.

1 Introduction

Over the past few years, I have been working on a theory of meaning and information content, one that aims at being rich enough to give a semantics for English that can account for the way language users handle information. We[1] call this Situation Theory, and the applications to natural language Situation Semantics. This project has led me to a number of unsuspected issues. In this paper, I discuss one of them, an issue in the foundations of set theory.

I take it that the basic tenet of set theory is that comprehensible families of objects can be considered as legitimate objects in their own right – that that is just what sets are. However, there are at least two other intuitions behind the various uses of sets. The first is that of sets are collections, things you get by acts of collecting. The metaphor is an empty box { } that can be used for collecting together some things $a, b, ...$ you already have on hand, and getting something *new*, namely the set $\{a, b, ...\}$.

The second intuition is that of sets arise from independently given structured situations by dropping the structure. Here the metaphor is an eraser, and operation S that takes a situation s, erases its structure, and leaves you with the underlying set $S(s)$ of constituents of s. Notice that this second conception encompasses the first, but not the other way around. That is, sets are presumably among the structured situations, and for any set s, one expects that $S(s) = s$. On the other hand, there is no *a priori* reason why the set $S(s)$ associated with some structured situation s should necessarily be able to be built up from priorly given objects $a, b, ...$

It is interesting to realize that while the first of these intuitions is what is used to generate the cumulative hierarchy, and so justifies the usual axioms of ZFC set theory, in particular, the axiom of foundation, it is the second which makes set theory so useful. Indeed, the fact that we can model most mathematical structures in the cumulative hierarchy in a natural way is what leads people to call (mistakenly, Kreisel has argued) ZFC a foundations for mathematics.

What I want to suggest in this paper is that if we take the second intuition as basic, thinking of sets as forgetful situations, then we get a different picture of the universe of sets, different from, and richer than, the cumulative picture. I am not sure, at this stage, just what the picture we get really is, but I am sure its different, and I am sure the alternative is interesting, useful, and worth pursuing.

There is an immediate difficulty with pursuing the second intuition, which accounts, perhaps, for why it has not been systematically pursued. One needs a prior theory of situations and their structure. But that is just what is needed for a theory of information content in any case, and what we have been trying to develop in Situation Theory. So, in this paper, I present part of situation theory and then try to apply it to study this notion of set.

I have been using, and will continue to use, the term "set" for the pre-theoretic notion, rather than for the cumulative conception that is now so prevalent. However, once I get to the point where the two conceptions really diverge, I will use the term "set" for the cumulative conception of wellfounded set, and the term "hyperset" for the broader notion of sets as forgetful situations. I suggest this terminology out of deference to the accepted notion, and out of similarity with the reals and hyperreals of non-standard analysis. As in that case, it turns out that every set is a hyperset, but not every hyperset is a set. And, at least under one conception of hypersets, the one that uses Aczel's axiom AFA discussed below, they can be approximated by sequences (of length ON) of wellfounded sets.

Even if one rejects hypersets and the intuition that gives rise to them, there is still a lesson to be learned, one I have learned the hard way. If you start with the cumulative picture of sets as collections, and then try to model various real world situations by imposing structure on sets, you may well get into trouble. John Perry and I fell into exactly this trap in our book *Situations and Attitudes*, [4], hereinafter referred to as S&A.[2] And I think we are not alone.

This paper is organized as follows. In the first section I present some basics of Situation Theory, since we need some idea of what the structure of a situation is before we can forget it. This part of the paper is an introduction the the rest of the paper, but also to a number of other papers that are planned as part of the development of Situation Theory. In the second section, I discuss the relation between sets and situations. In section three I turn to the issue of wellfounded situations and sets.

I have attempted to make this paper reasonably precise but still fairly simple, since I intend it as an introduction to Situation Theory that should be intelligible to non-mathematicians.

2 Rudiments of Situation Theory

In this section I collect together some of the most basic facts about situations and their structure. I have resisted the temptation to give the set of axioms listed below a name, since it is at best a tentative stab at a small fragment of a full-fledged theory of situations.

2.1 Everything is a first-class citizen. In computer science, it is common to contrast first and second class citizens, with respect to some language or theory. First class citizens are those things one can refer to; second class citizens are things one can use but not refer to. In Peano number theory, for example, numbers are first class citizens but the set of natural numbers and the numerical operations of addition and multiplication are second class citizens. In ZFC set theory, sets are first class citizens, but classes and the membership relation are second class citizens.

One of the assumptions of situation semantics is that human cognitive abilities make naturalization routine; everything is, or can easily become, upon reflection, a first class citizen. That is, anything humans systematically use is an invariant across situations so that they can step back and objectify it, and so treat it as a thing in its own right.[3] This is reflected in the ease with which we can nominalize in natural language – take the verb *to collect* and noun *collection*, for a relevant example.

To develop a theory of information that is adequate to cope with human language, this open-door policy has to be honored. Thus, in situation theory, anything we use can be objectified and talked about; this applies to situations themselves, relations, operations, conditions, parameters, whatever, and gives situation theory a rather different flavor from more traditional "closed" logical theories.

2.2 Relations and states of affairs. The place where this open-door policy makes it's first appearance is in the treatment of relations as first class citizens. Things relate in various ways. When we abstract across various instances of these relating, we get relations. Relations are basic in that they are not words, sets of n-tuples, ideas, or concepts. Relations are the glue that holds things together, the primary constituents of the facts that go to make up reality. Any relation we use in the theory can also be objectified and treated as an object of the theory.

Each relation comes with a set of argument places, roles that can be filled to get a basic state of affairs. Take the relation of holding, that is the relation that holds at some region of space-time if some object is holding another object there and then. There are three roles that need to be filled before we have a full-fledged holding: the spatio-temporal location l of the holding, the holder a, and the thing b that a holds.

Each relation R has a set of argument places that can be filled by appropriate objects. (A property is a 1-ary relation.) We take this notion of appropriateness between relations and the objects that can serve as their arguments as basic. In general, there is no reason to expect that we will be able to give a complete account of the conditions under which objects are appropriate to fill the argument places of a given relation, but in specifying axioms about relations, we will give as complete an account of appropriateness as possible. Note that each relation has some appropriate assignments, since the relation is an invariant across situations where things do or do not stand in the relation.

There is no reason to suppose the argument places of an arbitrary relation are intrinsically ordered. While it makes for notational convenience (at least in English) if we treat them that way, it also introduces a high degree of artificiality in some of the applications, especially to languages that use case-marking rather than order to indicate which objects are to fill which argument roles.

Thus, each relation R has a finite set $Arg(R)$ of argument places. By an *assignment* we mean a function $\mathbf{a}$ which assigns an object $\mathbf{a}_{arg}$ to some arguments arg. If $\mathbf{a}_{arg}$ is defined for some arguments arg in $Arg(R)$, and if the values are appropriate for R, then we call $\mathbf{a}$ an assignment for the relation R.

Note that we are not requiring that every assignment for R assign a value to every argument arg in $Arg(R)$, only to some, and that the assignment be appropriate for R.[4] Of course it *might* be that the only appropriate sequences for R are those that are defined on all its arguments, but there is no need to assume that this is the case in general. And there are good reasons for assuming it is not the case.

Example 1 Consider the relation E of eating. It seems reasonable to suppose that E has at least three argument roles, that of the eater, the eaten, and the space-time location of the eating. However, if only the first and last of these are filled, it seems we have enough for an appropriate assignment, since it either is or is not the case that the individual question is eating at that location. Of course if b is eating, then there will be something c that b is eating, and conversely. On the other hand, if b is not eating that b is not eating anything. These facts will have to be captured by constraints that express relationships between different states of affairs. In particular, there is a constraint that if b is not eating at l, then for each c, b is not eating c at l.

Example 2 Consider the activity of renting, wherein someone at some spatiotemporal location rents something to someone else, for some period of time, at some rate. Here we find quite a few arguments, at least six. However, there is not really any fixed order to these, even as we talk about them in English, since we can say that Bob rented his car to Tom, or Tom rented a car from Bob. Both of these statements have left some indeterminacy in the situation, in that they have not told us how to fill various other arguments, like the period of time, or the rent.

Example 3 In S&A, Perry and I found a need to have spatiotemporal locations arguments *loc* for properties denoted by common nouns – like the property of being a child. However, using the present approach, one can assume that this argument *loc* is optionally filled, and is usually determined by the situation being described. This makes the semantical rules associated with common nouns much simpler and provides a difference with verb phrases.

Given some n-ary relation R, we will often assume some arbitrary ordering of the argument places $arg_1,...,arg_n$ of R. Indeed, we will usually use the ordering suggested by the English expression that describes the relation in its active form, with the subject coming before the object of the relation, in the case of relations described by transitive verbs in a unique way. (This was not the case with renting.) In the case of two place relations with ordered arguments arg_1, arg_2, we often use infix notation and write "aRb" to describe $R(a,b)$ or $R(\mathbf{a})$, where $\mathbf{a}_1 = a, \mathbf{a}_2 = b$.

A relation R, and an appropriate assignment $\mathbf{a}$ of objects, determine two *basic states of affairs*, at most one of which is factual. We denote these states of affairs by $\langle\langle R, \mathbf{a}; 1\rangle\rangle$ and $\langle\langle R, \mathbf{a}; 0\rangle\rangle$, or, in case the arguments have been ordered, by $\langle\langle R, a_1, ..., a_n; 1\rangle\rangle$ and $\langle\langle R, a_1, ..., a_n; 0\rangle\rangle$.

Example 4 The state of affairs $\langle\langle eating, l, a; 1\rangle\rangle$, where a fills the role of eater and l that of the spatio-temporal location of the eating, obtains iff a is eating at l. The state of affairs $\langle\langle eating, b; 0\rangle\rangle$, where b is assigned to the role of eater, obtains if b is never eating. The state of affairs $\langle\langle renting, Bob, c; 1\rangle\rangle$, where Bob is assigned to the role of lessor, c to the role of the thing rented, obtains if Bob once rents c. On the other hand, the state of affairs $\langle\langle renting, Tom, c; 1\rangle\rangle$ where Tom is assigned to the role of lessee and c to that of the thing rented, obtains if Tom once rents c from someone.

We let σ, σ',... vary over states of affairs. If $\sigma = \langle\langle R, \mathbf{a}; i\rangle\rangle$ is a basic state of affairs, then R is called the *major constituent* of σ, and each $\mathbf{a}_{arg}$, for arg in $Arg(R)$, is called a *minor constituent* of σ. The value i ($= 1$ or 0) is called the *polarity* of σ. If $\sigma = \langle\langle R,\mathbf{a}; 1\rangle\rangle$ is a basic state of affairs with polarity 1, then we call it a *positive* basic state of affairs, and often write it without the polarity explicitly displayed, as $\sigma = \langle\langle R, \mathbf{a}\rangle\rangle$. Basic states with polarity 0 are called *negative*.

By a state of affairs, we mean a state that affairs may or may not be in. To emphasize this, we sometimes just call them states. We divide the basic states of affairs in to two classes, the facts, and the non-facts. (We use the term "fact" only for the *basic* states of affairs that obtain, not for more complex states of affairs that obtain.) If appropriate objects $\mathbf{a}$ do indeed stand in the relation R, which we write as $R(\mathbf{a})$, then the state $\langle\langle R, \mathbf{a}; 1\rangle\rangle$ with polarity 1 is a fact. That is, if there is at least one instance of R holding of $\mathbf{a}$, then $\langle\langle R, \mathbf{a}\rangle\rangle$ ($= \langle\langle R, \mathbf{a}; 1\rangle\rangle$) is a fact. If, on the contrary, appropriate objects $\mathbf{a}$ do not stand in the relation R, which we write $\neg R(\mathbf{a})$, then the state with polarity 0 is a fact. At most one of these states of affairs is a fact. If $\mathbf{a}$ is appropriate for R, then exactly one of the two states is a fact, since either $R(\mathbf{a})$ or $\neg R(\mathbf{a})$, but not both. Given a state of affairs σ, we write $\models \sigma$ to indicate that σ is obtains. Thus if σ is a basic state, then $\models \sigma$ iff σ is a fact. (Actually $\models$ is a binary relation. The other (optional) argument can be filled by situations, as we will see.) We summarize all of this with our first three axioms:

Axiom 1 *Every relation is the major constituent of some basic state of affairs, and everything is a minor constituent of some basic state of affairs.*

Axiom 2 *If* $\sigma = \langle\langle R, \mathbf{a}; i\rangle\rangle$ *and* $\sigma' = \langle\langle R', \mathbf{a}'; i'\rangle\rangle$ *are basic states of affairs, then* $\sigma = \sigma'$ *just in case they are equal item by item, that is, if* $R = R'$, $\mathbf{a}_{arg} = \mathbf{a}'_{arg}$ *for each argument arg of* R, *and* $i = i'$.

Axiom 3 *Given an n-ary relation R and an appropriate sequence* a *of arguments for R, either $R(\mathrm{a})$ or $\neg R(\mathrm{a})$, but not both. $R(\mathrm{a})$ if and only if $\models \langle\langle R, \mathrm{a}; 1\rangle\rangle$; $\neg R(\mathrm{a})$ if and only if $\models \langle\langle R, \mathrm{a}; 0\rangle\rangle$.*

It follows from Axiom 1 that everything is a constituent of some fact or other. So far we have only basic states of affairs and basic facts. There are more complicated states of affairs, but we will get to those later.

Many relations have a special argument over space-time locations, the location in space and time where the relation holds. Activities, for example, are usually located in space and time. So are physical properties of things. When an $(n+1)$-ary relation R has such a special argument, we call it an n-ary located relation, think of the special argument as being the first, and sometimes write the first minor constituent of a basic state of affairs $\langle\langle R, l, \mathrm{a}; i\rangle\rangle$ in front of the relation as follows: $\langle\langle at\, l : R, \mathrm{a}; i\rangle\rangle$.

Example 5 Since Claire is petting Jackie here at this space time location l, the state of affairs $\sigma = \langle\langle at\, l : petting, Claire, Jackie\rangle\rangle$ is a fact and so the state of affairs

$$\langle\langle at\, l : petting, Claire, Jackie; 0\rangle\rangle$$

is not a fact. Note that while the fact σ has l as a minor constituent, the fact $\langle\langle \models, \sigma\rangle\rangle$ doesn't have l (or any other space-time location) as a constituent.

There is a partial ordering $\sigma \rhd \sigma'$, read σ is as strong as σ', of a purely logical nature. Intuitively, $\sigma \rhd \sigma'$ just in case σ's being a fact entails σ''s being a fact, on purely logical grounds. A feeling for this ordering can be obtained from the next axioms.

Axiom 4 *The relation $\rhd$ is reflexive ($\sigma \rhd \sigma$) and transitive (if $\sigma \rhd \sigma'$ and $\sigma' \rhd \sigma''$ then $\sigma \rhd \sigma''$). If $\sigma \rhd \sigma'$ and σ is a fact, then so is σ'.*

Axiom 5 *Suppose* a *and* a$'$ *are both appropriate assignments for R, and that* a *is a sub-assignment of* a$'$. *For $i = 0, 1$, let $\sigma_i = \langle\langle R, \mathrm{a}; i\rangle\rangle$ and let $\sigma_i' = \langle\langle R, \mathrm{a}'; i\rangle\rangle$. Then $\sigma_1' \rhd \sigma_1$ and $\sigma_0 \rhd \sigma_0'$.*

Example 6 The state of affairs $\langle\langle atl, eating, b, c\rangle\rangle$ is stronger than the state of affairs $\langle\langle atl, eating$ since if b is eating c at l, b is eating at l. Similarly, the state of affairs $\langle\langle at\, l, eating, b; 0\rangle\rangle$ is stronger than the state of affairs $\langle\langle atl, eating, b, \mathrm{c}; 0\rangle\rangle$, since if b is not eating at l, then b is not eating c at l.

The next two axioms will not be needed for this paper, but are included (with little discussion) to prevent giving a false impression.

Axiom 6 *The states form a complete lattice under $\rhd$; that is, each set Σ of states has a least upper bound $\bigwedge\Sigma$ and a greatest lower bound $\bigvee\Sigma$. $\bigwedge\Sigma$ is a fact (in s) if and only if each state in Σ is a fact (in s), whereas $\bigvee\Sigma$ is a fact if and only if some state in Σ is a fact.*

There is another operation $\bigoplus$ on (basic) states of affairs It allows us to "unify" or "merge" two compatible states of affairs σ and σ' to obtain a state of affairs $\sigma \bigoplus \sigma'$. I won't go into this operation here, except to point out that it is different, and to give the basic axiom about it, to show how it differs from those just introduced.

Axiom 7 *The basic states of affairs $\sigma = \langle\langle R, f; i\rangle\rangle$ and $\sigma' = \langle\langle R, f'; i\rangle\rangle$ are compatible iff f and f' are compatible as functions, and in this case $\sigma \bigoplus \sigma' = \langle\langle R; f \cup f'; i\rangle\rangle$.*

It follows from this axiom and Axiom 5 that if $i = 1$, then $\sigma \bigoplus \sigma' \rhd \sigma \bigwedge \sigma'$, but if $i = 0$, then $\sigma \bigvee \sigma' \rhd \sigma \bigoplus \sigma'$. We can use this operation to fill missing arguments of a state of affairs by unifying it with one where that argument is filled.

2.3 Situations and facts. One of the starting points for situation semantics was the promotion of real situations from second class citizens to first class citizens. They have always been used in model-theoretic semantics, since they are parts of reality that correspond to the model-theoretic structures used to specify truth conditions. However, most previous semantic theories have stopped short of admitting they are first class citizens. The move of admitting situations as first class citizens in semantics is analogous (at least) to the admission of sets as first class citizens in mathematics. In fact, the latter might well be seen as a special case of the former.

By a situation, then, we mean a part of reality that can be comprehended as a whole in its own right – one that interacts with other things. By interacting with other things, we mean that they have properties or relate to other things. They can be causes and effects, for example, as when we see them or bring them about. Events are situations, but so are more static situations, even eternal situations involving mathematical objects.[5] We use $s, s', \ldots$ to range over real situations.

There is a binary relation $s \models \sigma$, read "σ holds in s", that holds between various situations s and states of affairs σ; that is, situations and states of affairs are the appropriate arguments for this relation of holding in. Our next axiom states the relation between facts and situations.

Axiom 8 *A state of affairs σ is a fact if and only if it holds in some situation; i.e. $\models \sigma$ iff there is a situation s such that $s \models \sigma$.*

If $s \models \sigma$, then the fact σ is called a fact of s, or, more explicitly, a fact about the internal structure of s. There are also other kinds of facts about s, facts external to s, so the difference between being a fact that holds in s and a fact about s more generally must be born in mind.

Note that for a basic state of affairs σ and its dual σ' (the state with the opposite polarity but everything else the same), either $\models \sigma$ or $\models \sigma'$ (and not both). However, for a particular situation, it may well not be the case that either of $s \models \sigma$ or $s \models \sigma'$, since s may not determine which of these is the fact of the matter.

We have characterized situations as those parts of all that is the case that can be comprehended as totalities in their own right. More generally, though, there is a relation $s \leq s'$ of one situation s being part of another situation s'.

Axiom 9 *(Partial ordering of situations:) (i) A situation s_1 is a part of a situation s_2 just in case every basic state of affairs that is a fact of s_1 is also a fact of s_2. (ii) If $s_1 \leq s_2$ and $s_2 \leq s_1$ then $s_1 = s_2$.*

The second part of the ordering axiom claims that each situation is determined (modulo an ontology) by the facts that hold in s. It provides, at one and the same time, an identity conditions on situations, but also a strong existence assumption on facts – there are enough of them to distinguish distinct situations. It follows from this axiom that the part-of relation between situations is a partial ordering:

- $s \leq s$
- $s_1 \leq s_2$ and $s_2 \leq s_3$ entails $s_1 \leq s_3$
- $s_1 \leq s_2$ and $s_2 \leq s_1$ entails $s_1 = s_2$.

Since situations themselves have properties and stand in relations, they are minor constituents of states of affairs, and so of facts. In particular, this applies to the relations of holding in and part-of discussed above. For example, since part-of is a relation we use, we call also objectify it and consider states of affairs with it as the major constituent. Then, as a special case of Axiom 1, we note that $s_1 \leq s_2$ if and only if the state of affairs $\langle\langle \leq, s_1, s_2 \rangle\rangle$ is a fact. Note that a fact of the form $\langle\langle \leq, s_1, s_2 \rangle\rangle$ is a fact about s_1 and s_2, but it need not be a fact that holds *in* either situation. It is rather a fact external to them.

By a *constituent* of a situation s, we mean a minor constituent of some basic state of affairs that holds in s. We write $b \in s$ to indicate that b is a constituent of s.

2.4 Reflecting on the axioms.

From the point of view of the theory of situations, there is something very misleading about the discussion above, and that to follow about hypersets. This subsection is included to correct this impression. It is irrelevant for the main point of this paper, though.

Our axioms are intended to describe facts about the world, facts that are necessary, given the way we carve the world up into relations, facts and situations. That being the case, we should be able to step back and objectify the facts described by these axioms. Thus, we think of each axiom not just as stating various relationships between things, but as describing facts involving the relationships as major constituents and the things as minor constituents. A full treatment of this must await the development of our theory of roles, conditions, and anchors, but we can at least hint at this development here.

Making this move requires us to objectify the properties and relations that constitute the major constituents of these facts. The most important one, the one that we have used most often, is the relation $\triangleright$ of one state entailing another, or the relation $\bowtie$ of two states each entailing the other. We cannot go into this in any detail here, but let's look at one example.

Let R be a binary relation with ordered arguments arg_1 and arg_2. Axiom 3 asserts, among other things, that $R(x,y)$ if and only if $\models \langle\langle R, x, y \rangle\rangle$. Just what fact or facts does this biconditional describe? As a first stab, we can objectify the fact described by this conditional by objectifying the state of affairs $\sigma = \langle\langle R, x, y \rangle\rangle$ and $\sigma' = \langle\langle \models, \sigma \rangle\rangle$ described by the antecedent and the consequents, respectively, and the relationship $\bowtie$ of mutual entailment that holds between them. This gives us, among other things, that σ entails and is entailed by σ': $\sigma \bowtie \sigma'$. We can, then, objectify this and assert that the state of affairs $\langle\langle \bowtie, \sigma, \sigma' \rangle\rangle$ is a fact.

Given what we have to work with so far, this is the best we can do at capturing the facts expressed by this statement. However, it is just not good enough. The above facts are not general enough, when compared with the facts expressed by the axioms in question. Consider, for example, the fact that

1 $R(x,y)$ iff $\models \langle\langle R, x, y; 1 \rangle\rangle$.

In order to interpret (1) as describing a fact, we had to interpret the uses of "x" and "y" as denoting specific things, x and y that gave us an appropriate sequence for R. However, it is more natural to interpret the assertion as expressing a general relationship, independent of what x and y happen to be (or to put it differently, independent of what "x" and "y" happen to denote) as long as they form an admissible assignment for R. To get at the general fact, we need to objectify as parameters the argument roles arg_1 and arg_2 introduced by our use of "x" and "y" in stating (1), rather than specify their values, as we did just now. The statement (1) is general in that it states a single fact about all such anchorings of the parameters. The importance of this move, for the general theory, can be seen from the role it plays in the treatment of conditions; see [3].

3 Situations and Sets

It is not really clear, at this stage, whether situation theory *needs* to be an extension of set theory, or whether it can be orthogonal to set theory. That is, it is not clear the extent to which one's ontology needs to have sets as a crucial ingredient to give an account of information. We could certainly have avoided the use of sets in everything we have done so far.

However, I am interested in the interaction of set theory with situation theory for three reasons. One is just that sets are mathematical objects so there has to at least be room for them in a theory like this, if it is to do the kind of job I want it to, in accounting form information in mathematics. The second reason is that I want to be able to use sets to model situations, since set-theoretic modelling has proven so extraordinarily powerful a method in logic. In so doing, I think the theory of situations can provide an alternative family of intuitions about sets, as I said earlier. The final reason is that an account of the relation between sets and situations is needed needed to explain the relation between the theory of situations and the set theoretic model Perry and I built of intuitive situation theory in S&A, and the ways it was right and where it went wrong.

3.1 Balancing sets with situations.

There is a strong interaction between our set-theoretic intuitions and our situation-theoretic intuitions. The basic tenet of set theory is that certain families of objects can be "comprehended" as a single object in their own right, an object that has properties and stands in relations to other objects. These collections are called sets. Other families cannot coherently be considered as completed objects – for example: the family of all sets.

We take a similar point of view with regard to situations. Reality, or "the world", is all there is. It consists, as Wittgenstein said, of all the facts and of their being all the facts. Situations, by contrast, are totalities in their own right, parts of reality that can enter into relations (e.g. causal) with other parts. But if situations are parts of reality that can be comprehended as completed totalities, then the internal structure of a situation should correspond to a set of facts, that is, a family of facts that can be comprehended as a completed totality.

This gives us a way to use sets to model real situations. Since facts can be individuated by their constituents and polarities, we can model facts with sequences and so model situations with sets of sequences. However, there is a very important assumption implicit in this. In claiming that we can model situations with sets of sequences, we are making a claim about two things, situations and sets. One can use it in two ways, to delimit the collection of situations, or to get new intuitions about what sets exist, or both. In this paper I want to do both. On the one hand, we will see that there is no largest situation. On the other, we will see that there are non-wellfounded sets.

Let us put the point more generally. Suppose we have any axiom of the form:

2 For every Tweedledee there is a Tweedledum R-related to it.

Axioms of the form (2) cut two ways. They can be construed as putting a limit on the Tweedledees: only those exist for which there is an R-related Tweedledum. On the other hand, they can be seen as postulating the existence of a rich collection of Tweedledums. There are so many of them that there is at least one R-related to every single Tweedledee. The "right" way to think of them is providing a balance between the Tweedledees and the Tweedledums.

As an example, let us consider Axiom 9. It entails that for every pair of distinct situations s and s' there is a fact σ that holds in one but not the other. If we have a limitation on the facts, this will impose a limitation on the pairs of distinct situations. On the other hand, it can be considered as an axiom asserting that there is a rich enough collection of facts to be able to distinguish between different situations.

Let's look at another example.

Axiom 10 *Every set F of facts determines a smallest situation s such that every fact in F holds in s.*

It is no more correct to think of this as positing a large collection of situations than it is to think of it as positing a limitation on sets.

Why is this axiom a reasonable assumption? It follows from the basic parity between situations and sets: if we have a comprehensible totality of facts, then we can find situations for each of them, and so a set of situations, which we should be able to bound by one situation that contains them all as parts. But then there should be a smallest such situation, since we humans find it so easy to carve situations out of the world.

Notice that we are not assuming that the smallest situation s has exactly the facts of F as its facts, though I am often tempted to do so. This leaves room for us to assume, if we want, that real situations are closed in various ways which sets of facts are not.

By using the same symbol for the membership relation and the constituent relation we are reflecting our prejudice that the former is just a special case of the latter. If we want to make explicit the intuition that sets are nothing but forgetful situations, we could do so in several ways. The simplest would be as follows.

Axiom 11 *Every set is the set of constituents of at least one situation.*

For example, one making this more definite is to consider a set as a situation giving one a set of facts about its members. Let us assume that $\in$ is a 2-ary relation, with arguments arg_1 and arg_2 for members and sets, and write $\langle\langle \in, a \rangle\rangle$ for the state of affairs of polarity 1 with the first argument filled by a. Thus a set b would give one a situation $s(b)$ with facts exactly those of the form $\langle\langle \in, a \rangle\rangle$ for $a \in b$. Then, using the terminology from the introduction, made precise below, $\mathcal{S}(s(b)) = b$.

3.2 Modelling situations with sets.

To make the modelling of situations with sets precise, we define an operation of M (for modelling). By a *state model* we mean an ordered triple $\langle R, \mathrm{a}, i \rangle$, where R is a relation, a is a function, and i is 0 or 1. (We called such triples constituent sequences in S&A, except that we required the assignments to be defined on all and only the arguments of R.) By a *situation model* we mean any set of state models. (We called these abstract situations in S&A.) However, since situations themselves are constituents of states of affairs, to successfully model situations with sets of state models we need to replace situations in state models by their models, and so on. Thus we suppose the following:

Axiom 12 *There is a unique operation $\mathcal{M}$ taking values in the (hyper)sets satisfying the following equations:*

- *if b is not a situation or state of affairs, then $\mathcal{M}(b) = b$;*
- *if $\sigma = \langle\langle R, \mathrm{a}; i \rangle\rangle$ then $\mathcal{M}(\sigma) = \langle R, \mathrm{b}, i \rangle$ (a state model), where b is the function on $dom(\mathrm{a})$ satisfying $\mathrm{b}(x) = \mathcal{M}(\mathrm{a}(x))$;*
- *if s is a situation, then $\mathcal{M}(s) = \{\mathcal{M}(\sigma) : s \models \sigma\}$.*

For any situation s, the set $\mathcal{M}(s)$ is called the *canonical model* of s. In S&A we called the canonical models of real situations "actual situations". We were implicitly assuming Axiom 12, since we were indeed assuming that these abstract actual situations were sufficient for modelling all real situations.

Given the guiding tenet of both situation theory and set theory, this seems like a reasonable assumption. But, as we have just seen, it amounts to a balancing act between situations and sets. We must leave open the possibility that the notion of set that this gives rise to may diverge from the notion that arises from other conceptions of sets, like the one that gives rise to the cumulative hierarchy.

In proving the propositions below, I am going to use the usual axioms of set theory, with the sole exception of the Axiom of Regularity, or Foundation, which is what is at issue.

Proposition 1 *For each situation s there is a set of all facts that hold in s, and hence a set of all constituents of s.*

Proof. The modelling operation $\mathcal{M}$ establishes a one-to-one correspondence between the facts of s and the elements of the model $\mathcal{M}(s)$. Thus, by replacement (which we assume to hold), the facts of s form a set. From the set of facts we can obtain the set of constituents in a routine manner. □

We write $\mathcal{F}(s)$ for the set of facts that hold in s, and $\mathcal{S}(s)$ for the set of constituents of s. By Axiom 9, $s_1 \leq s_2$ iff $\mathcal{F}(s_1) \subseteq \mathcal{F}(s_2)$, and $s_1 = s_2$ iff $\mathcal{F}(s_1) = \mathcal{F}(s_2)$.

Proposition 2 *Every situation s is a proper part of some other situation s'.*

Proof. Given s, we can form the set $\mathcal{S}(s)$ of constituents of s. This being a set, there are many objects not in it, since there is no universal set. (Contrary to popular opinion, the lack of a universal set has nothing to do with the axiom of foundation; it is provable from the axiom of separation alone, via the construction in Russell's paradox.) Let b be such an object, and let f be a fact which has b as a minor constituent. Then there is a situation s' such that the $\mathcal{F}(s')$ contains $\mathcal{F}(s) \cup \{f\}$. But then s' is a situation that has s as a proper part. □

Corollary 3 *There is no largest situation.*

One might find this a bit odd, since it seems to show that reality, all that is, is not a situation. However, all it really shows is that there is no possibility of having reality be a completed totality that can be treated as a first class citizen, for it always outstrips our attempts to comprehend it as a totality.

This is just a start at axiomatizing a theory of situations, and even this much is highly tentative. However, it is enough to let us ask, and answer, the foundational question that concerns us in this paper.

4 Is Reality Wellfounded?

A binary relation R is *wellfounded* if whenever a property P holds of some x in the range of R, there is something x that has P but such that no y with yRx has property P. When we ask whether reality is wellfounded, what we mean is: Is the constituent relation $\in$ on situations wellfounded?

Actually, there are two different ways in which reality may fail to be wellfounded. To see this, we can rephrase the notion as follows. Call a situation s *circular* if there is a finite sequence (of length ≥ 2) of situations, starting and ending with s, each of which is a constituent of the next:

$$s \in \ldots \in s.$$

Call s *ungroundable* if there is an infinite sequence of distinct situations, starting with s, each a constituent of the next:

$$\ldots \in s''' \in s'' \in s' \in s.$$

We will call s *wellfounded* if it is neither circular nor ungroundable. Then it is easy to see that reality is wellfounded if and only if every situation is wellfounded. So, is reality wellfounded? To answer this, let us look at the corresponding notion in set theory.

4.1 The axiom of foundation in set theory. Set theory, as it is usually described these days, actually rests on two assumptions. One is that a set is any family that can be comprehended as a totality. The other is that sets are built up in a cumulative hierarchy, that is, in stages, starting with some primitive non-sets, and iterating the operation of collecting "previously" constructed sets. It is this assumption that justifies the axiom of foundation in ZFC (and KPU), the axiom that asserts that any property that holds of some set must hold of some set that has no member which is itself a set with the property: that is, if $P(a)$ holds for some set a, then there is a set b such that $P(b)$ holds, but there is no set $b' \in b$ such that $P(b')$. The traditional justification for the axiom is this: assume some set a has property P. Among such sets, choose one b that occurs "earliest" in the "construction" of all sets. Then none of the members of b can be sets with the property in question.

This assumption is clearly capturing a kind of different intuition than the earlier ones, that of sets as totalities that have properties and stand in relations to other things, and that of sets as the sets of constituents of structured situations. Should we have a similar assumption in situation theory, with a corresponding axiom of foundation for situations?

Before arguing that there are real circular situations, and hence that reality is not wellfounded, let me make a couple of observations. First, if we assumed that situations were wellfounded, then we would not need to posit Axiom 12. It would be provable. But without such an axiom, it is not provable. Second, because of Axiom 12, the universe of sets is wellfounded if and only if the universe of situations is wellfounded. That is:

Proposition 4 *If the membership relation on sets is wellfounded, then the constituent relation on situations is wellfounded, and conversely.*

Proof. To prove the first half, we need the following apparent strengthening of foundation (which is well known to follow from it): if Q is any property that holds of some set a, then there is a set b such that $Q(b)$ but Q does not hold of any member of the transitive closure of b. Now suppose P is some property that holds of some situations. Consider the property Q of being a model $M = \mathcal{M}(s)$ of some situation s that has property P. Apply the above to Q to get a model $M = \mathcal{M}(s)$ with Q but such that no member of the transitive closure of M has Q. Now if s' were a constituent of s, and s' had property P, then $M' = \mathcal{M}(s')$ would be in the transitive closure of M and have property Q, which establishes the first half of the result. This half uses Axiom 12, but not 10 or 11. The converse follows immediately from Axiom 11.□

This result suggests the possibility that the intuition behind the notion of set and situation needed for an account of information might give rise to a concept of set that is at odds with (indeed richer than) the cumulative hierarchy concept of set. To someone (like me) who has literally grown up with the cumulative picture, this is a disquieting thought, and one I have strenuously resisted since first starting to think about situations. For example, in using my favorite theory KPU in S&A, Perry and I were working in a set theory that assumes the axiom of foundation, a fact that gave us trouble at various points. These problems have finally piled up to the point where I want to suggest that there are good reasons for supposing that the constituent relation on situations is in fact not wellfounded, and so rejecting the axiom of foundation for the notion of set that is commensurable with the notion of situation.

4.2 Some inherently circular situations. In this section I am going to list several sorts of situations that seem to me to be best thought of as involving an inherent circularity, and so as non-wellfounded situations. I won't try to justify these claims, except in the one case, that of stud poker, which I will go into in a bit more detail, and in a further paper.

Example 7 Suppose I am in a very embarrassing situation s, say I have forgotten to bring the main dish to a dinner given in John Perry's honor. If I say "This is an embarrassing situation", I am referring to s, so I stand in a certain relation to s, in virtue of referring to it with "this": $\langle\langle referring\ to, Barwise, s\rangle\rangle$. However, it seems that my utterance s_0 might also be part of the situation s referred to. But if so, that makes the situation s circular.

Example 8 I recently heard an airline announcement that concluded with "This announcement will not be repeated". If announcements are situations, then this one has itself as a constituent. In general, self-referential statements, and statements about other statements which are themselves about the first statement, are circular.[6]

Example 9 (Due to John Perry.) Imagine two parallel mirrors of the same size, facing each other, one A with an "X" painted on it, the other B with an "O". This is a simple finite physical situation which is, in some sense, circular. We can think of it as three situations, a situation s with two sub-situations s_A and s_B, the scenes reflected in A and B, respectively. In s_A we have the facts that B has an "X" and that B reflects s_B, while in s_B we have the symmetric facts about A. The facts of s are those of $s_A \cup s_B$ plus the facts that A and B are parallel and facing one another.

Example 10 (Due to Descartes.) Genuine self-awareness is circular. For example, what breaks the sceptical doubt in Descartes' "I think, therefor I am"? It is not the thinking alone, since Fermat's thinking did not convince Descartes that *anybody* existed, Descartes or Fermat. Rather, it is a mental act that comprehends itself as a constituent that breaks the chain of doubt.

Example 11 (Due to Grice) If there really are the kinds of M-intentions that Grice [7] uses in his account of speaker meaning, intentions to have others recognize various things include the very intention, then having such an intention must be a circular situation.

Example 12 (Due to the inventor of stud poker.) Consider five card stud, a poker game where each player receives one "down" card, and four "up" cards. A player's down card is one that she is entitled to see but to keep hidden from the other players, while the dealer displays the up cards for all to see. Any given hand can be viewed as one where some finite number of players $p_1, ..., p_n$ each have five out of 52 cards, 2♣, 3♣,..., K♠, A♠. If we take the spatio-temporal location as a fixed parameter, we can take the relation H of having that holds between players and cards, and write $H(p_2, 3♣)$ (or $p_2 H 3♣$) if p_2 has the 3♣, for example. Given a hand of the game, we can represent it with the atomic and negated atomic statements involving "H" and names for the players and cards. Call any propositional combination (using "and" and "or") of such statements a "level 0" statement.

What concerns us here, though, is not just what *cards* each player has, but what *information* each player has. In this regard, a central question becomes that of the informational difference between a player's up and down cards. How should we treat this?

One possibility would be to use different properties, that of being up, and of being down. That, however, is not really what is at stake here. Imagine a variant of five card stud where each player is required to show his last four cards only to the players sitting next to her. In this version, whether or not a card is "up" depends on a set of people. It is the information about values that count, now really whether it is face up or face down.

A standard approach to this would be to use an epistemic relation S, writing "$p_i S \phi$" to indicate that p_i sees that ϕ. (Here there are subtle variations possible, ranging from seeing, having the information that, knowing that, believing that, each of which has its own distinctive features, features that are logically significant.) Thus, if the 3♣ is one of p_2's up cards, then, for example, $p_1 S(p_2 H 3♣)$.

It is one thing to introduce new symbols into a language, quite another to say just what in the world it corresponds to, that is, to give its semantical interpretation. The standard modal logic interpretation for "S" would be as a relation between players and sets of possible worlds, the set of worlds where ϕ is true. Not surprisingly, I think that taking it to be a relation to a situation s where ϕ obtains is more reasonable (see Barwise [2] or Chapter 8 of S&A), but we can set this aside for now, since the problem I want to raise is more basic.

Consider those statements we can make that are propositional combinations of those of the form "$p_i S\phi$", where ϕ is of level 0. Call these the level one informational statements. While we can represent a lot about the information available to the different players by using level 1 statements, there is a lot we cannot represent, a lot that is relevant to the informational nature of the actual game.

To see this, contrast a hand of standard five card stud with exactly the same hand in terms of card distribution, except that all five cards are down cards, but where, as it happens, everyone just happens to see the last four cards dealt to the other players, or learn of their value is some other private way, so that the fact that they have seen these cards is not known to the other players.

Superficially, the situations look the same, in that everyone has seen exactly the same cards, and so seems to have the same information. Everyone has seen the last four cards of every other player in both cases. In fact, just the same level 0 and level 1 statements are true in the different situations. However, any card player realizes that the situations are as different as day and night. In the former, standard, situation there is a great deal more information. Everyone sees everyone else see the up cards, so, for example,

$$p_2 S(p_1 S(p_2 H 3♣)),$$

whereas this need not be the case in the second situation. So we need at least level 2 statements to capture the information present.

But then one can easily construct a situation with the same level two statements holding as in the standard hand, but which is informationally distinct from it. For example, in the standard hand, everyone sees everyone else see everyone else see the up cards, and so on up an infinite hierarchy of distinct informational facts. One can show that no finite level captures all the information. In fact, not even the whole infinite hierarchy of informational reports adds up the the true situation regarding the shared information.

Something seems to have gone radically wrong here, in that we seem driven to perceptual reports of arbitrary finite level (at least) in an attempt to capture the information present in a very simple, finite situation. To a model-theorist, it looks reminiscent of a number of results where second-order statements are give rise to an infinite hierarchy of first-order approximations. In a separate paper, I will show that the same thing is going on here, and that there is a simple account of the informational situation that (under various assumptions on the properties of the relation S in question) gives rise to the above hierarchy of levels. Here I will just hint at the results enough to show the importance of circular situations.

Let us shift the focus of attention from individual mental states, to something that people actually share, namely their environment. On this view, just what one knows is highly dependent on one's external circumstances – where they are, what they see, what is going on around them more generally, and what things are relatively stable in their environment. By focusing on the objective circumstances of individuals, and their role in the determination of what people know and believe, we can exploit shared circumstances in the explanation of shared understanding.

In the case of our poker players, obviously what best classifies their mental states is not some infinite hierarchy of mental states, but a public situation s, the one they are in, where they are all seeing what is there for them to see. What each player p_i sees is a certain situation s_i (which I will call the scene seen by p_i) which is part of the situation s. Each of these situations is determined by a set of facts. However, these situations s_i has an important sub-situations, the part s_i^p that is public (the superscript "p" being for "public part"), or shared with the other players. This includes the facts about who has what up cards, but also who sees who has what up cards. That is, part of what p_1 sees is p_2 seeing the cards face up on the table. Thus, for example, one of the facts that holds in s_1^p is that p_2 sees s_2^p.

Here are a few of the basic observational facts in $\mathcal{F}(s_i^p)$:

$$\langle\langle H, p_2, 3\clubsuit\rangle\rangle,$$

$$\langle\langle S, p_j, s_j^p\rangle\rangle,$$

for $j \neq i$. If we define the semantics of "$pS\phi$" in terms of there being a situation s where ϕ holds such that pSs, then out of the finite number of such basic observational facts all the desired consequences will follow. What is important to notice here is that all of these situations s_j^p are circular. Moreover, it is easy to see that no wellfounded situation will satisfy all of the desired informational statements.

In a separate paper, I will go into the model theory of shared information, and prove some approximation theorems that relate the two approaches just mention, along with a third, a fixed point approach. I hope that the brief discussion here at least makes it plausible that shared information may well arise from circular situations.

Example 13 Shared understanding in all its various guises (mutual belief, common knowledge, public information) rests on circular, or at least non-wellfounded situations. In as much as there are assumptions of such shared understanding throughout game theory, law, communication theory, and the like, we are constantly caught in non-wellfounded situations. The previous example is just one very special case of this.

There are basically three approaches to understanding shared understanding in its various forms: iterated attitudes, fixed points and a shared environment approach. The latter, suggested in Clark and Maxwell [6] characterizes the case where two agents a, b have common knowledge that f in terms of there being a situation s such that (using our framework):

- f is a fact of s,
- $\langle\langle knows, a, s\rangle\rangle$ is a fact of s, and
- $\langle\langle knows, b, s\rangle\rangle$ is a fact of s.

While this may not be the right characterization of common knowledge in general, in that there may well be other ways it can arise, it is certainly a common way for it to arise. Note, though, that such a situation s is circular. In the paper referred to above, I will go into the relation between the three approaches just mentioned.

I hope that some of these examples convince the reader that reality, unlike the cumulative hierarchy of sets, is not wellfounded, that it involves many sorts of inherently circular situations.

4.3 Aczel's AFA.

It is one thing to give up the axiom of foundation; it is quite another to know what to replace it with. What kind of non-wellfounded situations and sets (or hypersets, as I'll start calling them) exist? For example, how can we be sure our theory is consistent, given Axiom 12, and the assumption that there are non-wellfounded situations?

Peter Aczel, motivated by non-well founded situations in the theory of processes, has recently produced an elegant axiom AFA of anti-foundation, studied its consequences and consistency, and relations with the axiom of foundation and with other anti-foundation axioms due to Boffa, Finsler and Scott. I have found Aczel's axiom to be extremely useful in working with hypersets as models of non-wellfounded situations, so I will describe it briefly here. Actually, I will describe the variant of his axiom that allows atoms, not just pure sets; I call the resulting axiom AAFAA, for Aczel's Anti-Foundation Axiom with Atoms.

Quite briefly, and without much motivation, let me state AAFAA. A graph G is a set of nodes and directed edges, as usual. If there is an edge $x \to y$ from x to y, then y is said to be a child of x. A tagged graph is a graph such that if x has no children, then x can be tagged by an atom or by the empty set. A decoration for a graph is a function $\mathcal{D}$ from the nodes of the graph into the universe of hypersets such that for each node x, if x has no children, then $\mathcal{D}(x) = tag(x)$, whereas if x has children, then

$$\mathcal{D}(x) = \{\mathcal{D}(y) : y \textit{ is a child of } x\}.$$

A graph G is wellfounded if the relation child-of in G is wellfounded. It is easy to see that each wellfounded set can be obtained from many different wellfounded graphs as the set assigned to the top node. Mostowski's Collapsing Lemma tells us that, conversely, every wellfounded tagged graph has a unique decoration in the universe of wellfounded sets. Aczel's AAFAA, by contrast, asserts that *every tagged graph has a unique decoration.*

In [1], Aczel shows that AAFAA is consistent with the other axioms of set theory, and proves a number of remarkable properties of this universe of hypersets. Using Aczel's results, it is easy to see that our Axiom 12 becomes a consequence of the other axioms plus AAFAA. Using this, it is easy to see that our Axioms 1-12 are consistent, along with the assumption that there are circular situations, of course.

On the other hand, Axiom 12 does not imply AAFAA. Indeed, it is not clear to me that AAFAA is true under the conception of hyperset that we have been discussing here. There are really two sides to AAFAA. One side tells us that each tagged graph pictures some hyperset. The other side tells us that there is only one hyperset pictured by any graph. The motivations for these two parts are somewhat different, and I am less confident of the second than the first under the situation-theoretic conception of set.

In a more technical paper, I will use Aczel's work to prove a number of approximation theorems in the model theory of public information, along the lines mentioned in Example 11, above.

Footnotes

1. I have been involved in this project with John Perry, and more recently with a number of others as well. I wish to thank John and all the members of the STASS project (John Etchemendy, Joseph Goguen, Kris Halvorsen, David Israel, Jose Meseguer, John Perry, Stanley Peters and Brian Smith), and Robin Cooper, for the many discussions we have had on topics related to this paper, discussions that have resulted in numerous changes in the basic theory. This research was partially supported by an award from the System Development Foundation to the Center for the Study of Language and Information.

2. See Fenstad's Puzzle on page 223 of S&A, for example.

3. See the discussion of promiscuous realism in [5].

4. This particular feature of the account here grew out of discussion in the STASS group. It has been independently suggested in a draft of a paper called "Toward an anadic situation semantics," by Carl Pollard.

5. In S&A, we used the term "state of affairs" for the static sort of situation. Here we are using this term differently, since we are now taking them to be things that may or may not obtain.

6. John Etchemendy and I are working on a paper called "Truth, the Liar, and non-wellfounded situations", in which we give several situation semantics accounts of the Liar Paradox, accounts that differ depending on what conception of truth (and falsity) one has.

References

1. Aczel, P., *Notes on Non-well-founded Sets*, CSLI Lecture Notes, vol 3., 1985

2. Barwise, J., Scenes and other situations, *J. of Phil.*, 78 (1981) pp. 369-397

3. Barwise. J. Conditionals and conditional information, to appear in *On Conditionals*, ed. by E. Traugott et. al., Cambridge University Press

4. Barwise, J. and Perry, J., *Situations and Attitudes*, Bradford Books, M.I.T. Press, 1983

5. Barwise, J. and Perry, J., Shifting situations and shaken attitudes, *Linguistics and Philosophy*, 1985

6. Clark, H. and Marshall, C., Definite reference and mutual knowledge, *Elements of Discourse Understanding,* Joshi, Webber and Sag, ed., Cambridge University Press, 1981, pp 10-63

7. Grice, P., Meaning, *Phil Review*, 66 (1957), pp 377-388

LOGIC COLLOQUIUM '84
J.B. Paris, A.J. Wilkie, and G.M. Wilmers (Editors)
© Elsevier Science Publishers B.V. (North-Holland), 1986

ULTRAFILTERS ON DEFINABLE SETS IN ARITHMETIC

P. Clote *

Abstract. In this paper, we solve an outstanding problem due to G. Mills and J. Paris [MP] in the study of initial segments of models of arithmetic: an initial segment $I \subset_e M$ is n-Ramsey if and only if I is n-extendible. Along the way, we give an interesting characterization of n-extendible cuts which generalizes a result of L. Kirby [Ki].

Introduction

The relation between independence results in systems of arithmetic and rate of growth of recursive functions has been known for some time. For instance, from [LW] and [Pars] it follows that:

$I\Sigma_n \vdash$ "f is total" iff f is Kalmar elementary in f_α, for some $\alpha < \omega_n$.

Here $f_0(x) = x + 1,\ f_1(x) = (x+1)^2,\ f_{\alpha+1}(x) = f_\alpha^{(x)}(x)$

and $f_\lambda(x) = f_{\lambda_x}(x)$ for $\alpha, \lambda < \epsilon_0$.

In particular, it has been noted independently by Parsons, Minc, Takeuti that

$I\Sigma_1 \vdash$ "f is total" iff f is primitive recursive.

It is perhaps not widely known that for $k \geq 2$

$I\Sigma_0 + \forall x \exists y f_k(x) = y \vdash$ "f is total" iff f is in $\mathcal{E}_{k+1}$

(the $k+1^{st}$ Grzegorczyk class). This was pointed out by A. Wilkie and is easily verified by using a modification of the proof by J. Paris of a result of R. Parikh [Parikh]. Recently, weak fragments of arithmetic have been found by S. Buss [B] whose provably recursive functions are exactly those functions in certain well-known complexity classes. For instance, among other things, Buss has shown that

$S_2^n \vdash$ "f is total" if f is in the polynomial time closure of a set in $\sum_{n-1}^P$,

* **Part of this research was done while the author was visiting the Department of Computer Science of the University of Toronto and partially supported by an NSERC grant. I would like here to express my sincere thanks to S.A. Cook.**

the $n-1^{st}$ level of the Meyer-Stockmeyer polynomial time hierarchy, assuming that the graph of f is in $\sum_n^P$.

In particular,

$S_2^1 \vdash$ "f is total" iff f is a deterministic polynomial time computable function

where Buss' system S_2^1 is essentially (a version in arithmetic of) induction on notation for NP-formulas and S_2^n is essentially (a version in arithmetic of) induction on notation for $\sum_n^P$ formulas in the polynomial time hierarchy.

In [Pa-1], J. Paris studied the rate of growth of modified Ramsey functions and showed that

(1) $I\Sigma_n \vdash \forall x \exists y [x, y] \overset{\rightarrow}{*} (n+2)_m^{n+1}$, for all $m \in \mathbf{N}$, but

(2) $I\Sigma_n$ not $\vdash \forall x, m \exists y [x, y] \overset{\rightarrow}{*} (n+2)_m^{n+1}$.

To prove the first assertion, J. Paris made essential use of the *symbiocity* of initial segments satisfying $I\Sigma_n$ with n-extendible initial segments. (Two classes of initial segments are said to be *symbiotic* if each class is dense in the other.) Using symbiocity, in [Pa-1] Paris showed that for any Π_2 formula θ (θ can even be $\prod_{n+2}$):

$I\Sigma_n \vdash \phi$ iff ϕ holds in all n-extendible initial segments.

As n-extendibility is a model theoretic notion, there is a not too difficult model theoretic argument to show that

$$I \models \forall x \exists y [x, y] \overset{\rightarrow}{*} (n+2)_m^{n+1}$$

for any $m \in \mathbf{N}$ where I is an n-extendible initial segment. (See [Cl-2] for an alternative proof of (1) as well as of a stronger combinatorial statement.)

In this paper, we solve a long outstanding problem in the theory of initial segments of models of arithmetic, first raised by G. Mills and J. Paris [MP]: the notions of $B\Sigma_{n+1}^*$, n-Ramsey and n-extendible are not only symbiotic but equivalent for $n \geq 1$. Since this problem has resisted a direct approach in [MP], there had been some speculation that n-extendible was a stronger notion.

The plan of this paper is as follows: in section §1, we review some of the basic notation and background results and we give a generalization of a result of L. Kirby [Ki], yielding an interesting characterization of n-extendible and n-Ramsey initial segments in terms of existence of certain ultrafilters. In section §2, we present a proof of the MAIN THEOREM. The proofs of the results use formalized recursion theoretic methods: the LOW BASIS THEOREM of Jockusch-Soare and ideas about the leftmost branch in a tree. Since KÖNIG LEMMA type argument often fail in set theory, the study of set theoretic analogues of the results in this paper may require different methods.

§1. Notation and Background Results

We use standard notation as in [Pa-1], [Pa-2] and [Cl-1]. Throughout, all models are understood to satisfy bounded induction, denoted by $I\Sigma_0$, but models may be of arbitrary cardinality unless otherwise indicated. Collection for Σ_n-formulas, denoted by $B\Sigma_n$, is the scheme

$$\Big[\phi(0,u) \wedge \forall x\Big(\phi(x,u) \rightarrow \phi(x+1,u)\Big)\Big] \rightarrow \forall x\phi(x,u)$$

where ϕ is a Σ_n-formula. A model M is said to be an *end extension* of I, or I to be an *initial segment* of M, if I is closed under multiplication and M adds no new element below an element in I. This is denoted by $I \subseteq_e M$. A subset of A of I is *M-coded* or *coded in M* if A is the trace on I of a Σ_o definable subset of M: for some Σ_o formula ϕ and parameter u.

$x \in A \leftrightarrow x \in I$ and $\models \phi(x,u)$.

The collection of M-coded subsets of I is denoted by $\mathbf{R}_M I$. A collection $\mathfrak{U}$ of M-coded subsets of I^n is said to be a *filter* on the M-coded subsets of I^n if

(1) $I^n \in \mathfrak{U}$, $\phi \notin \mathfrak{U}$.

(2) $A, B \in \mathfrak{U}$ implies that $A \cap B \in \mathfrak{U}$.

(3) $A \in \mathfrak{U}$ and $B \supseteq A$ implies that $B \in \mathfrak{U}$, provided that B is M-coded.

(4) $A \in \mathfrak{U}$ implies that A is unbounded in I^n.

Here a subset A of I^n is *unbounded in* I^n means that

$\{a \in I: (A)_a$ is unbounded in $I^{n-1}\}$ is unbounded in I

where

$$(A)_a = \{(a_2, \ldots, a_n): (a, a_2, \ldots, a_n) \in A\}.$$

Such a filter $\mathfrak{U}$ is a *complete ultrafilter* if additionally

(5) for any $X \subseteq I^{n+1}$ which is M-coded and any $a \in I$, if

$$(X)_i \in \mathfrak{U} \text{ for all } i < a$$

then

$$\bigcap_{i<a} (X)_i \in \mathfrak{U}.$$

To be precise, one should say rather an ultrafilter complete with respect to coded collections, but we prefer the more succinct term. When working with initial segments, one must consider the "boldface" version of proof theoretic schemes. Thus for instance, an initial segment I of M satisfied $B\Sigma_n^*$ in M if I satisfies the scheme of Σ_n-collection allowing M-coded subsets as parameters.

The definable version of Ramsey's Theorem

$$M \vec{\Delta}_n (M)^k_{<M}$$

means that for any Δ_n (with parameters) definable partition of the collection of all increasing k-tuples of elements of M into boundedly many classes, there is an unbounded homogeneous set.

In the proof of Theorem 1 (see [Cl-1] Theorem 11), strong use was made of the notion of a low-Δ_n set. A subset A of M is *low-Δ_n* if any Σ_1 or $\prod_1$ formula in the enriched language $\{+, \times\ 0, 1, <, A\}$ (whose A is interpreted in M by A) is equivalent in M to both a Σ_n and a $\prod_n$-formula.

An initial segment I of M is *n-Ramsey* (in M), a notion due to L. Kirby, if for any M-coded partition F mapping I^n into I-boundedly many classes, there exists an i for which $F^{-1}(i)$ is unbounded in I^n. An initial segment I of M is *1-extendible* (in M) if there exists an extension K of M satisfying:

(1) for any bounded formula ϕ and element u of M,

$$M \models \phi(u) \text{ iff } K \models \phi(u).$$

(2) $\forall a \in K - M \forall b \in I\ K \models a > b$.

(3) $\exists a \in K - M \forall b \in M - I\ K \models b > a$.

Such an extension is called an *I-extension* of M and denoted by $M \prec_I K$.

An initial segment I of M is *$(n+1)$-extendible* (in M) if there exists an I-extension K of M in which I is n-extendible. This notion is due to J. Paris.

Finally, K is said to be an *n-elementary extension* of M, denoted by $M \overset{\prec}{_n} K$, if for any Σ_n-formula ϕ and u belonging to M,

$$M \models \phi(u) \text{ iff } K \models \phi(u).$$

Recall the

Theorem 1. (Cl-1) For integers $n, k \geq 1$,

$$M \models B\Sigma_{n+k} \text{ iff } M \overrightarrow{\Delta_n} (M)^k_{<M}.$$

Implicit in the proof of Theorem 1 is the equivalence also with the combinatorial property:

for any Δ_n partition $F\colon M^k \to \{0, \ldots, a\}$ with $a \in M$, there exists $i \leq a$ for which $F^{-1}(i)$ is unbounded in M^k. As well, the "boldface" version holds, which will be used later:

An initial segment $I \subset_e M$ satisfies $B\Sigma^*_{n+1}$ if $I \overrightarrow{\Delta^*_1} (I)^n_{<I}$ iff for any coded partition $F\colon I^n \to \{0, \ldots, a\}$ there exists $i \leq a$ such that $F^{-1}(i)$ is unbounded in I^n.

The same proof yields the boldface version of Theorem 1: for $n, k \geq 1$ and any initial segment I of M

$$I \models B\Sigma^*_{n+k} \text{ iff } I \underset{\Delta^*_n}{\longrightarrow} (I)^k_{<I}.$$

J. Paris kindly pointed about the above remark and its immediate

Corollary 2. (Cl-1) ($n \geq 1$). For any initial segment I of M, I is n-Ramsey if and only if $I \models B\Sigma^*_{n+1}$.

Concerning partial elementary end extensions and collection, we have

Theorem 3. (Cl-3) ($n \geq 1$) If M satisfies $B\Sigma_{n+1}$ then there exists a Σ_n-complete Σ_n-ultrafilter on M.

Using this, we can obtain a cardinality free version of a result due to Kirby-Paris [KP].

Corollary 4. (Cl-3) ($n \geq 1$) M satisfies $B\Sigma_{n+1}$ if and only if M admits a proper $(n+1)$-elementary end extension. The MacDowell-Specker Theorem on proper elementary end extensions of models of Peano arithmetic is cardinality free, so this corollary was to be expected.

Theorem 5. (Cl-3) ($n \geq 1$) For countable models, $M \models B\Sigma_{n+1}$ if and only if there exists the following tower

$$M \prec_{n,e} M_1 \prec_{n-1,e} M_2 \prec_{n-2,e} \cdots \prec_{1,e} M_n.$$

For $n \geq 2$ and for countable models, $M \models B\Sigma_{n+1}$ if and only if there exists a proper n-elementary end extension K of M satisfying $B\Sigma_n$.

Compare with

Theorem. ([PR]) ($n, k \geq 1$) If there exists a Δ_n-based Δ_n-ultrafilter on M^k, then M admits a k tower of proper $(n+1)$-elementary end extensions

$$M \prec_{n+1,e} M_2 \prec_{n+1,e} \cdots \prec_{n+1,e} M_k,$$

Theorem 6. (Cl-2) ($k \geq 1$) For countable models, the following are equivalent:

(1) $M \models B\Sigma_{k+1}$.

(2) There exists a Δ_0-complete Δ_0-ultrafilter $\mathfrak{U}$ on M^k.

By using the technique of Theorem 3, perhaps the countability assumption may be removed. Theorem 6 suggests the following nice characterization of n-extendible cuts.

Theorem 7. For an initial segment I of a countable model M of $I\Sigma_0$, I is n-extendible if any only if there exists a complete ultrafilter $\mathfrak{U}$ on the coded subsets of I^n.

Remark. This is a generalization of a result by L. Kirby in [Ki]: an initial segment of a countable model is regular if and only if there is a complete ultrafilter on the collection of coded subsets of I.

Proof.

($\Rightarrow$)

For $1 \le i \le n$, let a_i be in "gap" of $M_i - M_{i-1}$ where $M_o = M$.

For $A \subseteq I^n$ M-coded, suppose θ is a bounded formula and $c \in M$ so that for all $x_1, \ldots, x_n$ in I,

$$(x_1, \ldots, x_n) \in A \leftrightarrow M \models \theta(x_1, \ldots, x_n, c).$$

Then let

$$A \in \mathfrak{U} \overset{\leftrightarrow}{\text{def}} M_n \models \theta\Big(a_n, a_{n-1}, \ldots a_1, c\Big).$$

The only non-obvious property to verify is non-triviality of $\mathfrak{U}$. Here is a picture when $n = 2$:

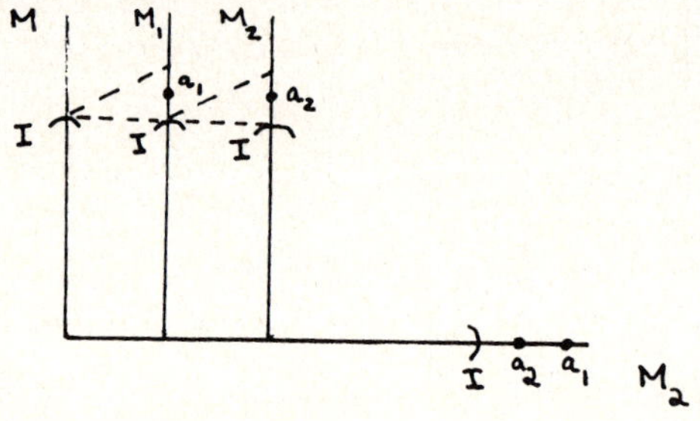

if A is not unbounded in I^2 then

$$\exists d \in I \forall x \in I, x \ge d \exists y \in I \forall z \ge y M \models \neg\theta(x, z, c).$$

For such a, d, and for all u in I,

$$\forall x \in I, x \ge d \exists y \in I\ M \models u \ge y \to \forall z \le u\Big(z \ge y \to \neg\theta(x, z, c)\Big).$$

As $M \models I\Sigma_o$, by overspill there exists $a \in M - I$ such that

$$\forall x \ge d, x \in I \exists y \in I\ M \models \forall z \le a\Big(z \ge y \to \neg\theta(x, z, c)\Big).$$

So as $M \prec_0 M_1$

$$\forall x \ge d,\ x \in I\ M \models \neg\theta(x, a_1, c).$$

So for all $e \in I$ $M_1 \models \forall x \le e\Big(d \le x \to \neg\theta(x, a_1 . c)\Big)$
and by overspill in M_1 there exists $e \in M_1 - M$ for which

$$M_1 \models \forall x \le e\Big(d \le x \to \neg\theta(x, a_1 . c)\Big)$$

and as $M_1 \prec_0 M_2$

$$M_2 \models \forall x \leq e\Big(d \leq x \rightarrow \neg\theta(x, a_1, c)\Big)$$

and so $M_2 \models \neg\theta(a_2, a_1, c)$.

But this is a contradiction.

In the general case if A is not unbounded in I^n, then

$$\exists d_1 \in I \forall x_1 \in I, x_1 \geq d_1 \ldots \exists d_n \in I \forall x_n \in I, x_n \geq d_n M \models \neg\theta(x_1, \ldots, x_n, c).$$

In a similar fashion, applying overspill successively to $M_1, M_2, \ldots, M_n$ yields the result.

$(\Leftarrow) n = 2$. Resolve the ultrafilter $\mathfrak{U}$ on M-coded subsets of I^2 into a product $\mathfrak{U}_2 \times \mathfrak{U}_1$, where $\mathfrak{U}_1$ is an ultrafilter on M-coded subsets of I and $\mathfrak{U}_2$ is an ultrafilter on K-coded subsets of I, where as in [Ki] the extension

$$K = \{f : f \text{ is M-coded mapping from } I \text{ into} M\}/\mathfrak{U}.$$

If A is an M-coded subset of I then let

$$A \in \mathfrak{U}_1 \overset{\leftrightarrow}{df} I \times A \in \mathfrak{U}.$$

Clearly. $\mathfrak{U}_1$ is a complete ultrafilter on the M-coded subsets of I, so as in [Ki], the model K is an I-extension of M.

If B is a K-coded subset of I, say

$$B = \{x \in I : K \models \theta(x, c)\},$$

where θ is bounded then let

$$B \in \mathfrak{U}_2 \overset{\leftrightarrow}{df} \{(x, i) \in I^2 :\ M \models \theta(x, c(i))\} \in \mathfrak{U}.$$

Notice that

$$\begin{aligned} B \in \mathfrak{U}_2 &\leftrightarrow \{x \in I : K \models \theta(x, c)\} \in \mathfrak{U}_2 \\ &\leftrightarrow \{x \in I : \{i \in I :\ M \models \theta(x, c(i))\} \in \mathfrak{U}_1\} \in \mathfrak{U}_2 \\ &\leftrightarrow \{(x, i) \in I :\ M \models \theta(x, c(i))\} \in \mathfrak{U}. \end{aligned}$$

Fact. For an M-coded subset A of I, say

$$A = \{x \in I :\ M \models \theta(x, c)\}$$

then

$$A \in \mathfrak{U}_1 \text{ iff } K \models \theta(id, c).$$

Proof. $K \models \theta(id, c)$ iff $\{i \in I : M \models \theta(id(i), c)\} \in \mathfrak{U}_1$ iff $A \in \mathfrak{U}_1$. ∎

Thus we have "resolved" $\mathfrak{U}$ into $\mathfrak{U}_2 \times \mathfrak{U}_1$: for any M-coded subset A of I^2, say

$$A = \{(x,i) \in I^2 \colon M \models \theta(x,i,c)\}$$

we have

$$A \in \mathfrak{U} \text{ iff } \{(x,i) \in I^2 \colon M \models \theta(x,i,c)\} \in \mathfrak{U}$$
$$\text{iff } \{x \in I \colon K \models \theta(x, id, c)\} \in \mathfrak{U}_2 \text{ by def of } \mathfrak{U}_2$$
$$\text{iff } \{x \in I \colon \{i \in I \colon M \models \theta(x, id(i), c)\} \in \mathfrak{U}_1\} \in \mathfrak{U}_2$$

So indeed $\mathfrak{U} = \mathfrak{U}_2 \times \mathfrak{U}_1$. ∎

Claim. $\mathfrak{U}_2$ is a complete ultrafilter on K-coded subsets of I.

Proof. Check completeness. Suppose X is a subset of I^2 coded in K, say

$$(x,y) \in X \leftrightarrow (x,y) \in I^2 \text{ and } K \models \theta(x,y,c).$$

Suppose $a \in I$ and $\forall j < a(X)_j \in \mathfrak{U}_2$. Thus

$$\forall j < a\{x \in I \colon K \models \theta(x,j,c)\} \in \mathfrak{U}_2.$$

$$\forall j < a\{x \in I \colon \{i \in I \colon M \models \theta(x,y,c(i))\} \in \mathfrak{U}_1\} \in \mathfrak{U}_2.$$
$$\forall j < a\{(x,i) \in I^2 \colon M \models \theta(x,j,c(i))\} \in \mathfrak{U}.$$

By completeness of $\mathfrak{U}$,

$$\{(x,i) \in I^2 \colon M \models \forall j < a\theta(x,j,c(i))\} \in \mathfrak{U}.$$

By the resolution of $\mathfrak{U}$ into $\mathfrak{U}_2 \times \mathfrak{U}_1$

$$\{x \in I \colon \{i \in I \colon M \models \forall j < a\theta(x,j,c(i))\} \in \mathfrak{U}_1\} \in \mathfrak{U}_2$$

so by Lôs' Lemma,

$$\{x \in I \colon K \models \forall j < a\theta(x,y,c)\} \in \mathfrak{U}_2$$

which means

$$\bigcap_{j<a} (X)_j \in \mathfrak{U}_2$$

∎

The general case is analogous. Let $M_o = M$. For $i = 0, \ldots, n-1$ let

$$M_{i+1} = \{f \colon f \text{ is an } M_i - \text{coded mapping from } I \text{ into } M_i\}/\mathfrak{U}_i.$$

Then M_{i+1} is an I-extension of M_i.

For an M-coded subset A_1 of I let

$$A_1 \in \mathfrak{U}_1 \leftrightarrow I^{n-1} \times A_1 \in \mathfrak{U}.$$

For an M_1-coded subset A_2 of I, say

$$x \in A_2 \longrightarrow x \in I \text{ and } M_1 \models \theta(x, c).$$

Then let

$$A_2 \in \mathfrak{U}_2 \overset{\leftrightarrow}{df} I^{n-2} \times \{(x,i) \in I^2 \colon M \models \theta(x, c(i))\} \in \mathfrak{U}$$

$$\vdots$$

For an M_{n-1}-coded subset A_n of I, say

$$x \in A_n \leftrightarrow x \in I \text{ and } M_{n-1} \models \theta(x, c).$$

Then let $A_n \in \mathfrak{U}_n$

$$\overset{\leftrightarrow}{df}$$

$$\{(x_1, i_2, \ldots, i_n) \in I^n \colon M \models \theta(x, c(i_2)(i_3)\ldots(i_n))\} \in \mathfrak{U}.$$

An inductive proof shows that $\mathfrak{U}$ can be resolved into $\mathfrak{U}_n \times \ldots \times \mathfrak{U}_1$ and further that for $i = 1, \ldots, n$. $\mathfrak{U}_i$ is a complete ultrafilter on the M_{i-1}-coded subsets of I. This completes the proof of Theorem 7. ∎

The referee has remarked that this may be much more easily proved by taking

$$M_k = \{f \colon I^n \to M | f(\vec{x}) \text{ depends only on } x_{n-k+1}, \ldots, x_n\}/\mathfrak{U}.$$

We have retained the original proof, feeling that the "resolution" of an ultrafilter may find other applications.

It is not hard to see

Theorem 8. For an initial segment I of a countable model M of bounded induction, I is n-Ramsey if and only if there is a complete ultrafilter $\mathfrak{U}$ on the collection Δ_n^* subsets of I, the latter meaning Δ_n with an M-coded parameter.

Proof.

($\Rightarrow$) By a theorem in [Pa-1], $I \models B\Sigma_{n+1}^*$ (an alternate proof can be given using the technique of proof of Theorem 11 of [Cl-1]). Now an obvious variant of the argument in [Ki],[Ka] or [Kr] suffices to obtain the ultrafilter. In the interests of self-containment, we repeat the well-known argument. Let

$$\{f_i \colon i \in \mathbf{N}\}$$

be an infinitely repetitive list of all Δ_n^* subsets of I. Construct $X_0 \supset X_1 \supset X_2 \supset \cdots$ so that each X_i is M-coded and unbounded in I, subject to:

$$X_o = I$$

$X_o = I$ and given X_m, if $f_m|X_m$ is unbounded in I, then set $X_{m+1} = X_m$; otherwise if the range of f_m on X_m is bounded by $a \in I$ then choose some $i < A$ for which

$$f_m^{-1}(i) \cap X_m \text{ is bounded in } I$$

and set

$$X_{m+1} = f_m^{-1}(i).$$

Such an i exists by $B\Sigma_{n+1}^*$.

Now set

$$\mathfrak{U} = \{A \subseteq I\colon\ A \text{ is } \Delta_n^* \text{ and } \exists m \in \mathbb{N}(X_m \subseteq A)\}.$$

Clearly, $\mathfrak{U}$ is a complete ultrafilter on the Δ_n^* subsets of I.

($\Leftarrow$) Suppose there exists a complete ultrafilter $\mathfrak{U}$ on the Δ_n^* subsets of I. Let

$$F\colon\ I \to \{0,\dots,a\} \text{ with } a \in I$$

be a Δ_n^*-partition. By completeness of $\mathfrak{U}$

$$\exists i \leq a\left[F^{-1}(i) \text{ is unbounded in } I\right],$$

thus we have

$$(*) \qquad I \overrightarrow{\Delta_n^*} (I)^1_{<I}.$$

Recall Theorem 4 of [Cl-1] which states that

$$I \overrightarrow{\Delta_n} (2)^1_{<I} \text{ iff } I \models I\Sigma_n.$$

The boldface version of this result holds by the same proof, so by $(*)$ I satisfies $I\Sigma_n^*$.

Now if an instance of $B\Sigma_{n+1}^*$ fails in I, say A is an M-coded subset of I and for some Σ_{n-1}^* formula ϕ

$$(I, A) \models \forall i < a\exists x\forall y\phi(i, x, y, A).$$

but $(I, A) \models \forall b\exists i < a\forall x < b\exists y\neg\phi(i, x, y, A)$.

By $I\Sigma_{n-1}^*$

$$(I, A) \models \forall b\exists i < a\forall x < b\exists y\left[\neg\phi(i, x, y, A) \wedge Ay' < y\phi(i, x, y, A)\right]$$

Again using $I\Sigma^*_{n-1}$ and the Hay-Manaster-Rosenstein trick (see [Cl-1] Lemma A after Theorem 8) there exists a $\prod^*_{n-1}$ formula ψ such that

$$(I, A) \models \forall i, x \left[\exists y \psi(i, x, y, A) \leftrightarrow \exists y(\neg\phi(i, x, y, A) \wedge \forall y' < y\phi(i, x, y, A))\right]$$

Thus $(I, A) \models \forall b \exists i < a \forall x < b \exists! y \psi(i, x, y, A)$. Using $I\Sigma^*_n$ and the same trick, there exists a Σ^*_n formula θ with

$$(I, A) \models \forall b \exists! i < a\theta(i, b, A).$$

Now define a Δ^*_n function

$$F\colon I \to \{0, \ldots, a-1\}$$

by

$$F(b) = i \leftrightarrow \theta(i, b, A)$$

$$F(b) \neq i \leftrightarrow \exists j(j \neq i \text{ and } \theta(j, b, A)).$$

But then an application of

$$I \overrightarrow{\Delta^*_n} (I)^1_{<I}$$

yields an $i_o < a$ for which $F^{-1}(i_o)$ is unbounded in I, hence

$$(I, A) \models \forall x \exists y \neg\phi(i_o, x, y, A)$$

contradicting the hypothesis that

$$(I, A) \models \forall i < a \exists x \forall y \phi(i, x, y, A).$$

∎

Remark. Realizing that any cut I closed under multiplication satisfies the following:

$$(I, \text{ coded subsets of } I) \models \Delta_1 \text{ comprehension.}$$

a more suggestive rendering of Theorem 7 is the result that I is n-extendible if and only if there exists a complete ultrafilter $\mathfrak{U}$ on the collection of Δ^*_1 subsets of I^n. At present, we are unable to directly construct a complete ultrafilter on the collection of Δ^*_1 subsets of I^n from a complete ultrafilter on the collection of Δ^*_n subsets of I, so this characterization of n-Ramsey and n-extendible cuts is not apparently useful in proving the equivalence of the two notions. However, by a very different proof, we do have the

MAIN THEOREM. For an initial segment I of a countable model M of bounded induction, I is n-Ramsey if and only if I is n-extendible.

Before going into the details of the MAIN THEOREM, we begin by some intuitive discussion. The most direct attempt at constructing a complete ultrafilter $\mathfrak{U}$ on the M-coded subsets of I^2 might proceed as follows: Enumerate all functions $\{g_n : n \in \mathbf{N}\}$ which are

M-coded maps from I^2 having range bounded in I. Let $X_0 = I^2$. Given X_n, consider $g_n|X_n$.

Pseudo-claim. If a is a bound on the range of g_n, then there exists $i_0 < a$ for which $g_n^{-1}(i_0) \cap X_n$ is unbounded in I^2.

Now, assuming the pseudo-claim, let X_{n+1} be $g_n^{-1}(i_0) \cap X_n$. Define the desired ultrafilter by

$$\mathfrak{U} = \{A \subset I^2\colon\ A \text{ is } M - \text{coded and } \exists n \in \mathbf{N}(X_n \subseteq A)\}.$$

Then clearly $\mathfrak{U}$ is a complete ultrafilter on the M-coded subsets of I^2.

However, G. Mills and J. Paris [MP] have shown that it requires in general $B\Sigma_4^*$ to obtain: For any M-coded unbounded subset X of I^2 and any M-coded partition $g\colon X \to \{0,\ldots,a\}$ with a belonging to I, there exists $i \le a$ with $g^{-1}(i)$ unbounded in I^2.

In fact,

Theorem 9. ([MP]) ($k \ge 1$) The following are equivalent:

(1) $M \models B\Sigma_{2k}$.

(2) In partitioning a Δ_1 set unbounded in M^k in a Δ_1 fashion into boundedly many pieces, one piece contains a set unbounded in M^k.

Thus $B\Sigma_4^*$ instead of $B\Sigma_3^*$ is necessary to prove the pseudo-claim and hence the most natural attempt to construct an appropriate ultrafilter is blocked. This observation led in fact to a feeling that 2-Ramsey and 2-extendible were symbiotic yet distinct notions. This is the reason for our rather roundabout proof of the MAIN THEOREM.

The crucial new idea in proving for instance that any 2-Ramsey cut I is 2-extendible is the following:

(∗) for any unbounded subset X of I coded in M and any M-coded partition $F\colon I \times X \to \{0,\ldots,a\}$ with $a \in I$, there exists an M-*coded* unbounded subset Y of X and a *not necessarily coded* unbounded subset Z of I and $i_0 \le a$ such that

$$\forall x \in Z \exists m \in I \forall n \in I[n \ge m \text{ and } n \in Y \to F(x,n) = i_0].$$

Using this property (∗), an appropriate ultrafilter $\mathfrak{U}$ may be constructed for which I is 1-Ramsey, hence 1-extendible, in

$$K = \{f\colon f \text{ is an } M - \text{coded map from } I \text{ into } M\}/\mathfrak{U}.$$

The proof of (∗), which uses $B\Sigma_3^*$, involves a leftmost branch in a modification of the Erdös-Rado tree associated with the partition. The idea of investigating the property (∗) was imminent upon hearing the yet unpublished result of J. Paris:

For $I \subset_e M \models I\Sigma_0, I$ is strong if and only if for any coded partition $F\colon I^3 \to \{0,\ldots,a\}$ with $a \in I$, there exist coded unbounded subsets X, Y, Z of I such that

$$F|X \times Y \times Z \text{ is constant.}$$

Proof of MAIN THEOREM.

($\Leftarrow$) This direction is due to L. Kirby and J. Paris and found in [Pa-1]. Using Theorem 7, the following is another proof: Let

$$F: I^n \to \{0, \ldots, a\} \text{ with } a \in I$$

be an M-coded partition. By n-extendibility let $\mathfrak{U}$ be a complete ultrafilter on the M-coded subsets of I^n. Then

$$\bigcup_{i<a} F^{-1}(i) = I^n \in \mathfrak{U}$$

so by completeness of $\mathfrak{U}$, there exists $i_0 \leq a$ with

$$F^{-1}(i_0) \in \mathfrak{U}.$$

But $\mathfrak{U}$ contains only sets which are unbounded in I^n, so I is n-Ramsey.

($\Rightarrow$) Now suppose we wish to show that any 2-Ramsey initial segment is 2-extendible. This will be obtained if we can construct a complete ultrafilter $\mathfrak{U}$ on the M-coded subsets of I which satisfies the property:

($*$) for any unbounded M-coded subset X of I and any M-coded partition $F: I \times X \to \{0, \ldots, a\}$ with $a \in I$, there exists an unbounded M-*coded* subset Y of X and an unbounded *but not necessarily coded* subset Z of I and $I_0 \leq a$ such that

$$\forall x \in Z \exists m \in I \forall n \in I \Big[n \geq m \text{ and } n \in Y \to F(x, n) = i_0 \Big].$$

(The idea is that $F|Z \times Y$ is almost constant.)

Let $\mathfrak{U} = \{A \subseteq I\colon A \text{ is } M\text{-coded and } \exists n \in \mathrm{N}(X_n \subset A)\}$, where the sequence $X_0 \supset X_1 \supset X_2 \supset \cdots$ is defined as follows: Using countability of M, let $\{g_n : n \in \mathrm{N}\}$ enumerate infinitely often all M-coded partitions of I^2 having range bounded in I. Let $X_0 = I$. Given X_n, suppose that

$$g_n|I \times X_n : I \times X_n \to \{0, \ldots, a\}.$$

By the property ($*$) let $X_{n+1} \subset X_n$ be M-coded and unbounded for which there exists $i_0 \leq a$ and an unbounded but not necessarily M-coded set Z satisfying

$$\forall x \in Z \exists r \in I \forall s \in I \Big[s \geq r \text{ and } S \in X_{n+1} \to g_n(x, s) = i_0 \Big].$$

This determines X_{n-1} and terminates the description of $\mathfrak{U}$. Clearly $\mathfrak{U}$ is a complete ultrafilter on M-coded subsets of I.

Now letting

$$K = \{f : f \text{ is an } M - \text{coded mapping from } I \text{ into } M\}/\mathfrak{U}$$

we claim that I is 1-Ramsey in K.

This latter suffices, since it is clear that 1-Ramsey and 1-extendible are synonomous. For I to be 1-Ramsey in K, consider any K-coded partition $G: I \to \{0, \ldots, a\}$ with $a \in I$. So for $x, y \in I$

$$G(x) = y \leftrightarrow K \models \theta(x, y, c)$$

where θ is a Σ_o formula and by Łôš' Lemma,

$$G(x) = y \leftrightarrow \{i \in I: M \models \theta(x, y, c(i))\} \in \mathfrak{U}.$$

Define the M-coded partition

$$F: I^2 \to \{0, \ldots, a+1\}$$

by

$$F(x, i) = y \leftrightarrow M \models \theta(x, y, c(i))$$

and

$$F(x, i) = a + 1 \text{ otherwise.}$$

By construction of $\mathfrak{U}$, there exists an unbounded M*-coded* subset Y of I belonging to $\mathfrak{U}$ and an unbounded *but not necessarily coded* subset Z of I and $i_0 \leq a + 1$ such that

$$\forall x \in Z \exists m \in I \forall n \in I \Big[n \geq m \text{ and } n \in Y \to F(x, n) = i_0 \Big].$$

Fix any $x \in Z$ and let $m \in I$ be such that

$$\forall x \in I \Big[n \geq m \text{ and } n \in Y \to F(x, n) = i_o \Big].$$

Then by completeness of $\mathfrak{U}$,

$$\{i \in Y: i \geq m\} \in \mathfrak{U}$$

so

$$\{i \in I: M \models \theta(x, i_o, c(i)\} \in \mathfrak{U}$$

and

$$K \models \theta(x, i_o, c)$$

so

$$F(x) = i_0.$$

By hypothesis that the range of F is contained in $\{0, \ldots, a\}$, it follows that $i_0 \leq a$ and we have shown the existence of an unbounded subset Y of I on which F is constant. Thus I is 1-Ramsey in K, hence 1-extendible in K and hence I is 2-extendible in M.

Our idea for proving the property (*) is rather roundabout.

Suppose the partition

$$F: I^2 \to \{0, \ldots, a\}$$

is coded in M. Let $T \subseteq I^{<I}$ be the usual M-coded Erdös-Rado tree corresponding to the partition F (see Theorem 11 of [Cl-1] for definition) and let $Y = \{\sigma \in I^{<I} : \sigma$ is the leftmost or lexicographic least node in T at level lh $(\sigma)\}$. Clearly Y is recursive in the code of the partition F and so coded in M (recall that (I, coded subsets of I) $\models \Delta_1$ comprehension). Also clearly, Y is unbounded. Let Z be the leftmost unbounded branch in T ($B\Sigma_3^*$ is sufficient to construct such a branch). The set Z is *pseudo-homogeneous* in the sense that $\forall x, y, z \in Z(x < y < z \rightarrow F(x,y) = F(x,z))$. Now plainly,

$$\forall x \in Z \exists y \in I \forall z, z' \in I \left[y < z, z' \text{ and } z, z' \in Y \rightarrow F(x,z) = F(x,z') \right]$$

By pseudo-homogeneity of Z, the partition F induces a partition

$$H\colon Z \rightarrow \{0, \ldots, a\}.$$

defined by

$$H(x) = F(x,y)$$

where y is any element of Z larger than x.

Now

(§) If one can extract an unbounded homogeneous set Z_o from Z, or in other words if $\exists i_0 \leq a \left[H^{-1}(i_0) \text{ is unbounded in } I \right]$, *then*

(§§) $\forall x \in Z_o \exists y \in I \forall z \geq y \left[z \in Y \rightarrow F(x,z) = i_0 \right]$. The property (§§) is exactly the desired condition used in constructing the ultrafilter $\mathfrak{U}$. By a formalized basis theorem, working in $B\Sigma_3^*$, the set Z can be taken to be Δ_3^*. If property (§) fails then

$$(I, \mathbf{R}_M I) \models \forall i \leq a \exists y \forall z, z' \left[y \leq z < z' \text{ and } z, z' \in Z \rightarrow F(z,z') \neq i \right].$$

We wish to obtain a contradiction by deducing

$$(I, \mathbf{R}_M I) \models \exists b \forall i \leq a \exists y < b \forall z, z' \left[\cdots \right].$$

However, the set Z is Δ_3^* and the apparent amount of collection necessary would seem to be $B\Sigma_4^*$. If Z were low $-\Delta_3^*$ then indeed $B\Sigma_3^*$ would suffice. However, an easy argument ([Cl-4] Proposition 4) produces a recursive finite branching infinite tree $T \subseteq \omega^{<\omega}$ whose leftmost infinite branch is Turing equivalent with $0''$, so in general the leftmost unbounded branch of a recursive tree will *not* be low Δ_3^*. To push through this argument, we modify the original Erdös-Rado tree T into a Δ_1^*-tree $\tilde{T}$ so that the leftmost unbounded branch f in $\tilde{T}$ will be Δ_3^* and will provide a low-Δ_3^* branch g in T. Furthermore the leftmodes nodes at each level of $\tilde{T}$ will provide a Δ_1^* set Y as in the previous discussion, so as to permit the argument to go through.

DETAILS OF THE TREE

Let T be the usual Erdös-Rado tree associated with the M-coded partition

$$F:[I]^2 \to \{0,\ldots,a\}$$

with $a \in I$. Then T is an M-coded tree contained in $I^{<I}$ which is finite branching in the sense of I. Define the M-coded tree $\tilde{T}$ inductively:

$$\langle\langle\ \ \rangle\rangle \in \tilde{T} \text{ and } (0) \in \tilde{T}.$$

Case 1. $\sigma \in \tilde{T}$ and $lh(\sigma) = 2n+2$.

At this stage we work on controlling the Σ_1-consequences or jump. Let $\sigma_E = \langle\sigma(0),\sigma(2),\ldots,\sigma(2n)\rangle$

$$\sigma_0 = \langle\sigma(1),\sigma(3),\ldots,\sigma(2n+1)\rangle$$

Then let

$$\sigma \frown (0) \in \tilde{T} \leftrightarrow \neg\exists y < 2n+2\Big[T\Big(\sigma_E,n,n,y\Big)\Big]$$

and

$$\forall ezn\Big[\sigma_0(e) = 0 \to \neg\exists y < 2n+2\Big[T\Big(\sigma_E,n,n,y\Big)\Big]\Big].$$

$$\sigma \frown (1) \in \tilde{T} \leftrightarrow \forall e < n\Big[\sigma_0(e) = 0 \to \neg\exists y < 2n+2\Big[T\Big(\sigma_E,n,n,y\Big)\Big]\Big]$$

Here $T(r,n,n,y)$ is Kleene's T-predicate whose interpretation is: The nth Turing machine given argument n and oracle τ halts in $\leq y$ steps.

Case 2. $\sigma \in \tilde{T}$ and $lh(\sigma) = 2n+3$.

At this stage we ensure that the "even part" of any node in $\tilde{T}$ is a node in T.

Let $\sigma_E = \langle\sigma(0),\ldots,\sigma(2n+2)\rangle$ and $\sigma_0 = \langle\sigma(1),\ldots,\sigma(2n+1)\rangle$.

For any $x \in I$, let

$$\sigma \frown \langle x\rangle \in \tilde{T} \leftrightarrow \sigma_E \frown \langle x\rangle \in T.$$

This concludes the construction of $\tilde{T}$.

If f is any unbounded branch of $\tilde{T}$ then it is clear that for all $e \in I$,

$$f_0(e) = 0 \to \neg\exists y T(f_E,e,e,y),$$

where in an analogous fashion f_E resp. f_0 is the even resp. odd part of f.

Now let f be the leftmost unbounded branch of $\tilde{T}$. Then f is Δ_3^* and for all $e \in I$,

$$f_0(e) = 0 \leftrightarrow \neg\exists y T(f_E,e,e,y).$$

Thus the jump f'_E of f_E is Δ_1 in f, hence Δ_3^*. Thus f_E is low Δ_3^*. (The formalization of this argument would follow the lines of that in [Cl-1].)

Let $\tilde{Y}$ be the collection of leftmost nodes in $\tilde{T}$:

$$\tilde{Y} = \{\sigma \in \tilde{T}: \sigma \text{ is leftmost node at level } lh(\sigma)\}.$$

Now let

$$Y = \{\sigma_E : \sigma \in \tilde{Y}\}.$$

Clearly Y is Δ_1 in $\tilde{Y}$, hence Δ_1^* and thus coded in M.

Let Z be the image of f_E, so as f_E is a strictly increasing function, it follows that Z is low-Δ_3^*. Then clearly the following holds in I:

$$(\$)\forall x \in Z \exists y \in I \forall z, z' \in I\Big[y \leq z \leq z' \text{ and } z, z' \in Y$$

$$\rightarrow F(x,z) = F(x.z')\Big].$$

Since Z is low-Δ_3^*, by the preceding discussion, there exists $i_0 \leq a$ and an unbounded subset Z_o of Z such that the following holds in I:

$$\forall x \in Z_o \exists y \in I \forall z\Big[y \leq z \text{ and } z \in Y \rightarrow F(x,z) = i_0\Big].$$

This is the desired result $(*)$. Before going on, notice that Z_o is low-Δ_3^* as well.

Notice that $B\Sigma_3^*$ was indeed invoked to prove ($):

$$I \models \forall \text{ levels } n \exists m \forall \sigma\Big[lh(\sigma) \geq m \text{ and } \sigma \in \tilde{T} \rightarrow$$

$$\rightarrow \sigma \text{ is to the right or above } f|m\Big].$$

Indeed, for if the above statement were not true, then using a formalized basis theorem ([Cl-1]) one could show that f was not the leftmost unbounded branch in $\tilde{T}$. ∎

Now suppose $n \geq 2$ and that n-Ramsey and n-extendible are equivalent. Suppose that I is $(n+1)$-Ramsey in M.

Claim: For any unbounded M-coded subset X of I and any M-coded partition

$$F: I^n \times X \rightarrow \{0.\ldots,a\}$$

with $a \in I$, there exists an unbounded M-*coded* subset Y of X and an unbounded low Δ_3^*-subset Z_o of I and $i_o \leq a$ such that

$$\forall x_1,\ldots,x_n \in Z_o \exists y \in I \forall z \in I\Big[x \geq y \text{ and } z \in Y \rightarrow F(x_1,\ldots,x_n,z) = i_o\Big].$$

Proof of claim. Analogous to the case $n = 1$, define the *modified* Erdös-Rado tree $\tilde{T}$ corresponding to the partition F. As before, obtain an unbounded low-Δ_3^* subset Z of I such that

$$\forall x_1, \ldots, x_n \in Z \exists y \in I \forall z, z' \in I \Big[y < z, z' \to F(x_1, \ldots, x_n, z) = F(x_1, \ldots, x_n, z') \Big].$$

Let Y be as before

Let $H: [Z]^n \to \{0, \ldots, a\}$ be the induced partition defined by

$$H(x_1, \ldots, x_n) = F(x_1, \ldots, x_n, z)$$

for the least z belonging to Z and greater than $x_1, \ldots, x_n$. The partition H is Δ_1 in Z and thus low-Δ_3^*.

Now as I is $(n+1)$-Ramsey, $I \models B\Sigma_{n+2}^*$ so the proof of Corollary 17 of [Cl-1], we have

$$I \underset{low - \Delta_3^*}{\to} (I)^n_{<I}.$$

Thus there exists an unbounded subset Z_o of Z (Z_o can be taken to be low-Δ_{n+2}^*) and $i_0 \leq a$ for which $H''[Z_o]^n = \{i_o\}$. Thus an argument similar to that establishing ($) yields

$$\forall x_1, \ldots, x_n \in Z_o \exists y \in I \forall z \in I \Big[y \leq z \text{ and } z \in Y \to$$

$$\to F(x_1, \ldots, x_n, z) = i_o \Big].$$

This completes the proof of the claim. ∎

Now construct an ultrafilter $\mathfrak{U}$ on M-coded subsets of I as follows. By countability of M, let $\{g_m : m \in \mathrm{N}\}$ enumerate infinitely often all M-coded partitions of I^{n+1} having range bounded in I. Let $X_o = I$. Given X_m, suppose that

$$g_m | I^n \times X_m : I^m \times X_m \to \{0, \ldots, a\}.$$

By the property $(*)$, let $X_{m+1} \subset X_m$ be M-*coded* and unbounded and for which there exists $i_o \leq a$ and an unbounded but not necessarily M-coded set Z_o satisfying

$$\forall x_1, \ldots, x_n \in Z_o \exists y \in I \forall z \in I \Big[y \leq z \text{ and } z \in X_{m+1} \to$$

$$\to g_m(x_1, \ldots, x_n, z) = i_o \Big].$$

Let $\mathfrak{U} = \{A \subseteq I : A \text{ is } M\text{-coded and } \exists m \in \mathrm{N}(X_m \subseteq A)\}$. Clearly $\mathfrak{U}$ is a complete ultrafilter on M-coded subsets of I. Let

$$K = \{f : f \text{ is an } M - \text{coded mapping from } I \text{ into } M\}/\mathfrak{U}.$$

Then

$$M \prec_I K$$

and we claim that I is n-Ramsey, hence by the induction hypothesis n-extendible, in K. Consider any K-coded partition $G: I^n \to \{0, \ldots, a\}$ with $a \in I$.
So for $x_1, \ldots, x_n, y$ in I

$$G(x_1, \ldots, x_n) = y \leftrightarrow K \models \theta(x_1, \ldots, x_n, y, c(i))$$

where θ is a Σ_o-formula and by Lôs' Lemma,

$$G(x_1, \ldots, x_n) = y \leftrightarrow \{i \in I: M \models \theta(x_1, \ldots, x_n, y, c)\} \in \mathfrak{U}.$$

Define the M-coded partition

$$F: I^{n+1} \to \{0, \ldots, a+1\}$$

by

$$F(x_1, \ldots, x_n, i) = y \leftrightarrow M \models \theta(x_1, \ldots, x_n, y, c(i))$$

and

$$F(x_1, \ldots, x_n, i) = a + 1 \text{ otherwise.}$$

By construction of $\mathfrak{U}$, there exists an unbounded *M-coded* subset Y of I belonging to $\mathfrak{U}$ and an unbounded *but not necessarily M-coded* subset Z of I and $i_o \leq a + 1$ such that

$$\forall x_1, \ldots, x_n \in Z \exists y \in I \forall z \in I \Big[z \geq y \text{ and } z \in Y \to$$

$$\to F(x_1, \ldots, x_n, z) = i_o \Big]$$

It follows that

$$G^{-1}(i_o) \text{ is unbounded in } I^n$$

and thus that I is n-Ramsey in K. By the induction hypothesis, I is n-extendible in K, so that I is $(n+1)$-extendible in M.

This concludes the proof of the MAIN THEOREM. ∎

We have the following Corollary which answers a question raised in [Cl-2].

Theorem 10. $(n, k \geq 1)$ For countable models M, the following are equivalent:

(1) $M \models B\Sigma_{n+k}$.

(2) There exists a Δ_n-complete Δ_n-ultrafilter $\mathfrak{U}$ on M^k.

Proof. By Theorem 2 of [Pa-2], there exists an end extension K of M satisfying $B\Sigma_{n+k}$ in which $M \models B\Sigma^*_{n+k}$. By the MAIN THEOREM, M is $(n+k-1)$-extendible in K, so by Theorem 7, there is a complete ultrafilter $\mathfrak{U}$ on the K-coded subsets of M^{n+k-1}. By the LIMIT LEMMA ([Cl-1]), a Δ_n-subset A of M^k can be approximated by a Δ_o-function F in the sense that for all $\vec{x} \in M^k$,

$$\vec{x} \in A \leftrightarrow \lim_{s_1} \cdots \lim_{s_{n-1}} F(\vec{x}, s_1, \ldots, s_{n-1}) = 1$$

$$\vec{x} \notin A \leftrightarrow \lim_{s_1} \cdots \lim_{s_{n-1}} F(\vec{x}, s_1, \ldots, s_{n-1}) = 0$$

Hence define the desired ultrafilter $\mathfrak{U}^*$ on Δ_n-subsets of M^k by

$$A \in \mathfrak{U}^*$$

$$\underset{df}{\leftrightarrow}$$

$$\{(x_1, \ldots, x_k, s_1, \ldots, s_{n-1}) \colon F(x_1, \ldots, x_k, s_1, \ldots, s_{n-1}) = 1\} \in \mathfrak{U}.$$

Recalling that any Δ_o-subset of M^{n+k-1} is K-coded, it is easy to check that $\mathfrak{U}^*$ is the desired ultrafilter. ∎

Remark. Using the techniques of this paper, we have recently provided a solution to a problem asked on p. 261 of [Pa-2]: an initial segment I of M satisfies $I\Sigma_n^*$ if and only if I is $(n - \frac{1}{2})$-extendible in M. We hope to report the details elsewhere.

REFERENCES

[B] S. Buss, *Bounded Arithmetic*, Preliminary Draft of Ph.D. Dissertation, Princeton University, October 1984.

[Cl-1] P. Clote, "Partitions in arithmetic," Springer Lecture Notes in Mathematics 1130, *Methods in Mathematical Logic*, Proceedings Caracas 1983, ed. C.A. DiPrisco (1985), pp. 32-68.

[Cl-2] P. Clote, "Applications of the low basis theorem in arithmetic," Springer Lecture Notes in Mathematics 1141, *Recursion Theory Week*, Proceedings Oberwolfach 1984, ed. H.D. Ebbinghaus, G.H. Müller, G.E. Sacks (1985), pp. 65-88.

[Cl-3] P. Clote, "A note on the MacDowell-Specker theorem," submitted to *Fundamenta Mathematicae*.

[Cl-4] P. Clote, "A note on the leftmost infinite branch of a recursive tree," *Open Days in Model Theory and Set Theory*, Proceedings of a Conference held in September 1981 at Jadwisin, near Warsaw, Poland, Ed. W. Guzicki, W. Marek, A. Pelc, and C. Rauszer, pp. 93-102, published by Leeds University (1984).

[Jo] C.G. Jockusch, Jr., "Ramsey's theorem and recursion theory," *Journal of Symbolic Logic*, **Vol. 37** (1972), pp. 268-280.

[Ka] M. Kaufmann, "On existence of Σ_n-end extensions," *Logic Year 1979-80*, Springer Lecture Notes in Mathematics, **Vol. 859** (1980), pp. 92-103.

[Ki] L. Kirby, "Flipping properties in arithmetic," *Journal of Symbolic Logic*, **Vol. 47** (1982), pp. 416-422.

[KiPa] L. Kirby and J. Paris, "Σ_n-collection schemas in arithmetic," *Logic Colloquium '77*, North Holland, Amsterdam (1978), pp. 199-209.

[Kr] E. Kranakis, "Partition and reflection properties of admissible ordinals," *Annals of Mathematical Logic*, **Vol. 22** (1982), pp. 213-242.

[LöWa] M. Löb and S. Wainer, "Hierarchies of number theoretic functions-Part I," *Archiv für Mathematische Logik und Grundlagenforschung* **Vol. 13** (1970), pp. 39-51, Part II in pp. 97-113.

[MP] G. Mills and J. Paris, "Regularity in models of arithmetic," *Journal of Symbolic Logic*, **Vol. 49** (1984), pp. 272-280.

[Parikh] R. Parikh, "Existence and Feasibility in Arithmetic," *Journal of Symbolic Logic*, **Vol. 36** (1971), pp. 494-508.

[Pa-1] J. Paris, "A hierarchy of cuts and models of arithmetic," *Model theory of algebra and arithmetic* (Proceedings, Karpacz, 1979), Lecture Notes in Mathematics, vol. 834, Springer-Verlag (1980), pp. 312-337.

[Pa-2] J. Paris, "Some conservation results for fragments of arithmetic," *Model theory and arithmetic*, Springer Lecture Notes in Mathematics, vol. 890, Springer-Verlag (1980), pp. 251-262.

[Pars] C. Parsons, Abstract in *Notices of the American Mathematical Society*, **Vol. 13** (1966), pp. 857-858.

[Pi] R. Pino, "$\prod_n$-collection, indicatrices et ultrafiltres dèfinissables," These de 3° cycle, Universitè Paris VII (1983).

[PR] R. Pino and J.-P. Ressayre, "Definable Ultrafilters and Elementary End Extensions," Springer Lecture Notes in Mathematics 1130, ed. C.A. Di Prisco (1985), pp. 341-350.

LOGIC COLLOQUIUM '84
J.B. Paris, A.J. Wilkie, and G.M. Wilmers (Editors)

TARSKI'S PROBLEM AND PFAFFIAN FUNCTIONS

Lou van den Dries*
Department of Mathematics
Stanford University
Stanford, California
U.S.A.

Contents. The introduction outlines the subject in precise but nontechnical terms. In the remainder of the paper we go into details on some points, prove various partial results, and show how our main goal, the Decomposition Conjecture, reduces to a technical question on sets of asymptotes, cf.(3.13). Many open problems and observations not directly related to the Decomposition Conjecture are scattered throughout the text.

Complete proofs of the facts discussed will appear elsewhere.

Introduction. (Divided in Sections I - IX.)

I. Tarski Systems

This notion is useful because it allows concise formulation of several theorems.

Definition. A Tarski system is a sequence $(\mathcal{S}_n)_{n\in\mathbf{N}}$, where each $\mathcal{S}_n$ is a boolean algebra of subsets of $\mathbf{R}^n$, such that for all n:

(T_1) $A \in \mathcal{S}_n \Rightarrow \mathbf{R} \times A,\ A \times \mathbf{R} \in \mathcal{S}_{n+1}$,

(T_2) $\{(x_1,\ldots,x_n) \mid x_1 = x_n\} \in \mathcal{S}_n$,

(T_3) $A \in \mathcal{S}_{n+1} \Rightarrow \pi(A) \in \mathcal{S}_n$, where $\pi\colon \mathbf{R}^{n+1} \longrightarrow \mathbf{R}^n$ is given by $\pi(x_1,\ldots,x_n,x_{n+1}) = (x_1,\ldots,x_n)$,

(T_4) $\{(x,y) \mid x < y\} \in \mathcal{S}_2$, and $\{r\} \in \mathcal{S}_1$ for each $r \in \mathbf{R}$.

It is an easy exercise to show that the smallest Tarski system containing a given family of finite place relations on $\mathbf{R}$ consists exactly of the subsets of $\mathbf{R}, \mathbf{R}^2, \mathbf{R}^3,\ldots$ that are (parametrically) definable from $<$ and the relations in the family. Sometimes it happens that we can give a more informative description of the Tarski system generated by given relations, and the most striking example of this phenomenon is the system of semi-algebraic sets (SA_n).

Tarski's Theorem. Let SA_n consist of the boolean combinations of sets $\{x \in \mathbf{R}^n \mid p(x) > 0\}$, where $p \in \mathbf{R}[X_1,\ldots,X_n]$. Then (SA_n) is the Tarski system generated by the relations $x + y = z$ and $x \cdot y = z$ on $\mathbf{R}$.

* Partially supported by NSF.

We shall not discuss here the many surprising applications of Tarski's result: I refer to [G] and [vdD4] for more information. Property (T_3) is of course the crucial "elimination of quantifiers" which Tarski accomplished by generalizing Sturm's theorem on the real roots of a polynomial equation.

What makes Tarski's theorem so remarkable is the link that it establishes between the algebraic - analytic structure of the real field and its logical properties: Let an m-parameter family[1)] of "elementary" statements $\{\Phi(x)\}_{x \in \mathbf{R}^m}$ about $(\mathbf{R},<,+,\cdot)$ be given.

How does the truth value of $\Phi(x)$ change as x runs through $\mathbf{R}^m$? Tarski's theorem tells us that $\{x \mid \Phi(x) \text{ is true}\}$ is a semi-algebraic set, and for $m = 1$ this means that $\Phi(x)$ is true exactly on a finite union of intervals and points. (Intervals are always open in this article.) When $m > 1$ there are similar finiteness results for semi-algebraic subsets of $\mathbf{R}^m$. (See the part of this introduction on 0-minimality.) These finiteness properties of definable sets are at least as striking as the decidability and completeness of $\mathrm{Th}(\mathbf{R})$ stressed by Tarski.

II. Tarski's Problem

The Renaissance gave us the isomorphism between addition and multiplication: multiplying (positive) reals amounts to adding their logarithms. The (graph of) the exponential function, however, is not semi-algebraic, so the system of semi-algebraic sets is missing a fundamental connection between its two generating operations. The following problem [T, p.45] is now rather natural.

Give an informative description of the Tarski system generated by the relations $x + y = z$, $x \cdot y = z$ and $\exp(x) = y$. (My paraphrase)

The following negative result shows that an obvious analogue of Tarski's theorem fails. (See [vdD3, p.99] for the proof.)

There is no Tarski system $(\mathcal{S}_n)$ properly extending the system (SA_n) and such that each set in $\mathcal{S}_n$ is a boolean combination of sets $\{x \in \mathbf{R}^n \mid f(x) > 0\}$, f a real analytic function on (all of) $\mathbf{R}^n$. (Special case, due to Osgood: the set $\{(x,y) \mid x > 0 \text{ and } \exp(\frac{1}{x}) = y\}$ is not "semi-analytic".)

---

1) In conventional terms: let $\Phi(x_1,\dots,x_m)$ be a formula in the language of $(\mathbf{R},<,+,\cdot)$.

III. 0-minimality

What feature of the system of semi-algebraic sets is worth preserving in larger Tarski systems that also contain the exponential function? The remarkable finiteness properties of semi-algebraic sets are clearly desirable. These finiteness phenomena are consequences of 0-minimality.

Definition. A Tarski system $(\mathcal{S}_n)$ is called 0-minimal if each set in $\mathcal{S}_1$ is a finite union of intervals and points.

(The term "0-minimal", due to Pillay and Steinhorn [P-S], is suggested by the fact that each Tarski system must contain at least each finite union of intervals and points. A seemingly more restricted notion than 0-minimality was already discussed in [vdD3] but recent work by J. Knight, A. Pillay and C. Steinhorn [K-P-S] has shown its equivalence with 0-minimality. What follows is an update on [vdD3] in view of this equivalence.)

Although it is not known whether there is any 0-minimal Tarski system properly extending the system of semi-algebraic sets, there is a lot of evidence for the following.

Conjecture. The Tarski system generated by the relations $x + y = z$, $x \cdot y = z$ and $\exp(x) = y$ is 0-minimal.

Before discussing our program for proving the conjecture let us mention some attractive consequences of 0-minimality.

Finiteness Theorem. If $(\mathcal{S}_n)$ is an 0-minimal Tarski system then each set in $\mathcal{S}_n$ has only finitely many connected components, and these components are also in $\mathcal{S}_n$.

To make this result more explicit we have to introduce the notion of cell.

Terminology

1. A continuous function on $A \subset \mathbf{R}^m$ is either real valued, or one of the constant functions $+\infty$, $-\infty$ on A.

2. The graph of a function $f: A \longrightarrow \mathbf{R}$, $A \subset \mathbf{R}^m$, is the set

$$\Gamma(f) = \{(a,f(a)) \mid a \in A\} \subset \mathbf{R}^{m+1}.$$

3. Given two continuous functions f,g on $A \subset \mathbf{R}^m$ such that $f < g$ on A, put

$$(f,g) = \{(a,r) \in A \times \mathbf{R} \mid f(a) < r < g(a)\} \subset \mathbf{R}^{m+1}$$

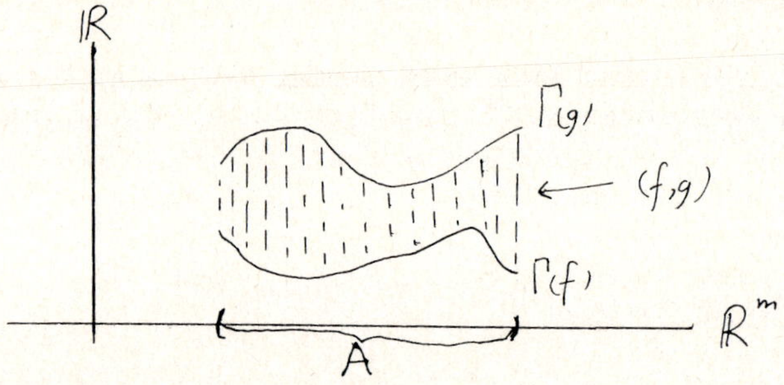

4. Cells are the sets obtained inductively as follows:

(i) the cells in $\mathbf{R} = \mathbf{R}^1$ are the intervals (open as always) and the points (= singletons),

(ii) the cells in $\mathbf{R}^{m+1}$ are the sets $\Gamma(f)$, where f is a real valued continuous function on a cell $A \subset \mathbf{R}^m$, and the sets (f,g) where f and g are continuous functions on a cell $A \subset \mathbf{R}^m$, such that $f < g$ on A.

Note that a cell in $\mathbf{R}^m$ is homeomorphic to $\mathbf{R}^k$ for some $k \leq m$. This notion of cell and the following theorem were inspired by Łojasiewicz's proof [Ł, p.110] of Tarski's theorem, and by Collins' method of cylindrical decomposition [C]. It implies the finiteness theorem on the preceding page.

Cell Decomposition Theorem

Let $(\mathcal{S}_n)$ be an 0-minimal Tarski system. Given any sets $S_1,\ldots,S_k$ in $\mathcal{S}_n$ there is a finite partition of $\mathbf{R}^n$ into cells belonging to $\mathcal{S}_n$ such that each S_i is a union of cells in the partition.

Instead of finitely many sets, let us now consider an infinite parametrized family of sets. Precisely, let $S \subset \mathbf{R}^{m+n}$ determine the family of sets $(S_a)_{a \in \mathbf{R}^m}$, where

$S_a = \{x \in \mathbf{R}^n \mid (a,x) \in X\}$.

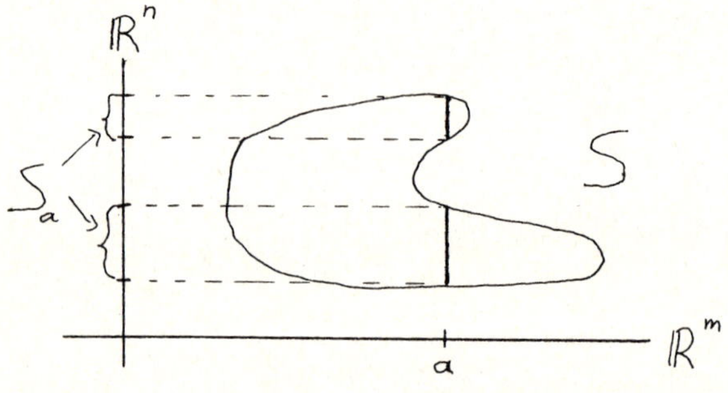

Cellwise Triviality Theorem ([Hardt], for (SA_n))

Let $(\mathcal{S}_n)$ *be an 0-minimal Tarski system extending* (SA_n), *and let* $S \in \mathcal{S}_{m+n}$. *Then there is a finite partition of* R^m *into cells of* $\mathcal{S}_m$ *such that for each cell* A *in the partition there is a* (so called trivializing) *homeomorphism* $h: \pi^{-1}(A) \xrightarrow{\sim} A \times S_a$ (a some point in A) *that has its graph in* $\mathcal{S}_{2(m+n)}$ *and makes the following diagram commutative*:

$$\begin{array}{ccc} \pi^{-1}A & \xrightarrow{\sim} & A \times S_a \\ & \searrow \quad \swarrow & \\ & A & \end{array}$$

($\pi: S \to R^m$ is given by $\pi(x_1,\dots,x_{m+n}) = (x_1,\dots,x_m)$.)

(In [vdD5] I give an "0-minimal" proof of this theorem.)

Comment. If $b \in A$ then S_b is homeomorphic to S_a, so the family $(S_c)_{c \in R^m}$ contains only finitely many homeomorphism types. All together there are only countably many homeomorphism types among the sets in $(\mathcal{S}_n)$.

IV. *Pfaffian functions*

The main point about these functions in connection with $(R,+,\cdot,\exp)$ is their usefulness as "Skolem functions" *in eliminating quantifiers. Pfaffian functions* are, roughly speaking, real analytic functions defined on certain cells, called *Pfaffian* cells, and they satisfy differential equations of a certain form.

Examples:

$\frac{dy}{dx} = y$ has the right form, and its solution $y = e^x$ is therefore a Pfaffian function;

$\frac{d^2y}{dx^2} = -y$ does not have the right form, and its solution $y = \sin x$ is not Pfaffian;

however, the restriction of sin to $(-\frac{1}{2}\pi, \frac{1}{2}\pi)$ is Pfaffian since on this interval it satisfies $\frac{dy}{dx} = (1-y^2)^{\frac{1}{2}}$ which has the right form.

The following inductive schema generates all Pfaffian functions on all open Pfaffian cells.

(P_1) *Intervals in* R *are* (open) *Pfaffian cells*.

(P_2) *If* $A \subset R^m$ *is an open Pfaffian cell and* f,g *are Pfaffian functions on* A, *or* $f = -\infty$, *or* $g = +\infty$, *such that* $f < g$ *on* A, *then* $(f,g) \subset R^{m+1}$ *is an open Pfaffian cell*.

(P_3) The constant functions and the coordinate functions $X_1,\dots,X_m$ are Pfaffian functions on each open Pfaffian cell in $\mathbf{R}^m$.

(P_4) If A is an open Pfaffian cell, then sums and products of Pfaffian functions on A are Pfaffian, and each real analytic function g on A satisfying an equation $f_0g^n + f_1g^{n-1} +\dots+ f_n = 0$, where $f_0,\dots,f_n$ are Pfaffian on A, $f_0 \neq 0$, is a Pfaffian function on A.

(P_5) If $f\colon A \longrightarrow \mathbf{R}$ and $F\colon B \longrightarrow \mathbf{R}$ are Pfaffian, where $A \subset \mathbf{R}^m$ and $B \subset \mathbf{R}^{m+1}$ are open Pfaffian cells such that $\Gamma(f) \subset B$, then the function $x \mapsto F(x,f(x))$ on A is Pfaffian.

(P_6) If $f\colon A \longrightarrow \mathbf{R}$ is analytic on an open Pfaffian cell $A \subset \mathbf{R}^m$ and $y = f(x_1,\dots,x_m)$ satisfies

$$\frac{\partial y}{\partial x_1} = F_1(x_1,\dots,x_m,y)$$
$$\vdots \qquad \vdots$$
$$\frac{\partial y}{\partial x_m} = F_m(x_1,\dots,x_m,y)$$

where $F_1,\dots,F_m$ are Pfaffian functions on an open Pfaffian cell in $\mathbf{R}^{m+1}$ containing $\Gamma(f)$, then f is Pfaffian.

(The last clause is the essential one; without it we would only obtain semi-algebraic functions.)

Besides open Pfaffian cells there are non-open Pfaffian cells and Pfaffian functions on them. We generate these as follows: if $A \subset \mathbf{R}^m$ is a Pfaffian cell and $f\colon A \longrightarrow \mathbf{R}$ is Pfaffian, then $\Gamma(f) \subset \mathbf{R}^{m+1}$ is a (necessarily non-open) Pfaffian cell; the Pfaffian functions on $\Gamma(f)$ correspond to those on A under the bijection $a \longleftrightarrow (a,f(a))$ between A and $\Gamma(f)$. Further, we let the singletons $\{r\} \subset \mathbf{R}$ and the constant functions on them be Pfaffian and extend clause (P_2) to non-open Pfaffian cells.

It can be shown, see (2.12), (2.19), that the partial derivatives of a Pfaffian function on an open cell are Pfaffian, that the restriction of a Pfaffian function to a Pfaffian subcell of its domain is Pfaffian, and that "superposition" of Pfaffian functions gives Pfaffian functions. (Clause (P_5) is a special case of superposition.)

The following decomposition conjecture is the ultimate goal. It implies our earlier conjecture that $+,\cdot$ and exp generate an 0-minimal Tarski system, but it goes much beyond it in precision, and is more manageable.

Decomposition Conjecture

The zeroset of a Pfaffian function is a finite (disjoint) union of Pfaffian cells.

A consequence would be that the finite (disjoint) unions of Pfaffian cells (in $\mathbf{R}^n$, for $n = 1,2,\ldots$) form a (necessarily 0-minimal) Tarski system. The decomposition conjecture is likely to be more general than the 0-minimality conjecture in III: many Pfaffian functions do not seem to be definable from $+,\cdot$ and exp. (However, I do not know how to prove this for even a single Pfaffian function, like arctg.)

V. Pfaffian functions in one and two variables

Suppose we introduce a new Pfaffian function $f: (a,b) \longrightarrow \mathbf{R}$ by clause (6), so $y = f(x)$ satisfies $\frac{dy}{dx} = F(x,y)$, where F is a Pfaffian function introduced "earlier", with domain, say, $(a,b) \times \mathbf{R} \subset \mathbf{R}^2$.

To verify the decomposition conjecture for f means to show that f has only finitely many zeros (unless f vanishes identically). Of course we may assume as an inductive hypothesis that the function $F(x,0)$ has only finitely many zeros. This assumption becomes relevant because of the following (Rolle-type) lemma which shows that the function $F(x,0)$ serves as a 'fake' derivative of f.

Key Lemma. If $f: (a,b) \longrightarrow \mathbf{R}$ is analytic and $y = f(x)$ satisfies $\frac{dy}{dx} = F(x,y)$, where F is analytic on an open set in $\mathbf{R}^2$ containing $\Gamma(f)$ and $(a,b) \times \{0\}$, then between any two zeros of f there is a zero of the function $F(x,0)$. (Consequently, if $F(x,0)$ has only finitely many zeros, so does f.)

To see why this is true, take successive zeros $c < d$ of f, and assume for simplicity that these zeros are simple, say $f'(c) > 0$ and (hence) $f'(d) < 0$. But $F(c,0) = F(c,f(c)) = f'(c) > 0$, and $F(d,0) = F(d,f(d)) = f'(d) < 0$, so by the intermediate value theorem $F(x,0)$ vanishes at some point between c and d.

What about the zeroset of a two-variable Pfaffian function f? To simplify the discussion, assume f is defined on all of $\mathbf{R}^2$, does not vanish identically, and that $y = f(x_1,x_2)$ satisfies

$$\frac{\partial y}{\partial x_1} = F_1(x_1,x_2,y)$$

$$\frac{\partial y}{\partial x_2} = F_2(x_1,x_2,y)$$

where F_1 and F_2 are Pfaffian functions on $\mathbf{R}^2$ introduced "earlier". Let us assume inductively that the decomposition conjecture holds for Pfaffian functions introduced "earlier". (The relevant functions are $F_1(x_1,x_2,0)$ and $F_2(x_1,x_2,0)$.) Let C be the "curve" $f(x_1,x_2) = 0$. It is not difficult to prove the following three properties:

(1) Only finitely many vertical lines $x_1 = r_1,\dots,x_1 = r_k$ are tangent to C. (If (r,s) is a singular point of C we count $x_1 = r$ as tangent to C.)

(2) There is a natural number M such that for $r \neq r_1,\dots,r_k$ the vertical line $x_1 = r$ intersects C in at most M points.

(3) Outside of the vertical lines $x_1 = r_1,\dots,x_1 = r_k$ the curve C consists of disjoint pieces of the form $x_2 = \alpha(x_1)$, where α is an analytic function on an interval I_α, each of whose two endpoints a is either in $\{r_1,\dots,r_k,\pm\infty\}$, or satisfies $\lim_{x_1 \to a} \alpha(x_1) = \pm\infty$. In the latter case we say that $x_1 = a$ is a vertical asymptote to C. (See picture, where C is decomposed in graphs of functions α as above.)

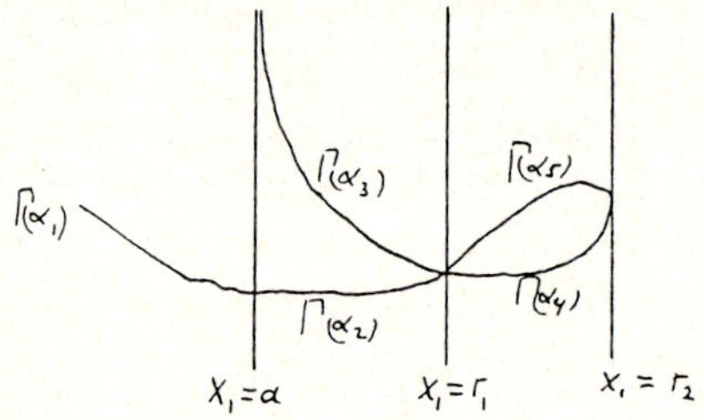

$x_1 = a$
is a vertical asymptote.

To verify the conjecture for the function f now amounts to proving:

(4) Each of the functions α is (piecewise) Pfaffian.

(5) There are only finitely many functions α as in (3).

Now (4) is easy: if $x_2 = \alpha(x_1)$ is a piece of C, then $\frac{dx_2}{dx_1} = \frac{F_1}{F_2}(x_1,x_2,0)$, as follows from differentiating the identity $f(x_1,\alpha(x_1)) = 0$. Hence α is piecewise Pfaffian. (We ignore here a technicality: one has to split up $\mathbf{R}^2$ into finitely many Pfaffian cells such that $\frac{F_1}{F_2}(x_1,x_2,0)$ is well defined and Pfaffian on the open cells.)

We now turn to (5). In view of (1), (2), (3) this property will follow from:

(6) The curve C has only finitely many vertical asymptotes.

The discussion of (6) merits a section of its own: it leads to the new concept of tameness, which, I hope, will be the key to proving the conjecture.

VI. Tameness: a use of infinite numbers

To detect the points $a \in \mathbf{R}$ for which $x_1 = a$ is a vertical asymptote to the curve $C: f(x_1,x_2) = 0$, we take a positive infinite hyper real $b \in {}^*\mathbf{R}$, and note that the equation $f(x_1,b) = 0$, or the equation $f(x_1,-b) = 0$, must have a root a' infinitely close to a. So it is enough to show that the function $f(x_1,b)$ has only finitely many zeros in $^*\mathbf{R}$. But $f(x_1,b)$ is Pfaffian on $^*\mathbf{R}$, in an obvious sense. Provided we extend our inductive hypothesis to such (nonstandard) Pfaffian functions - keeping however the standard meaning of finite - we may conclude that $f(x_1,b)$ has only finitely many zeros. (The inductive hypothesis applies, since the one variable function $f(x_1,b)$ comes "earlier" than the two variable function $f(x_1,x_2)$.)

Our attempt to verify the conjecture inductively must therefore be carried out simultaneously over $\mathbf{R}$ and $^*\mathbf{R}$.

Let us now consider a three variable Pfaffian function $f: \mathbf{R}^3 \longrightarrow \mathbf{R}$. Here a new aspect comes into play, and the role of $^*\mathbf{R}$ becomes more prominent. The technicalities (domains of definition, etc.) are more severe than in the two variable case, but, grosso modo, the verification of the conjecture for f reduces to proving the analogue of point (6) above: Let $S \subset \mathbf{R}^3$ be the surface $f(x_1,x_2,x_3) = 0$. Think of the x_3-axis as the vertical axis. Define As(S) to be the set of points $(a_1,a_2) \in \mathbf{R}^2$ such that the vertical line $x_1 = a_1$, $x_2 = a_2$ is an asymptote to S. The analogue of (6) is as follows:

As(S) is contained in a Pfaffian curve, i.e., in the union of finitely many non-open Pfaffian cells in $\mathbf{R}^2$.

To prove this, we take an infinite hyperreal b and consider the equation $f(x_1,x_2,b) = 0$ on $(^*\mathbf{R})^2$. Since the two variable function $f(x_1,x_2,b)$ comes before the three variable function $f(x_1,x_2,x_3)$, the inductive hypothesis applies and tells us that the "curve" $f(x_1,x_2,b) = 0$ in $(^*\mathbf{R})^2$ is, modulo finitely many lines, a union of graphs $\alpha(x_1) = x_2$, α ranging over finitely many Pfaffian functions on intervals in $^*\mathbf{R}$. Let us consider such a function α, and assume for simplicity that its domain is all of $^*\mathbf{R}$. Keep in mind that we are not interested so much in the subset $\Gamma(\alpha) \subset (^*\mathbf{R})^2$ but rather in the set of points of $\mathbf{R}^2$ that are infinitely close to $\Gamma(\alpha)$ since it is these points that are in As(S). So it suffices to establish the following tameness property of α:

(T) There are $a_1 <\ldots< a_k$ in $\mathbf{R}$ such that on the hull[1)] of each interval $(a_i,a_{i+1}) \subset \mathbf{R}$ $(i = 0,\ldots,k,\ a_0 = -\infty, a_{k+1} = +\infty)$ the function α is either positive infinite, or negative infinite, or infinitely close to $^*\alpha_i$, where α_i is a (standard) Pfaffian function on (a_i,a_{i+1}).

[1)] The hull of a subset of $\mathbf{R}$ consists of all $x \in {}^*\mathbf{R}$ which are infinitely close to a point in that set. Note: $0 \notin$ hull of (0,1).

I have verified this tameness property in many cases and under various inductive assumptions, which are too complicated to state here. (But see VIII.) The result of all this is a proof of the "three-variable" case of the decomposition conjecture. More precisely, consider the generating clauses (P_1) - (P_6) for Pfaffian cells and functions in IV. Remember that clause (P_6) is the one with the differential equations. Define $\mathcal{P}_M$ as the collection of Pfaffian functions and Pfaffian cells obtained by applying clauses (P_1) - (P_5) without restriction, and clause (P_6) with the restriction that $1 \leq m \leq M$. (So clause (P_6) can only introduce functions in at most M variables.) The non-open Pfaffian cells in $\mathcal{P}_M$, and the functions on them that belong to $\mathcal{P}_M$ are obtained as usual, see IV. So $\mathcal{P}_1 \subset \mathcal{P}_2 \subset \mathcal{P}_3 \subset \mathcal{P}_4 \subset \ldots$, and their union is the collection of all Pfaffian functions and Pfaffian cells.

Theorem. _The zeroset of a Pfaffian function in $\mathcal{P}_3$ is a finite disjoint union of Pfaffian cells in $\mathcal{P}_3$._

How large is $\mathcal{P}_3$? The following is a rough indication: when $A \in \mathcal{P}_3$ is a cell in $\mathbf{R}$, $\mathbf{R}^2$, or $\mathbf{R}^3$, then the functions on A which belong to $\mathcal{P}_3$ form a ring containing all polynomial functions on A, and this ring is also closed under many transcendental operations, such as $f \mapsto \exp(f)$, $f \mapsto \operatorname{arctg}(f)$, and, for $f > 0$ on A, under $f \mapsto f^r$ $(r \in \mathbf{R})$ and $f \mapsto \log(f)$.

The proof of the theorem will be given in [vdD6] and is along the lines sketched above: the main point is the (inductive) verification of the tameness property (T).

The theorem implies a considerable amount of "0-minimality". Expressed in logic jargon: _Let the set $S \subset \mathbf{R}^m$ be defined by a formula_

$$\Phi(x_1,\ldots,x_m) := (Q_1 x_{m+1})\ldots(Q_n\, x_{m+n})\ \psi(x_1,\ldots,x_{m+n}),$$

where $Q_1,\ldots,Q_n$ are quantifiers $\exists$, $\forall$ and ψ is a quantifier free formula in the language $\{<,+,\cdot,\exp,\text{real constants}\}$ in which $\exp$ is only applied to subterms $\tau(x_1,x_2,x_3)$ not involving $x_4,x_5,\ldots$. Then S is a finite disjoint union of Pfaffian cells.

There are some special features of $\mathcal{P}_3$ that enable me to prove the theorem above, and not (yet) the analogue for $\mathcal{P}_4$, $\mathcal{P}_5$, etc., which would settle the conjecture. Roughly speaking, the difficulty in verifying the tameness property (T) above, occurs when the one variable function α on $^*\mathbf{R}$ is obtained from earlier functions by "_superposition_", i.e., by clause (P_5). It turns out that this clause is _not needed_ for producing α, as long as α arises from a 3-variable function f in $\mathcal{P}_3$ in the manner described above. For details, see [vdD6].

The principal difficulty in proving the full decomposition conjecture along the lines sketched in this paper is to get rid of clause (P_5).

VII. Relation to Khovanskii's Work

I should acknowledge here that the idea of using Pfaffian functions to study Tarski's problem was inspired by Khovanskii's paper [Kh1]. (I thank M. Boshernitzan and G. Cherlin for calling this article to my attention. The methods I had available previously were mainly restricted to exponential functions in two variables, cf. [vdD2].) In particular, the Key Lemma in V above is hidden in Proposition 1 of [Kh1]. At the International Congress of Mathematicians, Warsaw 1983, Khovanskii introduced the class of Pfaff manifolds and Pfaff functions, and showed that this class shares many finiteness properties with the smaller class of real algebraic manifolds and their Nash functions. The property of being a Pfaff submanifold of $\mathbf{R}^m$ is invariant under coordinate permutations, in contrast to our notion of Pfaffian cell in $\mathbf{R}^m$. (We fixed in advance coordinate systems on $\mathbf{R}, \mathbf{R}^2, \mathbf{R}^3, \ldots$, and compatible projection maps $\ldots \longrightarrow \mathbf{R}^3 \longrightarrow \mathbf{R}^2 \longrightarrow \mathbf{R}$.) It follows from the results mentioned in [Kh2] that our Pfaffian cells and Pfaffian functions are special cases of Khovanskii's Pfaff manifolds and Pfaff functions. Fortunately, our decomposition conjecture would imply that the two notions are essentially equivalent: a Pfaff submanifold of $\mathbf{R}^m$ would be a finite disjoint union of Pfaffian cells in $\mathbf{R}^m$.

One of the open problems in [Kh2] asks whether there is an analogue of the Seidenberg-Tarski theorem for Pfaff manifolds. Presumably, this question can be made precise as follows:

If $f: M \longrightarrow N$ is a Pfaff map between Pfaff manifolds M and N, does it follow that $f(M)$ is a finite (disjoint) union of Pfaff submanifolds of N?

Our decomposition conjecture would imply a positive answer.

Apart from the Key Lemma in V we do not use Khovanskii's results or his differential -topological methods (Sard's lemma, Morse theory) in our proof of the $\mathcal{P}_3$-theorem in VI. (Our analytical tools are analytic continuation, Weierstrass preparation, and, most of all, tameness.) There is, however, one finiteness theorem in his recent paper [Kh3] which overlaps with our $\mathcal{P}_3$-result: it implies that the decomposition conjecture holds for *all* Pfaffian functions on Pfaffian cells in $\mathbf{R}$ and $\mathbf{R}^2$, also for those not in $\mathcal{P}_3$. I shall state Khovanskii's finiteness theorem in IX and indicate why the decomposition conjecture for Pfaffian functions in one and two variables follows from it.

Khovanskii's theorems imply new (effective) bounds in real and complex algebraic geometry, plane dynamical systems, and related areas, cf. [Kh1], [Kh2], [Kh3], [Ri]. It would be very desirable to complete his theory of Pfaff manifolds by establishing the Tarski-Seidenberg property.

VIII. More on Tameness

Tameness phenomena are established inductively. One starts with semi-algebraic functions.

Theorem. Let $f: ({}^*R)^m \to {}^*R$ be a semi-algebraic function, in the sense of the real closed field $^*R$. Then there is a finite partition of R^m into semi-algebraic cells such that on the hull[1)] C^h of each open cell C of the partition:

either f is positive infinite on C^h,

or f is negative infinite on C^h,

or there is a continuous semi-algebraic function $g: C \to R$ such that f is infinitely close on C^h to $^*g$.

(Note: $^*g$ is defined on $^*C$, which contains C^h, since C is open.)

This theorem has surprising standard consequences. To describe these, let $\bar{R} = R \cup \{\pm\infty\}$ be the two point compactification of R. Let $f: R^{m+n} \to \bar{R}$ be semi-algebraic, not necessarily continuous. For each $y \in R^n$ define $f_y: R^m \to \bar{R}$ by $f_y(x) = f(x,y)$. Let us call $\mathcal{F} = \{f_y \mid y \in R^n\}$ a semi-algebraic collection (of functions); consider $\mathcal{F}$ as a subset of the (compact) product space $(\bar{R})^{R^m}$. The theorem now implies:

(*) The closure $\mathcal{F}^{cl}$ of $\mathcal{F}$ in $(\bar{R})^{R^m}$ consists entirely of semi-algebraic functions.

(Open problem: is $\mathcal{F}^{cl}$ a semi-algebraic collection?) For example, if a function $\psi: R^m \to \bar{R}$ is the pointwise limit of a sequence of f_y's, then ψ is semi-algebraic.

The theorem actually implies a stronger form of convergence which seems hard to express in terms of a topology on the function space $(\bar{R})^{R^m}$, namely:

(**) For each sequence in $\mathcal{F}$ there is a subsequence and a finite partition of R^m into semi-algebraic cells such that the subsequence converges uniformly on each compact subset of each of the cells.

The program I have sketched for proving the Decomposition Conjecture requires a simultaneous proof of the Pfaffian analogue of the semi-algebraic tameness theorem. (The Pfaffian analogues of (*) and (**) are then of course consequences.) It would be nice to know in advance whether "Pfaffian tameness" is implied by the Decomposition Conjecture. A certain amount of tameness is present in any 0-minimal Tarski system:

If $\mathcal{S}$ is an 0-minimal Tarski system and $f: R^m \times R \to R$ is a function in $\mathcal{S}$ then for each $x \in R^m$ $\lim_{y\to\infty} f(x,y)$, exists in $\bar{R}$, and, consequently, the limit function $\psi: R^m \to \bar{R}$ defined by $\psi(x) = \lim_{y\to\infty} f(x,y)$, belongs to $\mathcal{S}$.

This general result is for the system of semi-algebraic sets a special case of (*) and (**) above. Another special case of the semi-algebraic tameness theorem says:

---

1) The hull C^h of a set $C \subset R^m$ consists of all $x \in ({}^*R)^m$ which are infinitely close to a point in C.

If $S \subset (^*\mathbf{R})^m$ is semi-algebraic in the sense of the real closed field $^*\mathbf{R}$, then $S \cap \mathbf{R}^m$ is a semi-algebraic subset of $\mathbf{R}^m$. Since $^*\mathbf{R}$ contains an isomorphic copy of each ordered field extension of finite transcendence degree over $\mathbf{R}$, we obtain from this the following model theoretic equivalent, as observed by Cherlin:

Each n-type over $\mathbf{R}$ is definable. (Note that this is false for any other real closed field.) It would be nice to know whether each n-type over the p-adic field $\mathbf{Q}_p$ is definable.

Going beyond semi-algebraic tameness, let me just state the following result, a special case of theorems in [vdD6].

Define on each connected open set $U \subset \mathbf{R}^m$ the ring of logarithmico-exponential functions LE(U) as the smallest ring of real-analytic functions on U which contains the polynomial functions, contains each real analytic function g on U satisfying an equation $f_0 g^n + \ldots + f_n = 0$, where $f_0, \ldots, f_n \in$ LE(U), $f_0 \neq 0$, and which is closed under $f \mapsto \exp(f)$, and under $f \mapsto \log(f)$ if $f > 0$ on U.

Theorem. Let $f \in \mathrm{LE}(\mathbf{R} \times \mathbf{R}^n)$. Let $y \in (^*\mathbf{R})^n$, and define f_y: $^*\mathbf{R} \longrightarrow {}^*\mathbf{R}$ by: $f_y(x) = {}^*f(x,y)$. Then there are real numbers $a_1 < \ldots < a_k$ such that on the hull of each interval (a_i, a_{i+1}) $(i = 0,\ldots,k.$ $a_0 = -\infty$, $a_{k+1} = +\infty)$ the function f_y is either positive infinite, or negative infinite, or infinitely close to $^*\alpha_i$, where α_i is a (standard) Pfaffian function on (a_i, a_{i+1}).

Even for $n = 1$ this is of interest:

Let $f \in \mathrm{LE}(\mathbf{R}^2)$, and define g: $\mathbf{R} \longrightarrow \mathbf{R}$ by: $g(x) = \lim_{y \to \infty} f(x,y)$. Then there are real numbers $a_1 < \ldots < a_k$ such that on each interval (a_i, a_{i+1}) $(i = 0,\ldots,k,$ $a_0 = -\infty$, $a_{k+1} = +\infty)$ the function g is either identically $+\infty$, or identically $-\infty$, or equal to a Pfaffian function. (And the "complexity" of this Pfaffian function is "bounded" in terms of the complexity of f.)

This result eliminates some hairy possibilities first raised by Angus Macintyre: could the limit function g have an infinite set of discontinuities? According to Macintyre such nightmarish pathology can also be eliminated on the basis of recent work by Dahn [D]. See also Wolter's contribution to this volume.

IX. Miscellaneous

Theorem. Let A be a Pfaffian cell and $f = (f_1,\ldots,f_k)$: $A \longrightarrow \mathbf{R}^k$ be a map whose coordinate functions $f_1,\ldots,f_k$ are Pfaffian. Then the number of connected components of $f^{-1}(c)$ admits a uniform finite bound independent of $c \in \mathbf{R}^k$.

This result is essentially Khovanskii's Theorem 4 in [Kh3]. We shall use this theorem in combination with two other propositions:

Proposition 1. Each Pfaffian function on an open cell is differentially algebraic. (See (2.15).)

For a Pfaffian function f on an interval $I \subset \mathbf{R}$ this means that f satisfies an equation $P(f,f',\ldots,f^{(n)}) = 0$, for some polynomial $P(X_0,\ldots,X_n) \in \mathbf{R}[X_0,\ldots,X_n] \setminus \{0\}$. Now each nonzero Pfaffian function on an interval (a,∞) has only finitely many zeros according to Khovanskii's theorem, so the germs at ∞ of these functions form, what is called a Hardy field. We can therefore apply the following result of M. Singer [S] which was rediscovered by M. Boshernitzan [B] and by myself [vdD1]; see also M. Rosenlicht [Ro] for more precise information.

Proposition 2. If f is a real analytic function on an interval (a,∞) whose germ at ∞ belongs to a Hardy field and which satisfies $P(f,f',\ldots,f^{(n)}) = 0$, where $P(X_0,\ldots,X_n) \in \mathbf{R}[X_0,\ldots,X_n] \setminus \{0\}$, then: $f(x) = 0(e_n(x^m))$ as $x \to \infty$, for some $m \in \mathbf{N}$. (Here e_n is the n-times iterated exponential function, i.e., $e_1(x) = e^x$, $e_2(x) = e_1(e_1x))$, etc.)

Corollary. If $f: (a,\infty) \to \mathbf{R}$ is Pfaffian, then there is n such that $f(x) = 0(e_n(x))$ as $x \to \infty$.

Another indication that Pfaffian functions behave like polynomials is the following: If f is a nonzero Pfaffian function on an open Pfaffian cell $A \subset \mathbf{R}^m$ then there is N such that for each point $x \in A$ at least one of the partials

$$\frac{\partial^{i_1+\cdots+i_m}}{\partial x_1^{i_1}\cdots\partial x_m^{i_m}} f, \text{ with } i_1+\ldots+i_m \le N \text{ does not vanish at } x.$$

This can be proved by induction on "complexity", see [vdD6] for details. Similarly we can show, for f and A as above, that there is M such that if the first M partials $\dfrac{\partial^i f}{\partial x_m^i}$, $i < M$, vanish at x, then all partials $\dfrac{\partial^i f}{\partial x_m^i}$, $i \in \mathbf{N}$, vanish at x. (With M independent of $x \in A$.)

This last result can be used for $m = 2$ in combination with Khovanskii's finiteness theorem and Weierstrass' Preparation Theorem to obtain:

Theorem. The zeroset of a Pfaffian function on an open Pfaffian cell in $\mathbf{R}^2$ is a finite disjoint union of Pfaffian cells. (Again, we refer to [vdD6] for details.)

This theorem is only mentioned here to inspire confidence in the Decomposition Conjecture. The methods used in its proof do not seem to generalize beyond the two variable case.

I thank Greg Cherlin, George Kreisel and Angus Macintyre for useful discussions.

§1. Cell Systems and Pfaffian Functions

We start with some terminological conventions concerning analytic functions, manifolds and cells. Then we define important notions such as: decompositions of $\mathbf{R}^m$, functions that are Pfaffian relative to a ring of functions, and cell systems. These cell systems allow us to generate the Pfaffian cells and functions in a carefully controlled manner.

(1.1) A function $f: A \longrightarrow \mathbf{R}$, $A \subset \mathbf{R}^m$, is called analytic if for each point $a = (a_1,\ldots,a_m) \in A$ there is a power series with real coefficients

$$f_a(x) = \Sigma\, c_{(i_1,\ldots,i_m)} (x_1 - a_1)^{i_1} \ldots (x_m - a_m)^{i_m}$$

converging on an open set $U \ni a$, such that $f(x) = f_a(x)$ for $x \in U \cap A$.

(1.2) A map $f = (f_1,\ldots,f_n): A \longrightarrow B$ where $A \subset \mathbf{R}^m$, $B \subset \mathbf{R}^n$, is called analytic if each of the component functions $f_i: A \longrightarrow \mathbf{R}$ is analytic.

If $f: A \longrightarrow B$ is a bijection and f and f^{-1} are both analytic we call f an analytic diffeomorphism from A onto B. If $f: A \longrightarrow B$ and $g: B \longrightarrow C$ are analytic, then $g \circ f: A \longrightarrow C$ is analytic. Any inclusion map $A_0 \hookrightarrow A$ is analytic.

(1.3) We call a set $A \subset \mathbf{R}^m$ an analytic manifold of dimension k if each $a \in A$ has an open neighbornood U_a in A that is analytically diffeomorphic to an open set V_a in $\mathbf{R}^k$. (For $k = 0$ this means that A is discrete, since $\mathbf{R}^0$ is a one-point set.) An open subset of $\mathbf{R}^m$ is an analytic manifold of dimension m.

(1.4) An extended analytic function on an analytic manifold $A \subset \mathbf{R}^m$ is either an analytic function $f: A \longrightarrow \mathbf{R}$, or one of the constant functions $-\infty$, $+\infty$ on A.

(1.5) Let $A \subset \mathbf{R}^m$ be an analytic manifold. Then the graph $\Gamma_A(f)$ of an analytic function $f: A \longrightarrow \mathbf{R}$ is an analytic manifold in $\mathbf{R}^{m+1}$. Given two extended analytic functions f and g on A such that $f < g$ on A we put

$$(f,g)_A = \{(a,r) \in A \times \mathbf{R} \mid f(a) < r < g(a)\}.$$

Note that $(f,g)_A$ is an analytic manifold in $\mathbf{R}^{m+1}$. If A is clear from context we write also $\Gamma(f)$ and (f,g) for $\Gamma_A(f)$ and $(f,g)_A$.

(1.6) We introduce $(i_1,\ldots,i_m)$-cells as certain analytic manifolds in $\mathbf{R}^m$ of dimension $i_1 + \ldots + i_m$, by induction on m:

A (0) - cell is a one-point set $\{a\} \subset \mathbf{R}$.

A (1) - cell is an interval $(a,b) \subset \mathbf{R}$, $-\infty \leq a < b \leq +\infty$.

An $(i_1,\ldots,i_m,0)$-cell is the graph $\Gamma(f)$ of an analytic function f on an $(i_1,\ldots,i_m)$-cell $A \subset \mathbf{R}^m$.

An $(i_1,\ldots,i_m,1)$-cell is a set (f,g) where f,g are extended analytic functions on an $(i_1,\ldots,i_m)$-cell A such that $f < g$ on A.

(1.7) Since the cells from (1.6) are the only cells we need, from now on cell means: $(i_1,\ldots,i_m)$-cell, for some sequence $(i_1,\ldots,i_m)$. The $(\underbrace{1,\ldots,1}_{m\text{-times}})$-cells are the cells which are open in their ambient space $\mathbf{R}^m$, and therefore we call them open cells (in $\mathbf{R}^m$). (They are of course analytically diffeomorphic to $\mathbf{R}^m$.) It is also convenient to consider the one-point set $\mathbf{R}^0$ as an open cell (in $\mathbf{R}^0$).

(1.8) Certain questions on general cells reduce to questions on open cells as follows: let $(i_1,\ldots,i_m)$ be a sequence of zeros and ones and let $\lambda(1) <\ldots< \lambda(k)$ be the indices $\lambda \in \{1,\ldots,m\}$ for which $i_\lambda = 1$. (So $i_1 +\ldots+ i_m = k$.) Consider the projection

$$h\colon (x_1,\ldots,x_m) \mapsto (x_{\lambda(1)},\ldots,x_{\lambda(k)})\colon \mathbf{R}^m \longrightarrow \mathbf{R}^k.$$

Then h restricts on each $(i_1,\ldots,i_m)$-cell $A \subset \mathbf{R}^m$ to an analytic diffeomorphism

$$h_A\colon A \longrightarrow h(A),$$

with image an open cell $h(A) \subset \mathbf{R}^k$. (This follows easily by induction.)

(1.9) In the following we shall use the notations h_A and $h(A)$ established in (1.8). Note that if A is an open cell then $h(A) = A$ and h_A is the identity.

(1.10) A decomposition of $\mathbf{R}^m$ is a certain kind of finite partition of $\mathbf{R}^m$ into cells. The definition is by induction on m.

(D_1) A decomposition $\mathcal{D}$ of $\mathbf{R}^1 = \mathbf{R}$ is given by real numbers $a_1 <\ldots< a_k$ as follows: put $a_0 = -\infty$, $a_{k+1} = +\infty$, and

$$\mathcal{D} = \{(a_i,a_{i+1}) \mid 0 \leq i \leq k\} \cup \{\{a_i\} \mid 1 \leq i \leq k\}$$

(D_{m+1}) A decomposition of $\mathbf{R}^{m+1}$ is a finite partition of $\mathbf{R}^{m+1}$ into cells such that the projections of these cells in $\mathbf{R}^m$ form a decomposition of $\mathbf{R}^m$. (The projections are the images under the map $(x_1,\ldots,x_m,x_{m+1}) \mapsto (x_1,\ldots,x_m)$.)

(1.11) Let A be a cell and S a ring of analytic functions on A. (We always assume S contains the constant function 1 which is then the unit element of S; since A is connected S is an integral domain.)

(1) We call an analytic function f on A Nash over S, if f satisfies an equation $p_n f^n + p_{n-1} f^{n-1} + \ldots + p_0 = 0$, where $p_i \in S$, $p_n \neq 0$. (Special cases: If f is rational over S, i.e., $pf = q$ for certain $p,q \in S$, $p \neq 0$, then f is Nash over S, and in the fraction field of the ring of all analytic functions on A we have $f = q/p$.)

(2) We call S Nash-closed if S contains each analytic function on A which is Nash over S.

(1.12) Let A and B be open cells in $\mathbf{R}^m$ and $\mathbf{R}^{m+1}$ such that $\pi(B) = A$ where $\pi\colon \mathbf{R}^{m+1} \longrightarrow \mathbf{R}^m$ is the projection on the first m coordinates. Let T be a ring of analytic functions on B.

(1) We call an analytic function f on A Pfaffian relative to T if $\Gamma(f) \subset B$ and there are functions $F_1,\ldots,F_m$ in T such that $y = f(x_1,\ldots,x_m)$ is a solution of the system

$$\partial y/\partial x_1 = F_1\,(x_1,\ldots,x_m,y)$$

$$\partial y/\partial x_m = F_m\,(x_1,\ldots,x_m,y)$$

[Remark: a system of partial differential equations of this form is sometimes called a Pfaffian equation in the literature.]

(2) If S is a ring of analytic functions on A then we call S Pfaff-closed in T if S contains each analytic function on A which is Pfaffian relative to T.

(1.13) Example. With $A \subset \mathbf{R}^m$ and $B \subset \mathbf{R}^{m+1}$ as above, let S be a ring of analytic functions on A closed under the operators $\partial/\partial x_i$, $i = 1,\ldots,m$. Define $S[X_{m+1}]$ as the ring of all polynomial functions

$$P = p_d X_{m+1}^d + \ldots + p_0\colon A \times \mathbf{R} \longrightarrow \mathbf{R}, \text{ where } p_i \in S.$$

($P(a,r) = p_d(a)r^d + \ldots + p_0(a)$ for $(a,r) \in A \times \mathbf{R}$.) Put $T = \{P|_B\colon P \in S[X_{m+1}]\}$. Then the exponential $\exp(g) = e^g$ of each function $g \in S$ is Pfaffian relative to T: $y = \exp(g)$ satisfies $\partial y/\partial x_i = (\partial g/\partial x_i)\cdot y$ for $i = 1,\ldots,m$.

(1.14) A cell system $\mathcal{S}$ on $\mathbf{R}^M$ is a map defined on a cell collection $\bigcup_{m=1}^{M} \mathcal{S}[m]$, each $\mathcal{S}[m]$ consisting of open cells in $\mathbf{R}^m$. The map $\mathcal{S}$ assigns to each cell $A \in \mathcal{S}[m]$ a ring $\mathcal{S}(A)$ of analytic functions on A, such that the following conditions are satisfied:

(CS_1) All intervals belong to $\mathcal{S}[1]$.
(CS_2) The cells in $\mathcal{S}[m+1]$ are exactly the sets (f,g) where $f,g \in \mathcal{S}(A) \cup \{+\infty,-\infty\}$, $f < g$ on $A \in \mathcal{S}[m]$. (Here $m < M$.)
(CS_3) For each cell $A \in \mathcal{S}[m]$ the ring $\mathcal{S}(A)$ contains all constant functions and the coordinate functions $X_1,\ldots,X_m$.
(CS_4)For each $A \in \mathcal{S}[m]$ and $B \in \mathcal{S}[m+1]$ with $\pi(B) = A$, and each function $f \in \mathcal{S}(A)$ the function $(a,r) \mapsto f(a)\colon B \longrightarrow \mathbf{R}$, belongs to $\mathcal{S}(B)$. (Here $\pi\colon \mathbf{R}^{m+1} \longrightarrow \mathbf{R}^m$ is the usual projection map, and $m < M$.)

It is also convenient to define $\mathcal{S}[0] = \{\mathbf{R}^0\}$, and $\mathcal{S}(\mathbf{R}^0) = \mathbf{R}$, each real number being considered as the corresponding constant function on $\mathbf{R}^0$. We do allow $M = \infty$; in that case m ranges over all natural numbers.

(1.15) Given two cell systems $\mathcal{S}_1$ and $\mathcal{S}_2$ on $\mathbf{R}^M$ we write $\mathcal{S}_1 \subset \mathcal{S}_2$ if $\mathcal{S}_1[m] \subset \mathcal{S}_2[m]$ for each m and $\mathcal{S}_1(A) \subset \mathcal{S}_2(A)$ for each $A \in \mathcal{S}_1[m]$. Each collection $\mathcal{C}$ of cell systems on $\mathbf{R}^M$ has a greatest lower bound $\inf \mathcal{C}$ with respect to this ordering:

$$(\inf \mathcal{C})[m] = \cap \{\mathcal{S}[m] \mid \mathcal{S} \in \mathcal{C}\}, \text{ and}$$
$$(\inf \mathcal{C})\ (A) = \cap \{\mathcal{S}(A) \mid \mathcal{S} \in \mathcal{C}\}, \text{ for each } A \in (\inf \mathcal{C})[m].$$

(1.16) Definition. The cell system $\mathcal{P}$ is defined as the smallest cell system on $\mathbf{R}^\infty$ such that for each pair of open cells $A \in \mathcal{P}[m]$ and $B \in \mathcal{P}[m+1]$ with $\pi(B) = A$ the following closure conditions hold:

(P_1) $\mathcal{P}(A)$ is Nash-closed,
(P_2) $\mathcal{P}(A)$ is Pfaff-closed in $\mathcal{P}(B)$,
(P_3) For each function $f \in \mathcal{P}(A)$ with $\Gamma(f) \subset B$ and each $F \in \mathcal{P}(B)$ the function $a \mapsto F(a,f(a))$ belongs to $\mathcal{P}(A)$. (Here $\pi\colon \mathbf{R}^{m+1} \longrightarrow \mathbf{R}^m$ is the usual projection map.)

(1.17) Note that clause (P_3) requires closure under a very limited form of substitution. It turns out that this is sufficient to guarantee closure under arbitrary substitutions. (See (2.19).)

Each cell system, in particular $\mathcal{P}$, contains all real polynomial functions on all its open cells.

(1.18) Lemma. Let $\mathcal{S}$ be a cell system on $\mathbf{R}^M$. To each finite sequence $(i_1,\ldots,i_m)$ of zeros and ones, with $m \leq M$, we can associate a collection $\mathcal{S}(i_1,\ldots,i_m)$ of $(i_1,\ldots,i_m)$-cells, and to each cell $A \in \mathcal{S}(i_1,\ldots,i_m)$ a ring $\mathcal{S}(A)$ of analytic functions on A, such that the following conditions hold:

(1) If $i_1 = .. = i_m = 1$, then $\mathcal{S}(i_1,\ldots,i_m) = \mathcal{S}[m]$, and $\mathcal{S}(A)$ has the usual meaning for $A \in \mathcal{S}(i_1,\ldots,i_m)$.

(2) $\mathcal{S}(i,\ldots,i_m,0)$ consists of all graphs $\Gamma(f)$ with $f \in \mathcal{S}(A)$, $A \in \mathcal{S}(i_1,\ldots,i_m)$.

(3) $\mathcal{S}(i_1,\ldots,i_m,1)$ consists of all sets (f,g) with $f,g \in \mathcal{S}(A) \cup \{-\infty,+\infty\}$, $f < g$ on $A \in \mathcal{S}(i_1,\ldots,i_m)$.

(4) If $A \in \mathcal{S}(i_1,\ldots,i_m)$ and $i_1 + \ldots + i_m = k$, then $h(A) \in \mathcal{S}[k]$ and the bijection $h_A \colon A \longrightarrow h(A)$ induces a ring isomorphism $\mathcal{S}(A) \simeq \mathcal{S}(h(A))$. (See (1.8) for the notation used here.)

(1.19) The proof is by induction. Note that properties (1)-(4) uniquely determine $\mathcal{S}(i_1,\ldots,i_m)$ and $\mathcal{S}(A)$ for $A \in (i,\ldots,i_m)$. A *cell* A is said to *belong to* $\mathcal{S}$ or to be *in* $\mathcal{S}$ if $A \in \mathcal{S}(i_1,\ldots,i_m)$ for some sequence $(i_1,\ldots,i_m)$. A *function* f is said to *belong to* $\mathcal{S}$ or to be *in* $\mathcal{S}$ if $f \in \mathcal{S}(A)$ for some cell A in $\mathcal{S}$. When $\mathcal{S} = \mathcal{P}$ we also write *Pfaffian cell* and *Pfaffian function* for a cell or function belonging to $\mathcal{P}$. Clearly this gives the same notion as in the Introduction. A more convenient formulation of the Decomposition Conjecture is as follows.

(1.20) *Decomposition Conjecture for* $\mathcal{P}$

If $f \in \mathcal{P}(A)$ *where* $A \in \mathcal{P}[m]$ *then there is decomposition* $\mathcal{D}$ *of* $\mathbf{R}^m$ *consisting of Pfaffian cells such that* A *is a union of cells in* $\mathcal{D}$, *and such that for each cell* $C \in \mathcal{D}$ *with* $C \subset A$ *we have:*

either $f > 0$ *on* C,
or $f = 0$ *on* C,
or $f < 0$ *on* C.

(1.21) Let us define $\mathcal{N}$ as the least cell system such that $\mathcal{N}(A)$ is Nash-closed for each open cell A in $\mathcal{N}$. Clearly $\mathcal{N} \subset \mathcal{P}$. The conjecture on $\mathcal{P}$ becomes a theorem when $\mathcal{P}$ is replaced everywhere by $\mathcal{N}$. This can be established along the lines of Łojasiewicz proof [Ł, p.110] of Tarski's theorem.

§2. *Generating* $\mathcal{P}$. *Elementary Properties of* $\mathcal{P}$.

(2.1) In this section we show how to build up $\mathcal{P}$ from below, which is the only way to obtain any results on Pfaffian functions. The first such result, (2.12), says that restrictions of Pfaffian functions to Pfaffian subcells of their domain are again Pfaffian functions.

(2.2) Some terminology.

(a) If $\mathcal{S}$ is a cell system on R^M, $1 < M < \infty$, then $\mathcal{S}| R^{M-1}$ is the cell system on R^{M-1} defined by:

$$(\mathcal{S}| R^{M-1})[m] = \mathcal{S}[m] \quad \text{for} \quad 1 \leq m \leq M-1,$$
$$(\mathcal{S}| R^{M-1})(A) = \mathcal{S}(A) \quad \text{for} \quad 1 \leq m \leq M-1,\ A \in \mathcal{S}[m].$$

(b) A cell system $\mathcal{S}$ is called coherent if $f|A$ belongs to $\mathcal{S}$ for each function f in $\mathcal{S}$ and each cell A in $\mathcal{S}$ which is a subset of the domain of f.

(2.3) The smallest cell system on R consists exactly of the restrictions of the polynomial functions $a_nX_1^n + \ldots + a_0$, $a_i \in R$, to intervals. We shall denote this cell system by $R[X_1]$. It is clearly coherent.

More generally, let $\mathcal{S}$ be a cell system on R^M, $M < \infty$. Then we define the cell system $\mathcal{S}' = \mathcal{S}[X_{M+1}]$ on R^{M+1} as follows:

(1) $\mathcal{S}'| R^M = \mathcal{S}$,

(2) $\mathcal{S}'[M+1]$ consists of all cells $A' = (\beta,\gamma)_A$ where $A \in \mathcal{S}[M]$ and $\beta, \gamma \in \mathcal{S}(A) \cup \{-\infty,+\infty\}$ $\beta < \gamma$ on A; for each such A' we put $\mathcal{S}'(A') = \{P|_{A'}: P \in \mathcal{S}(A)[X_{M+1}]\}$. (Here $\mathcal{S}(A)[X_{M+1}]$ is considered as a ring of functions on $A \times R$, see (1.13).)

(2.4) Lemma. If $\mathcal{S}$ is coherent, then $\mathcal{S}[X_{M+1}]$ is coherent. This follows in a straightforward way from the definitions.)

(2.5) The following operations on cell systems are only partially defined. First some terminological conventions: A cell system $\mathcal{S}$ on R^M is called rationally closed (Nash closed) if each ring $\mathcal{S}(A)$, A an open cell in $\mathcal{S}$, is rationally closed (Nash closed).

Definition. Let $\mathcal{S}$ be a cell system on R^M, $M < \infty$, such that $\mathcal{S}| R^{M-1}$ is rationally closed (Nash closed). (This condition holds by convention if $M = 1$.) Then the cell system $\mathcal{S}^{rat}$ ($\mathcal{S}^N$) on R^M is defined by:

(1) $\mathcal{S}^{rat}| R^{M-1} = \mathcal{S}| R^{M-1}$ ($\mathcal{S}^N| R^{M-1} = \mathcal{S}| R^{M-1}$).

(2) for each open cell $A \in \mathcal{S}[M]$:

$\mathcal{S}^{rat}(A) = \{f: A \longrightarrow R \mid f$ is analytic and rational over $\mathcal{S}(A)\}$

($\mathcal{S}^N(A) = \{f: A \longrightarrow R \mid f$ is analytic and Nash over $\mathcal{S}(A)\}$).

Clearly $\mathcal{S}^{rat}$ ($\mathcal{S}^N$) is rationally closed (Nash closed).

(2.6) Weierstrass prepared us, among other things, for the following;

Lemma. If $\mathcal{S}$ as above is coherent, then $\mathcal{S}^{rat}$ ($\mathcal{S}^N$) is coherent.

(2.7) Besides coherence there are other nice properties that the cell systems used in building $\mathcal{P}$ enjoy. One of these is "closure under partial derivatives."

Definition. A cell system $\mathcal{S}$ on R^M is called a differential cell system if for each open cell $A \in \mathcal{S}[m]$, $1 \leq m \leq M$, the ring $\mathcal{S}(A)$ is closed under the operators $\partial/\partial x_i$, $i = 1,\dots,m$.

Note that the operations $\mathcal{S} \mapsto \mathcal{S}^{rat}$, $\mathcal{S} \mapsto \mathcal{S}^N$ and $\mathcal{S} \mapsto \mathcal{S}[X_{M+1}]$ carry differential cell systems to differential cell systems.

(2.8) A very general adjunction operation is defined as follows.

Definition. Let $\mathcal{S}$ be a cell system on R^M, $M < \infty$, and $f\colon A \longrightarrow R$ an analytic function with domain $A \in \mathcal{S}[M]$. Then the cell system $\mathcal{S}\langle f\rangle$ on R^M is defined by the conditions:

(1) $\mathcal{S}\langle f\rangle | R^{M-1} = \mathcal{S} | R^{M-1}$. (It follows that necessarily $\mathcal{S}\langle f\rangle[M] = \mathcal{S}[M]$.)

(2a) If $B \in \mathcal{S}[M]$ and $B \not\subset A$, then $\mathcal{S}\langle f\rangle(B) = \mathcal{S}(B)$.

(2b) If $B \in \mathcal{S}[M]$ and $B \subset A$, then

$$\mathcal{S}\langle f\rangle(B) = \mathcal{S}(B)\left[\frac{\partial^{i_1+\dots+i_M}}{\partial x_1^{i_1}\dots\partial x_M^{i_M}} f|B : i_1 \geq 0,\dots,i_M \geq 0\right]$$

(the ring generated over $\mathcal{S}(B)$ by the restrictions to B of the partials of f).

Note that if $\mathcal{S}$ is a differential cell system on R^M then $\mathcal{S}\langle f\rangle$ is also a differential cell system on R^M. Of course, coherence is generally not inherited by $\mathcal{S}\langle f\rangle$ from $\mathcal{S}$. For that we need further assumptions on $\mathcal{S}$ and f which we discuss now.

(2.9) Definition. A cell system $\mathcal{S}$ on R^M, $M < \infty$, is called Pfaff closed in a cell system $\mathcal{S}'$ on R^{M+1} if for each pair of cells $A \in \mathcal{S}[M]$ and $B \in \mathcal{S}'[M+1]$, such that B maps onto A under the projection $R^{M+1} \longrightarrow R^M$, the ring $\mathcal{S}(A)$ is Pfaff closed in $\mathcal{S}'(B)$.

(2.10) Lemma. Let $\mathcal{S}$ be a cell system on R^M, $M < \infty$, and f an analytic function on an open cell $A \in \mathcal{S}[M]$. Then $\mathcal{S}\langle f\rangle$ is coherent if the following conditions are satisfied:

$\mathcal{S}|R^{M-1}$ is Pfaff closed in $\mathcal{S}$, and there is a coherent differential cell system $\mathcal{S}'$ on R^{M+1} such that $\mathcal{S}'|R^M = \mathcal{S}$ and f is Pfaffian relative to $\mathcal{S}(B)$ for some cell $B \in \mathcal{S}'[M+1]$ which maps onto A under the projection $R^{M+1} \longrightarrow R^M$.

(When $M = 1$ we call $\mathcal{S}|\mathbf{R}^{M-1}$ Pfaff closed in $\mathcal{S}$ by convention.)

The proof is a tedious verification. One has to use the following fact which is also important for other reasons.

Let $A_0 \in \mathcal{S}[M]$, $A_0 \subset A$. Then each function $g \in \mathcal{S}\langle f\rangle(A_0)$ can be written as

$$g(x) = G(x, f(x)) \quad (x \in A_0)$$

where $G \in \mathcal{S}'(B_0)$, B_0 the inverse image of A_0 under the projection $B \longrightarrow A$. ($\mathcal{S}'$ and B are as in the lemma.)

This fact implies that if the (old) systems $\mathcal{S}$ and $\mathcal{S}'$ consist of Pfaffian cells and functions, so does the (new) system $\mathcal{S}\langle f\rangle$. (By virtue of condition (P_3), see (1.16).)

(2.11) From the definition of $\mathcal{P}$ it should be fairly clear that $\mathcal{P}$ can be built up by starting with the system $\mathbf{R}[X_1]$, carrying out the operations $\mathcal{S} \mapsto \mathcal{S}^N$, $\mathcal{S} \mapsto \mathcal{S}[X_{M+1}]$ and $\mathcal{S} \mapsto \mathcal{S}\langle f\rangle$ when they apply, and taking unions when an increasing chain has been built. (Here it is understood that the operation $\mathcal{S} \mapsto \mathcal{S}\langle f\rangle$ is only applied when the hypothesis of lemma (2.10) is satisfied for an earlier constructed system $\mathcal{S}'$.)

Some basic properties of Pfaffian cells and functions follow from this way of generating $\mathcal{P}$.

(2.12) Proposition. $\mathcal{P}$ is a coherent differential cell system.

Remark. The proof of this result is typical for many of our arguments and therefore we give it in detail.

Proof. $\mathbf{R}[X_1]$ is a coherent differential subsystem of $\mathcal{P}|\mathbf{R}$. Take a maximal coherent differential subsystem $\mathcal{S}_1$ of $\mathcal{P}|\mathbf{R}$. Of course $\mathcal{S}_1$ is Nash closed since $\mathcal{S}_1 \subset \mathcal{S}_1^N \subset \mathcal{P}|\mathbf{R}$. Consider $\mathcal{S}_1[X_2]$. It is a coherent differential subsystem of $\mathcal{P}|\mathbf{R}^2$ and its restriction to $\mathbf{R}$ equals $\mathcal{S}_1$. Take a maximal coherent differential subsystem $\mathcal{S}_2$ of $\mathcal{P}|\mathbf{R}^2$ with $\mathcal{S}_2|\mathbf{R} = \mathcal{S}_1$. We claim that $\mathcal{S}_2$ is Nash closed and that $\mathcal{S}_1$ is Pfaff closed in $\mathcal{S}_2$. Nash closedness is clear from maximality. If $\mathcal{S}_1$ were not Pfaff closed in $\mathcal{S}_2$ there would be an analytic function f on a cell $A \in \mathcal{S}_2[1] = \mathcal{S}_1[1]$ such that $f \notin \mathcal{S}_1(A)$ and f is Pfaffian relative to $\mathcal{S}_2(B)$, for some cell $B \in \mathcal{S}_2[2]$ projecting onto A. Then $\mathcal{S}_2\langle f\rangle$ would be a coherent differential system properly extending $\mathcal{S}_1$ but still contained in $\mathcal{P}|\mathbf{R}$. (See lemma (2.10) and the comments following it.) This contradicts the maximality property of $\mathcal{S}_1$. The claim is proved. $\mathcal{S}_2[X_3]$ is a coherent differential subsystem of $\mathcal{P}|\mathbf{R}^3$ whose restriction to $\mathbf{R}^2$ equals $\mathcal{S}_2$. Take a maximal coherent differential subsystem $\mathcal{S}_3$ of $\mathcal{P}|\mathbf{R}^3$ with $\mathcal{S}_3|\mathbf{R}^2 = \mathcal{S}_2$. As with $\mathcal{S}_2$ it follows that $\mathcal{S}_3$ is Nash closed and $\mathcal{S}_2$ is Pfaff closed in $\mathcal{S}_3$.

Continuing in this fashion we obtain a sequence $(\mathcal{S}_n)$ such that each $\mathcal{S}_n$ is a Nash closed coherent differential subsystem of $\mathcal{P}|\mathbf{R}^n$, with $\mathcal{S}_{n+1}|\mathbf{R}^n = \mathcal{S}_n$ and $\mathcal{S}_n$ Pfaff closed in $\mathcal{S}_{n+1}$. The "union" $\mathcal{S}_\infty$ of the $\mathcal{S}_n$'s is then a cell system contained in $\mathcal{P}$ which satisfies the closure conditions (P_1), (P_2) and (P_3) of (1.16). ((P_3) follows from coherence of $\mathcal{S}_\infty$.) Since $\mathcal{P}$ is the least such system we must have $\mathcal{S}_\infty = \mathcal{P}$. Now use that $\mathcal{S}_\infty$ is a coherent differential system. ■

(2.13) Remark. It follows from the proof that an alternative definition of $\mathcal{P}$ is as the least differential cell system on $\mathbf{R}^\infty$ which satisfies (P_1) and (P_2). (I.e., condition (P_3) of (1.16) could have been replaced by the requirement that the rings $\mathcal{P}(A)$, $A \in \mathcal{P}[m]$, are closed under the operators $\partial/\partial x_i$, $i = 1,\ldots,m$.)

(2.14) To formulate the next result it is convenient to call an analytic function f on an open cell $A \subset \mathbf{R}^m$ a differentially algebraic function if the integral domain $\mathbf{R}\langle f\rangle$ has finite transcendence degree over $\mathbf{R}$. Here $\mathbf{R}\langle f\rangle$ is the $\mathbf{R}$-algebra generated by the partials $\dfrac{\partial^{i_1+\cdots+i_m}}{\partial x_1^{i_1}\cdots\partial x_m^{i_m}} f$ of f (including f itself). Alternatively, it is the least $\mathbf{R}$-algebra of analytic functions on A containing f and closed under the operators $\partial/\partial x_i$.

(2.15) Proposition. Each Pfaffian function on an open Pfaffian cell is differentially algebraic.

The proof is along the same lines as the proof of the previous proposition, using the two lemmas below. The first one is well known.

(2.16) Lemma. If $A \subset \mathbf{R}^m$ is an open cell and $\mathcal{F}$ a set of differentially algebraic functions on A, then $\mathbf{R}\langle\mathcal{F}\rangle$, the $\mathbf{R}$-algebra generated by the partials $\dfrac{\partial^{i_1+\cdots+i_m}}{\partial x_1^{i_1}\cdots\partial x_m^{i_m}} f$, with $f \in \mathcal{F}$, is closed under the operators $\partial/\partial x_i$, and all its members are differentially algebraic. Moreover, each analytic function on A which is Nash over $\mathbf{R}\langle\mathcal{F}\rangle$ is differentially algebraic.

(2.17) Lemma. Let $A \subset \mathbf{R}^m$ and $B \subset \mathbf{R}^{m+1}$ be open cells with B projecting onto A. Let the analytic function $f\colon A \longrightarrow \mathbf{R}$ be Pfaffian relative to an $\mathbf{R}$-algebra T of analytic functions on B. Suppose that T is closed under the operators $\partial/\partial x_i$, $i = 1,\ldots,m+1$, that all functions in T are differentially algebraic, and that T contains the coordinate function X_{m+1}.

Then f is differentially algebraic.

Proof. For each function G in T we put $G_f(x) = G(x,f(x))$, so the map $G \mapsto G_f$ is an $\mathbf{R}$-algebra morphism from T into the ring of analytic functions on A. Suppose now that $y = f(x)$ satisfies the partial differential equations $\partial y/\partial x_i = F_i(x,y)$, $i = 1,\ldots,m$ where $F_i \in T$. This can also be written as $\partial f/\partial x_i = (F_i)_f$. Since $\mathbf{R}\langle F_1,\ldots,F_m,X_{m+1}\rangle$ has finite transcendence degree over $\mathbf{R}$ it suffices to show that its image under the morphism $G \mapsto G_f$ contains f and is closed under the operators $\partial/\partial x_i$. The image contains f because $f = (X_{m+1})_f$. Closure under $\partial/\partial x_i$ is seen as follows: suppose $g = G_f$, $G \in T$. Then a simple computation gives:

$$\partial g/\partial x_i = (\partial G/\partial x_i)_f + (\partial G/\partial x_{m+1})_f \cdot (\partial f/\partial x_i)$$

$$= (\partial G/\partial x_i)_f + (\partial G/\partial x_{m+1})_f \cdot (F_i)_f \, .$$

(2.18) Question. When is a differentially algebraic function locally Pfaffian? (True for analytic $f\colon I \longrightarrow \mathbf{R}$, I an interval if f satisfies an equation $p(y,y^{(1)}, y^{(2)}) = 0$ where p is a nonzero real polynomial.)

(2.19) Several mathematicians asked me whether substituting Pfaffian functions in a Pfaffian function gives again a Pfaffian function. This is indeed the case but a precise formulation of the general result, let alone its proof, is messy, so we only indicate here a proof of the special case of composition with a one-variable Pfaffian function.

(2.20) Proposition. Let $f \in \mathcal{P}(I)$, I an interval, and $g \in \mathcal{P}(A)$, A an open Pfaffian cell, such that $g(A) \subset I$. Then $f \circ g \in \mathcal{P}(A)$.

Proof. The statement is certainly true if f belongs to $\underline{\mathbf{R}}[X_1]$. Further, if $\mathcal{S}$ is a cell system on $\mathbf{R}$, $\mathcal{S} \subset \mathcal{P}|\mathbf{R}$, such that the statement is true for all f in $\mathcal{S}$, then the statement is true for all f in $\mathcal{S}^N$. Take a maximal coherent subsystem $\mathcal{S}_1$ of $\mathcal{P}|\mathbf{R}$ such that the statement is true for f in $\mathcal{S}_1$. Then $\mathcal{S}_1$ is Nash closed.

Consider $\mathcal{S} = \mathcal{S}_1[X_2]$. It is easy to verify that $\mathcal{S}$ has the following property (*):

(*) If $F \in \mathcal{S}(B)$, $B = (\alpha,\beta)_I$ a cell in $\mathcal{S}[2]$, and $g \in \mathcal{P}(A)$, A an open Pfaffian cell, such that $g(A) \subset I$, then the function $(x,r) \mapsto F(g(x),r)$, with domain $(\alpha\circ g,\beta\circ g)_A$, belongs to $\mathcal{P}$. (Note: $\alpha\circ g, \beta\circ g$ are in fact (extended) functions in $\mathcal{P}$ since α,β are in $\mathcal{S}(I) \cup \{-\infty,+\infty\}$.)

Let us say that a cell system $\mathcal{S}$ on $\mathbf{R}^2$ is $\mathcal{S}_1$-good if $\mathcal{S}$ is a coherent subsystem of $\mathcal{P}|\mathbf{R}^2$ such that $\mathcal{S}|\mathbf{R} = \mathcal{S}_1$ and $\mathcal{S}$ satisfies property (*). It is easy to check that if $\mathcal{S}$ is $\mathcal{S}_1$-good then $\mathcal{S}^N$ is also $\mathcal{S}_1$-good. It is more work to show:

(**) Let $\mathcal{S}$ be $\mathcal{S}_1$-good. Then $\mathcal{S}_1$ is Pfaff closed in $\mathcal{S}_1$.

To see this, let $\Phi: I \longrightarrow \mathbf{R}$ be analytic, and Pfaffian relative to $\mathcal{S}(B)$ where $B = (\alpha,\beta)_I \in \mathcal{S}[2]$, say $\Phi'(t) = \psi(t,\Phi(t))$ $(t \in I)$, with $\psi \in \mathcal{S}(B)$. We claim:

(***) <u>The statement of the theorem holds for each function</u> f <u>in</u> $\mathcal{S}_1\langle\Phi\rangle$. (By maximality of $\mathcal{S}_1$, claim (***) implies $\mathcal{S}_1\langle\Phi\rangle = \mathcal{S}_1$, and therefore (**).)

To prove assertion (***), consider a function f in $\mathcal{S}_1\langle\Phi\rangle$ and a function $g \in \mathcal{P}(A)$, $A \in \mathcal{P}[m]$, such that $g(A) \subset$ domain (f). We have to show that $f \circ g \in \mathcal{P}(A)$.

We first consider the case $f = \Phi$. Then

$$\frac{\partial(\Phi\circ g)}{\partial x_i}(x) = \Phi'(g(x)) \cdot \frac{\partial g}{\partial x_i}(x) = \psi(g(x),(\Phi\circ g)(x)) \cdot \frac{\partial g}{\partial x_i}(x)$$

$$= \psi_i(x,(\Phi\circ g)(x)), \quad (x \in A),$$

where the function ψ_i is defined on the cell $(\alpha\circ g,\beta\circ)_A$ by $\psi_i(x,r) = \psi(g(x),r)\cdot\frac{\partial g}{\partial x_i}(x)$. By property (*) the functions ψ_i are Pfaffian, hence $\Phi\circ g$ is Pfaffian. This takes care of the case $f = \Phi$. For general f the case that domain $(f) \not\subset I$ is trivial since then f must belong to $\mathcal{S}_1$. (See (2.8), clause (2a).) So we assume that domain $(f) \subset I$. For simplicity of notation we assume even that domain $(f) = I$. Then, see (2.10), we have $f(r) = F(r,\Phi(r))$ $(r \in I)$, for some function $F \in \mathcal{S}(B)$. So $(f\circ g)(x) = F(g(x),(\Phi\circ g)(x))$. Define $H(x,r) = F(g(x),r)$ for $(x,r) \in (\alpha\circ g,\beta\circ g)_A$. Then H is Pfaffian by property (*). Since $\Phi\circ g$ is also Pfaffian it follows that the function $f\circ g$, which satisfies $(f\circ g)(x) = H(x,(\Phi\circ g)(x))$, is Pfaffian. This finishes the proof of claim (***), and therefore we have established (**).

Let $\mathcal{S}_2$ be a maximal cell system on $\mathbf{R}^2$ which is $\mathcal{S}_1$-good. (So $\mathcal{S}_2$ is Nash closed.) It is easy to verify that $\mathcal{S}_2[X_3]$ has a certain property relative to $\mathcal{S}_2$ which is analogous to the property (*) which $\mathcal{S}_2$ has relative to $\mathcal{S}_1$. As before we express this by saying that $\mathcal{S}_2[X_3]$ is $\mathcal{S}_2$-good. ($\mathcal{S}_2$-goodness is meant to include the property of being a coherent subsystem of $\mathcal{P}|\mathbf{R}^3$ whose restriction to $\mathbf{R}^2$ equals $\mathcal{S}_2$.) One can then prove, as we did for $\mathcal{S}_1$-good systems, that $\mathcal{S}_2$ is Pfaff closed in each $\mathcal{S}_2$-good system. Let $\mathcal{S}_3$ be a maximal $\mathcal{S}_2$-good cell system on $\mathbf{R}^3$. Continuing in this fashion we obtain a sequence $(\mathcal{S}_n)$ such that each $\mathcal{S}_n$ is a Nash closed coherent subsystem of $\mathcal{P}|\mathbf{R}^n$, with $\mathcal{S}_{n+1}|\mathbf{R}^n = \mathcal{S}_n$ and $\mathcal{S}_n$ Pfaff closed in $\mathcal{S}_{n+1}$. Then the "union" $\mathcal{S}_\infty$ of the $\mathcal{S}_n$'s is necessarily equal to $\mathcal{P}$, by definition of $\mathcal{P}$. This shows that $\mathcal{P}|\mathbf{R} = \mathcal{S}_1$, and proves closure of $\mathcal{P}$ under composition with functions in $\mathcal{P}|\mathbf{R}$. ∎

§3. <u>Zerosets of Pfaffian Functions.</u> <u>Completeness of Cell Systems.</u>

(3.1) <u>Definition</u>. A cell system $\mathcal{S}$ on $\mathbf{R}^M$ is called <u>complete</u> if $\mathcal{S}$ is coherent and the zeroset $Z(f)$ of each nonzero $f \in \mathcal{S}(A)$, A an open cell in $\mathcal{S}[m]$, $1 \leq m \leq M$, is contained in the union of finitely many cells in $\mathbf{R}^m$ of dimension $< m$ that belong to $\mathcal{S}$.

(Note: for $m = 1$ this means that each nonzero function $f \in \mathcal{S}(I)$, I an interval, has only finitely many zeros. In particular, $\underline{\mathbf{R}}[X_1]$ is complete.)

(3.2) One can show that if a cell system $\mathcal{S}$ on $\mathbf{R}^M$ is complete then there is for each function $f \in \mathcal{S}(A)$ (A an open cell in $\mathcal{S}[m]$, $1 \leq m \leq M$) a decomposition $\mathcal{D}$ of $\mathbf{R}^m$ consisting of cells in $\mathcal{S}$ such that A is a union of cells in $\mathcal{S}$ and such that f has constant sign (-1, 0, or 1) on each cell of $\mathcal{D}$ contained in A.

In particular, the Decomposition Conjecture states that the cell system $\mathcal{P}$ is complete. In view of the way we generate $\mathcal{P}$ it is therefore of vital importance to know under what conditions completeness is preserved by the various operations on cell systems. The algebraic operations are well behaved in this respect as the following two propositions show. The first one is essentially trivial.

(3.3) <u>Proposition</u>. If $\mathcal{S}$ is a complete cell system on $\mathbf{R}^M$, $M < \infty$, and $\mathcal{S}|\mathbf{R}^{M-1}$ is Nash closed then $\mathcal{S}^{rat}$ and $\mathcal{S}^N$ are complete.

(3.4) <u>Proposition</u>. If $\mathcal{S}$ is a complete Nash closed cell system on $\mathbf{R}^M$, $M < \infty$, then $\mathcal{S}[X_{M+1}]$ is complete.

(3.5) The proof of (3.4) uses Tarski's elimination theory and the following real analytic version of a topological lemma in Łojasiewicz [Ł, p. 107].

<u>Lemma</u>. Let C be a connected analytic manifold and $c_0,\ldots,c_d\colon C \longrightarrow \mathbf{R}$ analytic functions, not all zero, such that the number of <u>complex</u> zeros of the polynomial $c_d(x)\, Y^d + c_{d-1}(x)\, Y^{d-1} + \ldots + c_0(x)$ is constant, i.e., does not depend on $x \in C$. Then there are analytic functions $y_1 < \ldots < y_k\colon C \longrightarrow \mathbf{R}$ such that for each $x \in C$ the <u>real</u> zeros of the polynomial are exactly $y_1(x),\ldots,y_k(x)$.

(In this lemma zeros are counted <u>without</u> multiplicity.)

(3.6) <u>Remark</u>. From (3.3) and (3.4) we can obtain the decomposition theorem for $\mathcal{N}$ stated in (1.21), since $\mathcal{N}$ can be built up by means of the operations $\mathcal{S} \mapsto \mathcal{S}^N$ and $\mathcal{S} \mapsto \mathcal{S}[X_{M+1}]$, starting with $\underline{\mathbf{R}}[X_1]$.

(3.7) The next question is: when is $\mathcal{S}\langle f\rangle$ complete, under the hypotheses of (2.10), assuming moreover that $\mathcal{S}'$ is complete? A key ingredient in the analysis of this problem is the "Rolle" type result, mentioned in the Introduction, part V.

(3.8) <u>Proposition</u>. Let $\mathcal{S}$ be a cell system on $\mathbf{R}$, $f\colon I \longrightarrow \mathbf{R}$ an analytic function, and suppose there is a complete differential system $\mathcal{S}'$ on $\mathbf{R}^2$, such that $\mathcal{S}'|\mathbf{R} = \mathcal{S}$ and f is Pfaffian relative to $\mathcal{S}'(B)$ where $B \in \mathcal{S}'[2]$ projects onto I. Then $\mathcal{S}\langle f\rangle$ is complete.

Proof. From lemma (2.10) we obtain that $\mathcal{S}\langle f\rangle$ is coherent. We have to show that each nonzero function $g \in \mathcal{S}\langle f\rangle(J)$ has only finitely many zeros in its interval of definition J. To simplify notations let us assume J = I. Then we can write $g(x) = G(x,f(x))$, for some function $G \in \mathcal{S}'(B)$. (See (2.10).) Since $\mathcal{S}'$ is complete the zeroset Z(G) of G has a particularly simple form, and in fact we may reduce to the case that $Z(G) = \Gamma_I(h_1) \cup \ldots \cup \Gamma_I(h_k)$, where $h_1 < \ldots < h_k$ are functions in $\mathcal{S}'(I) = \mathcal{S}(I)$. Put $f_j = f - h_j$ for $j = 1,\ldots,k$. Then we have:

$$g(x) = 0 \iff G(x,f(x)) = 0 \iff f(x) = h_j(x) \text{ for some } j$$

$$\iff f_j(x) = 0 \text{ for some } j.$$

Hence to show Z(g) finite reduces to showing that each set $Z(f_j)$ is finite. Now $y = f_j(x)$ satisfies $\frac{dy}{dx} = F_j(x,y)$, where $F_j(x,y) = F(x,y + h_j(x)) - h_j'(x)$. We have no guarantee that F_j belongs to $\mathcal{S}'$, but fortunately the function $x \mapsto F_j(x,0) = F(x,h_j(x)) - h_j'(x)$ does belong to $\mathcal{S}'(I)$ and has, by completeness of $\mathcal{S}'$, only finitely many zeros. So we can apply the Key lemma, see Introduction, to conclude that $Z(f_j)$ is finite.

(3.9) Let us pause to consider a consequence of the last three propositions. For historical reasons we write from now on <u>Hardy system on</u> $\mathbf{R}^M$ for: <u>complete differential cell system on</u> $\mathbf{R}^M$. so a Hardy system on **R** is a map $\mathcal{H}$ which assigns to each interval $I \subset \mathbf{R}$ and **R**-algebra $\mathcal{H}(I)$ of analytic functions on I, such that:

(i) $X_1 \in \mathcal{H}(\mathbf{R})$,
(ii) $f \in \mathcal{H}(I)$, J a subinterval of $I \Rightarrow (f|_J) \in \mathcal{H}(J)$,
(iii) $f \in \mathcal{H}(I) \Rightarrow f' \in \mathcal{H}(I)$,
(iv) $f \in \mathcal{H}(I)$, $f \neq 0 \Rightarrow Z(f)$ is finite.

(G.H. Hardy, in his book [Hardy], proved that each "logarithmico-exponential" function has only finitely many real zeros. In our terminology this amounts to the following: for each interval I, put LE(I) = the smallest Nash closed ring of analytic functions on I which contains all real polynomial functions on I and is closed under $f \mapsto \exp(f)$, and under $f \mapsto \log(f)$ for $f > 0$ on I; then LE is a Hardy system on **R**.)

<u>Let $\mathcal{H}$ be a Hardy system on</u> **R**. If $\mathcal{H}$ is not Nash closed we can extend $\mathcal{H}$ to the larger Hardy system $\mathcal{H}^N$ on **R** which is Nash closed. So let us assume that $\mathcal{H}$ is already Nash closed. Then $\mathcal{H}[X_2]$ is a Hardy system on $\mathbf{R}^2$, hence $\mathcal{H}' = \mathcal{H}[X_2]^N$

is a Hardy system on $\mathbf{R}^2$, with $\mathcal{H}'|\mathbf{R} = \mathcal{H}$. If $\mathcal{H}$ is not Pfaff closed in $\mathcal{H}'$ then there is an analytic function $f\colon I \longrightarrow \mathbf{R}$, I an interval, which is Pfaffian relative to $\mathcal{H}'(B)$ for some $B \in \mathcal{H}'[2]$ projecting onto I, such that $f \notin \mathcal{H}(I)$. Then $\mathcal{H}\langle f\rangle$ is a Hardy system on $\mathbf{R}$, by (3.8), and is strictly larger than $\mathcal{H}$. Now the adjunction process just described can be started over again, with $\mathcal{H}\langle f\rangle$ in the role of $\mathcal{H}$.

These considerations lead to the following conclusion. For each Hardy system $\mathcal{H}$ on $\mathbf{R}$ there is a smallest Nash closed Hardy system $\mathcal{H}^P$ on $\mathbf{R}$ such that $\mathcal{H}^P \supset \mathcal{H}$ and $\mathcal{H}^P$ is Pfaff closed in $\mathcal{H}^P[X_2]^N$. (Note: if $\mathcal{H} \subset \mathcal{P}|\mathbf{R}$, then $\mathcal{H}^P \subset \mathcal{P}|\mathbf{R}$.)

The Hardy system $(\underline{\mathbf{R}}[X_1])^P$ is already quite large: it contains the functions exp, $\sin x$ $(-\frac{1}{2}\pi < x < \frac{1}{2}\pi)$, it contains each indefinite integral of each of its functions, and is closed under composition of functions. (The last statement can be proved along the same lines as proposition (2.20).)

(3.10) We now proceed to generalize these results to Hardy systems on $\mathbf{R}^M$ where $M > 1$. Here a new phenomenon, the existence of vertical asymptotes to the zeroset, may cause trouble.

Let $M > 1$ and let $A \subset \mathbf{R}^M$ be an open cell, say $A = (\alpha,\beta)_{\pi A}$. (Look at A as the union of vertical intervals parametrized by the points of πA; $\pi\colon \mathbf{R}^M \longrightarrow \mathbf{R}^{M-1}$ is the usual projection.) Let Z be a subset of A. (E.g., $Z = Z(f)$ for some analytic function f on A.)

Definition. (1) We say that a point $p \in \pi A$ lies on a vertical asymptote to Z in A, if there is a point $q \in \pi A$, $q \neq p$, such that $(q,p) \overset{\text{def.}}{=} \{(1-t)q + tp \mid 0 < t < 1\} \subset \pi A$, and an analytic function $\sigma\colon (q,p) \longrightarrow \mathbf{R}$, such that $(x,\sigma x) \in Z$ for all $x \in (q,p)$, and either $\lim_{x\to p} \sigma(x) = \alpha(p)$, or $\lim_{x\to p} \sigma(x) = \beta(p)$.
(N.B., it may be that $\alpha = -\infty$ or $\beta = +\infty$.)

(2) $\mathrm{As}(Z,A) \overset{\text{def.}}{=} \{p \in \pi A \mid p$ lies on a vertical asymptote to Z in $A\}$.

(If Z is the zeroset of an analytic function f on A, and $f \neq 0$, one naturally expects $\dim \mathrm{As}(Z,A) < \dim Z$.)

(3.17) Lemma. Let $\mathcal{H}$ be a Hardy system on $\mathbf{R}^M$, $1 < M < \infty$, such that $\mathcal{H}|\mathbf{R}^{M-1}$ is Pfaff closed in $\mathcal{H}$. Let $A \subset \mathbf{R}^M$ be an open cell in $\mathcal{H}$, and $f\colon A \longrightarrow \mathbf{R}$ an analytic function such that $y = f(x)$ satisfies $\partial y/\partial x_i = F_i(x,y)$, $i = 1,\dots,M$, where the F_i are analytic functions on an open set $B \subset \mathbf{R}^{M+1}$, with $\Gamma_A(f) \subset B$. Suppose the following three conditions are satisfied:

(1) $A \times \{0\} \subset B$ and each function $x \mapsto F_i(x,0)\colon A \longrightarrow \mathbf{R}$ belongs to $\mathcal{H}(A)$.
(2) For each cell $D \subset A$ in $\mathcal{H}$, of dimension $< M$, we have $(f|D) \in \mathcal{H}(D)$.
(3) $\mathrm{As}(Z(f),A) = \emptyset$.

Then $Z(f) \subset D_1 \cup .. \cup D_k$ for suitable cells $D_j \subset A$ in $\mathcal{H}$ of dimension $< M$.

(3.12) The proof of this important lemma is long, we do not give it here. (Weierstrass preparation, real analytic continuation, and a refinement of the Key Lemma in the Introduction are the principal ingredients.)

(3.13) The main consequence of lemma (3.11) is a generalization of proposition (3.8) to Hardy systems on $\mathbf{R}^M$:

Theorem. Let $\mathcal{H}$ be a Hardy system on $\mathbf{R}^M$, $1 < M < \infty$, such that $\mathcal{H} | \mathbf{R}^{M-1}$ is Pfaff closed in $\mathcal{H}$. Let f be an analytic function on a cell $A \in \mathcal{H}[M]$ such that f is Pfaffian relative to $\mathcal{H}'(B)$ where $\mathcal{H}'$ is a Hardy system on $\mathbf{R}^{M+1}$ with $\mathcal{H}' | \mathbf{R}^M = \mathcal{H}$ and $B \in \mathcal{H}'[M+1]$ projects onto A.

Then $\mathcal{H}\langle f\rangle$ is a Hardy system on $\mathbf{R}^M$ if and only if the following condition on asymptotes is satisfied: For each function $h \in \mathcal{H}(A_0)$, where $A_0 \subset A$, $A_0 \in \mathcal{H}[M]$, the set $As(Z(f_h), A_0)$, with $f_h = (f | A_0) - h$, is contained in the union of finitely many cells in $\mathbf{R}^{M-1}$ which belong to $\mathcal{H}$ and have dimension $< M - 1$.

Proof. (Sketch) The idea is the same as in the proof of proposition (3.8). Lemma (2.10) and completeness of $\mathcal{H}$ enable us to reduce to the case where lemma (3.11) can be applied. See also parts V, VI of the Introduction. ■

(3.14) Comment. The theorem is useful since it reduces an M-variable problem (on zerosets of M-variable functions) to an (M-1)-variable problem, i.e., the verification of the "condition on asymptotes." For example, if $M = 2$ we have to verify that certain plane curves have only finitely many vertical asymptotes. For $M = 3$ we must consider a certain collection of surfaces $S \subset \mathbf{R}^3 = \mathbf{R}^2 \times \mathbf{R}$, and check that for each surface S there is a curve $C \subset \mathbf{R}^2$ of a certain type such that all vertical asymptotes to S intersect $\mathbf{R}^2$ in C. I have been able to verify this condition on asymptotes in quite a few cases, by means of the notion of tameness. (See parts VI, VIII of the Introduction.)

To work with tameness it is essential to develop the theory of Hardy systems and Pfaffian functions also over $^*\mathbf{R}$ (without changing the notion of 'finite'). This development will be carried out in [vdD 6].

(3.15) Let us finish this paper by pointing out an algebraic consequence of completeness. If $\mathcal{S}$ is a complete rationally closed cell system on $\mathbf{R}$ then it is easy to see that each nonzero function $f \in \mathcal{S}(I)$, I an interval, is of the form

$$f = u \cdot (X_1 - r_1)^{e_1} \cdot \ldots \cdot (X_1 - r_k)^{e_k} \qquad (e_i > 0)$$

where $u \in \mathcal{S}(I)$ has no zeros and $r_1,\ldots,r_k$ are the distinct zeros of f. It follows that $\mathcal{S}(I)$ is a principal ideal domain and that $r \mapsto (X_1 - r) \cdot \mathcal{S}(I)$ is a bijection of I onto the set of maximal ideals.

More generally, if $\mathcal{S}$ is a complete rationally closed cell system on $\mathbf{R}^M$ one can show that for each open cell $A \in \mathcal{S}[m]$, $1 \leq m \leq M$, we have a bijection $a = (a_1,\ldots,a_m) \mapsto (X_1 - a_1,\ldots,X_m - a_m) \cdot \mathcal{S}(A)$ of A onto the set of maximal ideals of $\mathcal{S}(A)$.

Is it true that $\mathcal{S}(A)$ is noetherian? (Maybe one needs an extra hypothesis here.) In any case, completeness of $\mathcal{S}$ clearly implies that A is a noetherian space for the topology whose closed sets are the zerosets of functions in $\mathcal{S}(A)$.

References

[B] M. Boshernitzan, New "Orders of Infinity", Journal d'Analyse Mathematique, 41 (1982), 130-167.

[C] G. E. Collins, Quantifier Elimination for Real Closed Fields by Cylindrical Algebraic Decomposition, Automata Theory and Formal Language, 2nd G. I. Conference, Kaiserslautern, pp. 134-183, Berlin, Springer-Verlag, 1975.

[D] B. I. Dahn, The Limit Behaviour of Exponential Terms, to appear in Fund. Math.

[vdD1] L. van den Dries, Bounding the Rate of Growth of Solutions of Algebraic Differential Equations and Exponential Equations in Hardy Fields, Report, 23 pp., Stanford University, Jan. 1982 (unpublished).

[vdD2] L. van den Dries, Analytic Hardy Fields and Exponential Curves in the Real Plane, Amer. J. Math. 106 (1984), 149-167.

[vdD3] L. van den Dries, Remarks on Tarski's problem concerning (**R**,+,•,exp), Logic Colloquium 1982, pp. 97-121, Ed. by G. Lolli, G. Longo and A. Marcja, North-Holland, 1984.

[vdD4] L. van den Dries, Tarski's elimination theory for real closed fields, preprint.

[vdD5] L. van den Dries, Definable sets in 0-minimal structures, in preparation.

[vdD6] L. van den Dries, Elimination Theory for a class of transcendental equations, in preparation.

[G] E. A. Gorin, Asymptotic Properties of Polynomials and Algebraic Functions of Several Variables, Russian Math. Surveys 16.

[Hardt] R. Hardt, Semi-Algebraic Local-Triviality in Semi-Algebraic Mappings, Amer. J. Math., 102 (1980), 291-302.

[Hardy] G. H. Hardy, Orders of Infinity, Cambridge, 1910.

[Kh1] A. G. Khovanskii, On a class of systems of transcendental equations, Soviet Math. Dokl., 22 (1980), 762-765.

[Kh2] A. G. Khovanskii, Fewnomials and Pfaff Manifolds, Proc. Int. Congress of Mathematicians, Warsaw 1983.

[Kh3] A. G. Khovanskii, Real Analytic Varieties with the Finiteness Property and Complex Abelian Integrals, Functional analysis and its applications (Translated from Russian), 18 (1984), 119-127.

[K-P-S] J. Knight, A. Pillay, C. Steinhorn, Definable Sets in Ordered Structures II, preprint.

[Ł] S. Łojasiewicz, Ensembles Semi-Analytiques, mimeographed notes, IHES, 1965.

[P-S] A. Pillay and C. Steinhorn, Definable Sets in Ordered Structures I, preprint.

[Ri] J. J. Risler, Complexité et Géométrie Réelle (d'après A. Khovansky), Séminaire Bourbaki, no. 637, Nov. 1984.

[Ro] M. Rosenlicht, The Rank of a Hardy Field, Trans. AMS 280 (1983), 659-671.

[S] M. Singer, Asymptotic behavior of solutions of differential equations and Hardy fields, preliminary report, SUNY at Stony Brook, 1976 (unpublished).

[T] A. Tarski, A Decision method for Elementary Algebra and Geometry, 2nd ed. revised, Berkeley, and Los Angeles, 1951.

[W] H. Wolter, this volume.

LOGIC COLLOQUIUM '84
J.B. Paris, A.J. Wilkie, and G.M. Wilmers (Editors)

SITUATION SCHEMATA AND SYSTEMS OF LOGIC RELATED TO SITUATION SEMANTICS

Jens Erik Fenstad
University of Oslo

Logic and linguistics have over the ages lived in a sometimes uneasy relationship. From the strong medevial brew of metaphysics, logic and grammar characteristic of the modistae (eg. the Grammatica Speculativa og Thomas of Erfurt, 1315) we have gone to the modern extreme of the Americal structuralist school (L. Bloomfield, Z. Harris, 1930-50), where "theory" signified a scheme for discovery procedures rather than an attempt at explanation and understanding.

Such were the extremes; there were always well reasoned positions in the middle ground, eg. O. Jespersen's Philosophy of Grammar from 1924. But it seems fair to say that N. Chomsky's Syntactic Structures from 1956 marked a theoretical renewal of linguistic science. From a logician point of view we should note that with Chomsky one bond was forged between the two sciences centering around formal language theory, syntactic parsing and other aspects of "computational linguistics".

Meaning was always a recognized "box" in the Chomskian schema, but it was only with the work of R. Montague from 1967 that a technically adequate meaning component was joined to the syntactic part, and a richer picture emerged.

Of necessity there was in an initial phase sometimes confusion and misunderstanding - not always of the quite kind, and not always of substance. Sometimes matters of notation took precendence; sometimes grand empirical claims was read off arbitrary notational schemata.

Today we recognize the fundamental contributions of N. Chomsky and R. Montague, but we approach the current problems in the spirit of a cummulative science, exploiting the insights and results of the research community, which means both to accept and to reject. It is doubtful whether higher order intensional logic as developed by R. Montague is the right way to structure the "world"; what is beyond doubt is that Montague set an example for what it means for a theory of grammar to join linguistic form and meaning.

Remark. N. Chomsky and R. Montague are the "public figures" of the theoretical renewal. As a logician I would like to recall a "missed opportunity".

Around 1950 Y. Bar-Hillel resurrected the categorial grammar of S. Lesniewski (1929) and K. Ajdukiewicz (1936). Few have read Lesniewski's paper "Grundzüge eines neuen systems des Grundlagen der Mathematik"; his ideas were made more accessible through Ajdukiewicz's paper "Die Syntakische Konnexität". The 1950's was the time of great hopes for machine translation and Bar-Hillel intended categorial grammar as the proper theoretical frame-work for this enterprise.

But the enterprise failed, chiefly because there was no adequate theory of meaning. But something did exist at the time. Hans Reichenbach had included a chapter on "conversational languages" in his text book from 1947, Elements of Symbolic Logic, using as his technical tools higher order logic extended by certain "pragmatic" operators.

The two parts remained separate, no one saw or was at all interested in how to connect the categorial analysis of Ajdukiewicz with the semantical analysis of Reichenbach, which, perhaps, is a bit noteworthy since one of the explicit sources for Lesniewski's categories was Russel's theory of types, the very logical formalism Reichenbach built on.

However, independently of Bar-Hillel and Reichenbach, Haskell B. Curry read a paper in the linguistic seminar conducted by Z. Harris, Some logical aspects of grammatical structure. The paper was written in 1948, but was first published in 1961. In it Curry gives an analysis of the traditional parts of speech in terms of combinatory logic, thus combining both a syntactic and a semantic analysis. But no one took notice at the time.

Acknowledgement. In this talk I report on some joint work with P.K. Halvorsen, T. Langholm and J. van Benthem. This will be published in Equations, Schemata and Situations (to appear); henceforth referred to as ESS.

From linguistic form to meaning

The meaning or informational content of an utterance is determined by a number of contextual factors as well as the linguistic form in a strict sense (i.e. the phonology, morphology and syntax), and the interpretation must satisfy all the constraints imposed by all the relevant aspects of the larger "utterance situation". There seems to be no solid empirical evidence that one component has primacy over the others in arriving at the meaning content, thus there is no necessity in the orthodox point of view which requires one to channel the full informational content of the utterance and its context through a traditional syntactic structure in order to arrive at the meaning of the utterance.

On the contrary, experience has shown that one had to resort to a number of complicated devices in order to conserve strict compositionality in the passage from a standard syntactic structure to a relational semantic form. Combinatorial ingeniuity could overcome most difficulties, and the litterature on Montague grammar is full of pretty examples. But ingeniuity can be an artifice. And, more seriously, difficulties remained, e.g. with anaphoric reference.

Awareness of this led us in ESS to choose a different perspective. We would like to represent the constraints which are imposed on the interpretation of an utterance by its contextual and linguistic constituents through a cumulative system of constraint equations, reducing thereby the problem of systematically determining the meaning content to the problem of finding a consistent solution to the constraint equations.

In our actual system in ESS we are more modest. We have chosen a format which we call situation schemata as a theoretical notion convenient for summing up information from linguistic form and certain other aspects of the utterance situation. In ESS we present a grammar for a fragment of English in the style of lexical functional grammar (LFG), see Kaplan and Bresnan (1982), i.e. we have a simple context-free phrase structure augmented by constraint equations which are introduced in the main by the syntactic rules and in the lexicon, but could also possibly come from other features of the utterance situation. The first step toward the semantic interpretation is to seek a consistent solution to the constraint equations which, if it exists, can be represented in an array or tabular form. Such a representation, which is analogous to an f-structure in LFG-theory, is what we call a situation schema.

There is some similarity between the notion of a situation schema and the intuitive notion of "logical form", which should not be confused with the notion of a well-formed expression in a standard logical formalism. A simple declarative sentence has a main "semantic predicate" relating a number of "actors" playing various "rôles". Both the predicate and the rôle actors play can be modified (by adjectives, adverbs, prepositional phrases, relative clauses). This suggests the following basic format for situation schemata. Let ϕ be a simple declarative sentence.

Situation Schema of ϕ :

REL	-
ARG 1	-
⋮	⋮
ARG n	-
LOC	-

This means that SIT.ϕ, which we use to abbreviate the situation schema associated with ϕ, is a function with argument list REL, ARG 1,..., ARG n, LOC (and possibly others as we will see in the next section). The value of SIT.ϕ on the arguments is given by the predicate and rôles of the sentence ϕ. These values can either be simple, e.g. SIT.ϕ (ARG k) can be the name of an actor, or it can be a complex entity, i.e. SIT.ϕ (ARG k) can be some complex NP. In the latter case we get a new function or subordinate "box" as value determined by the structure of the NP (the determiner, the noun, the optional relative clause). The value of SIT.ϕ on LOC will be derived from the tense marker of ϕ.

Remark. We have presented the notion of situation schema as a refine-ment of the LFG notion of f-structure; recall also the "syntaxe structurale" of Tesnière (1959). We could also have started from the "discourse representation systems" of Kamp (1981), which gives a level of representation comparable to the "logical form" of situation schemata. Note, however, that we do not claim any psychological reality for our representional level. It would be interesting, even pleasing, but not necessary for our theoretical analysis. The notion of situation schema could also be abstracted out of various semantic theories, in particular, the format fits - and was partially designed to fit - the format of situation semantics, see Barwise and Perry (1983).

We shall recall some basic notions from situation semantics, see Barwise and Perry (1983), Barwise (these proceedings). Situation semantics is grounded in a set of primitives:

S	situations
Λ	locations
D	individuals
R	relations

For our immediate purposes we do not worry about the ontological status of the primitives; as mathematicians we assume that they come with some structure. A minimal requirement is that each relation in R is provided with an a-rity, i.e. a specification of the number of arguments slots or rôles of that relation. For the moment we impose no structure on the set of situations and individuals. The set Λ is or represents connected regions of space-time. Thus Λ may be endowed with a rich geometric structure, and should be if we were to give an analysis of seeing that which would correctly classify verb phrases describing spatio-temporal processes. Here we are much more modest and assume that Λ comes endowed with two structural relations:

$<$	temporally precedes
0	temporally overlaps,

to account for a simple-minded analysis of past and present tenses.

Primitives combine to form facts which are either located or unlocated. Let r be an n-ary relation, ℓ a location and $a_1,\ldots,a_n$ individuals. The format of a basic located fact is

$$\text{at } \ell : r,a_1,\ldots,a_n;1$$
$$\text{at } \ell : r,a_1,\ldots,a_n;0$$

where the first expresses that at location ℓ the relation r holds of the individuals $a_1,\ldots,a_n$; the second expresses that it does not hold. The basic format of unlocated facts is

$$r,a_1,\ldots,a_n;1$$
$$r,a_1,\ldots,a_n;0$$

In addition to basic facts we have "atomic" assertions concerning the location structure:

$\ell < \ell'$	ℓ wholly temporally precedes ℓ'
$\ell \circ \ell'$	ℓ and ℓ' temporally overlap.

A situation s determines a set of facts, but is not in the set-theoretic sense a set of facts. This distinction is not so important for our present discussion, but will be of crucial importance in the semantic analysis of attitude verbs (see Barwise and Perry (1983), p. 223). The "primitive" relation connecting situations, locations, individuals and relations is

$$\text{in } s : \text{at } \ell : r, a_1, \ldots, a_n; 1$$
$$\text{in } s : \text{at } \ell : r, a_1, \ldots, a_n; 0$$

the first expressing that in the situation s at location ℓ the relation r holds of $a_1, \ldots, a_n$. We have a corresponding reading of the second and of the unlocated versions.

The meaning of a (simple declarative) sentence ϕ is a relation between an utterance situation u and a described situation s. We shall use the situation scheme SIT.ϕ to spell out the connection between u and s, and we write the basic meaning relation as

$$u \, [\![\text{SIT.}\phi]\!] \, s.$$

We refer the reader to Barwise and Perry (1983), Chapter 6, for a full discussion and motivation for the chosen relational format of the meaning relation. In the next section we shall give a formal definition of situation schema associated to a fragment of English and use this to give a semantic interpretation of the fragment in a system of situation semantics.

Situation Schemata

Take a simple sentence such as

A girl handed the baby a toy

This sentence can be generated by a simple context free grammar. Indeed, any proposed grammatical analysis of such a specimen will be almost identical to the context free analysis:

S → NP VP
NP → Det N
VP → V NP NP
Det → a, the
N → girl, baby, toy
V → handed

This gives the following syntax tree for the sentence:

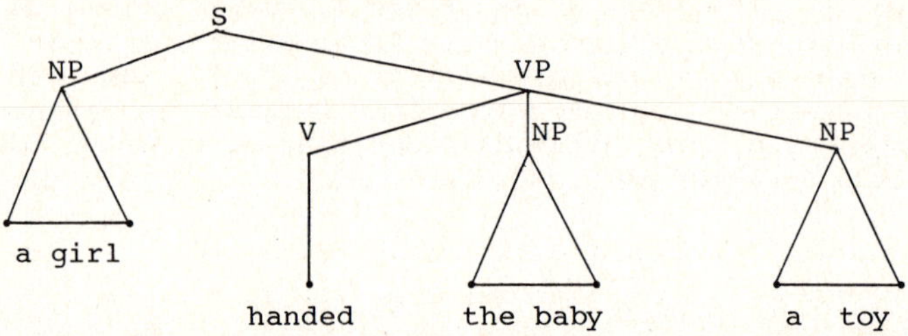

We want to convert this tree into a format better adapted for further semantic processing. From the point of view of "logical form" we have a main predicate <u>handed</u> and three rôles <u>a girl</u>, <u>the baby</u>, <u>a toy</u>. The standard contex free analysis gives us a tree where one of the rôles is separated from the predicate and the other actors. In this case it is not difficult to supplement the contex free structure with "constraint equations" leading to a functional form of the type

REL	hand
ARG 1	"a girl"
ARG 2	"the baby"
ARG 3	"a toy"
LOC	"past tense location"

We shall not enter into details of how to convert the syntax tree to schematic form; this is discussed in detail in part 1 of ESS. As remarked above we follow the pattern of LFG theory, and the reader is also urged to consult Kaplan and Bresnan (1982).

The schematic form above is only a first step toward a full unravelling of rôles actors play. In ESS we have proposed a formal definition of situation schema adapted to a simple fragment of English. The definition is given as a set of rewriting rules, using standard notational conventions including the Kleene ⋆-notation:

SIT.SCHEMA	→	(SIT)REL^n ARG1 ... ARGn LOC POL
SIT	→	<situation indeterminate>
REL^n	→	<n-ary relation constant>
ARGi	→	IND (SPEC COND (SIT.SCHEMA)*)
LOC	→	IND COND$_{loc}$
POL	→	{0|1}
IND	→	{<indeterminate>|<entity>}
SPEC	→	<quantifier>
COND	→	(SIT) REL ARG1 POL
COND$_{loc}$	→	REL$_{loc}$ ARG1 ARG2

Here entity stands for a proper name, quantifier for determiners such as a, the, some, most, ..., and REL$_{loc}$ is either $<$ or $\circ$ (see our discussion of the basic ideas of situation semantics above). The value of POL is either 1 or 0, compare the basic format of facts in situation semantics.

In ESS we have an algorithm which via the context free structure and the constraint equations converts a grammatically correct sentence of the fragment to a situation scheme, and every correctly generated situation schema is the schema derived from a sentence of the fragment. This algorithm has been implemented as an extension of the implementation of the LFG algorithm.

With our simple declarative sentence we arrive at the following situation schema:

REL	hand		
ARG 1	IND	IND 1	
	SPEC	a	
	COND	REL	girl
		ARG1	IND1
		POL	1
ARG 2	IND	IND 2	
	SPEC	the	
	COND	REL	baby
		ARG1	IND2
		POL	1
ARG 3	IND	IND 3	
	SPEC	a	
	COND	REL	toy
		ARG1	IND3
		POL	1
LOC	IND	IND4	
	$COND_{loc}$	REL_{loc}	$<$
		ARG1	IND4
		ARG2	1

The complete situation schema is a function of arguments REL, ARG1, ARG2, ARG3, LOC. The value on REL is a "constant", the 3-ary relation constant hand. The values on the other arguments are composite functional structures.

Let ϕ be our sample sentence, and let SIT.ϕ denote the above situation schema. Let u be an utterance situation and s a described situation. We have to explain the basic meaning relation u[[SIT.ϕ]]s. This we cannot do in full details in this talk, see ESS part 2 for the complete story. The reader will see that we have a fairly straight-forward recursion on the occurrences of SIT.SCHEMA within SIT.ϕ. There are a number of problems with quantifier scopes, anaphorie reference and definite descriptions which we evade in this brief exposition, again we refer the reader to ESS. Here we concentrate on a few simple points.

LOC is a function (or, rather, has a function as value) which anchors the described fact relative to the discourse location as given by u. Tense can be interpreted in many ways; taking a simpleminded referential and "unlocated" point of view we say that (the partial function) g anchors the location of SIT.ϕ, SIT.ϕ LOC, in the discourse situation u if

$$g(\underline{1}) = l_u$$
$$<\ ,\ g(\mathrm{IND.4}), l_u; 1,$$

where l_u is the discourse location determined by u and $<$ is the relation of temporally precede.

The functions corresponding to the three ARG's can in this case either be given singular NP-reading, or they can be interpreted as the standard generalized quantifiers, see ESS for details.

We could according to our definition of situation schema get a different box for ARG.2, viz.

ARG.2

<table>
<tr><td>IND</td><td>IND.2</td></tr>
<tr><td>SPEC</td><td><u>the</u></td></tr>
<tr><td>COND</td><td>
<table>
<tr><td>SIT</td><td>SIT.1</td></tr>
<tr><td>REL</td><td><u>baby</u></td></tr>
<tr><td>ARG1</td><td>IND.2</td></tr>
<tr><td>POL</td><td>1</td></tr>
</table>
</td></tr>
</table>

We see that ARG.2 COND has an extra argument which we can use to pick a "resource" situation for evaluating or anchoring <u>baby</u>; we have thus a mechanism for distinguishing between the so-called referential versus the attributive use of definite descriptions.

In this section we have described a map from linguistic form to a set of entities called situation schemata. The map is directed from linguistic form to the schemata, hence the primacy of the former over the latter. For us the situation schemata is a theoretical notion, a convenient way of summing up information relevant for the semantic interpretation. Two points follow:

- The notion of situation schemata is <u>open</u>; it can be amended or extended depending upon the chosen theory of grammar and upon how broadly one delinate the notion of linguistic form.
- Situation schemata can be adapted to various kinds of semantic or operational interpretation; it could be interpreted in higher order intensional logic as in the Montague tradition, or it could be given an operational interpretation in a suitable programming language.

In this section we have presented, or rather, hinted at an interpretation of $u[\![\mathrm{SIT}.\phi]\!]s$ in a system of situation semantics; we could as well have worked in a Montague style extending Halvorsen (1983). For an operational interpretation we could have based ourselves on the "instructable robot" developed by P. Suppes.

Systems of logic related to situation semantics

Sets with operations and relations form a (multisorted) algebraic structure. And we could pursue our study of

$$\mathfrak{A} = \langle S, \Lambda, D, R \rangle$$

in this spirit; see the contribution of J. Barwise to this volume. In ESS we proceed in a slightly different way. In the spirit of model-theoretic investigations we formulate several formal languages and use these to study structures of the type of $\mathfrak{A}$. Step by step we present a hierarchy, starting with a purely propositional language, continuing with a predicate logic system enriched with tense operators, ending up with rich many-sorted languages allowing direct reference to locations and, eventually, to situations themselves.

Our purpose is to use this as a kind of "semantical laboratory" for various notions proposed in the situation semantic literature, in particular, notions connected with inference and involvement. Our approach is somewhat conservative and not necessarily in harmony with the structure of natural language itself, where, for instance, quantifier mechanisms seem more basic than propositional connectives. However, in our scheme, linguistic form and model-theory interact through the level of situation schemata. Thus the formal tools we use to study the model-theoretic structure is not in any direct way to be thought of as the logical form of linguistic utterances.

We now turn to a brief exposition of some technical results. A situation semantic structure $\mathfrak{A} = \langle S, \Lambda, D, R \rangle$ has four non-empty domains. We impose no structure on S and D. We assume that each relation r in R comes with a specific "a-rity"; we thus take R to consist of two families $\{R_n^\ell\}_{n \geq 0}$, $\{R_n^u\}_{n \geq 0}$, where e.g. R_n^ℓ is the set of n-ary located relations in our model.

The location part Λ has one internal structural relation precede, i.e. Λ is of the form

$$\Lambda = \langle L, \underline{\text{precede}} \rangle$$

where L is a non-empty set and precede is a binary relation on L satisfying the structural conditions

(i) $\langle \lambda, \lambda \rangle \notin \underline{\text{precede}}$,

(ii) $\langle \lambda_1, \lambda_2 \rangle, \langle \lambda_3, \lambda_4 \rangle \in \underline{\text{precede}}$ implies that

either $\langle \lambda_1, \lambda_4 \rangle \in \underline{\text{precede}}$ or $\langle \lambda_3, \lambda_2 \rangle \in \underline{\text{precede}}$.

Overlap will be definable in terms of precede.

The four domains are connected by a "global" relation

$$\text{in } s : \text{at } \lambda : r, a_1, \ldots, a_n; \text{ pol},$$

which in the model will be represented by a set of tuples In of the form $\langle s, \lambda, r, a_1, \ldots, a_n, \text{pol} \rangle$, where the location λ is optional.

We impose one consistency constraint

$$\langle s,\vec{\alpha},1\rangle \in \text{In} \quad \text{implies} \quad \langle t,\vec{\alpha},0\rangle \notin \text{In}$$

(i.e. each situation is "actual"). We also require a compatability constraint

$$\langle s,\lambda,r,a_1,\ldots,a_n,pol\rangle \in \text{In} \quad \text{implies} \quad r \in R_n^{\ell},$$
$$\langle s,r,a_1,\ldots,a_n,pol\rangle \in \text{In} \quad \text{implies} \quad r \in R_n^{u}.$$

As a first stage we shall study a two-sorted language specific to one situation, i.e. we shall be interested in model-structures of the form

$$\langle D,R,\Lambda,\underline{in}\rangle$$

where D,R,Λ are as before and in is a set of tuples

$$\langle \lambda,r,a_1,\ldots,a_n,pol\rangle,$$

where the λ is optional, satisfying the consistency condition

$$\langle\vec{\alpha},1\rangle \in \underline{in} \quad \text{implies} \quad \langle\vec{\alpha},0\rangle \notin \underline{in},$$

and a compatability constraint as above. From In and an $s \in S$ we get a restricted structure by setting

$$\langle\vec{\alpha}\rangle \in \underline{in}_s \quad \text{iff} \quad \langle s,\vec{\alpha}\rangle \in \text{In}.$$

To study these structures we use a two-sorted language allowing both individual variables/constants and location variables/constants. We shall have both located and unlocated relation symbols and a special (unlocated) symbol $<$ for precede. We find it convenient to add the propositional constants t for truth and f for falsity. In addition to the quantifiers $\forall,\exists$ and the connnectives $\wedge,\vee$ we introduce both a strong negation $\neg$ and a weak negation $\sim$. Then notion of formula is entirely standard, we shall use $\phi \supset \psi$ as an abbreviation for $\sim\phi \vee \psi$ and $\phi \equiv \psi$ for $(\phi\supset\psi)\wedge(\psi\supset\phi)$. With this machinery we can write formulas such as

$$\forall\ell(\ell<\underline{\ell} \supset \exists x R(\ell,x,\underline{a})),$$

where $\underline{\ell}$ is a location constant and $\underline{a}$ an individual constant.

Let $\mathfrak{A} = \langle D,R,\Lambda,\underline{in}\rangle$ be a model-structure for our language and A a variable assignment, we shall use

$$[\![\phi]\!]^+_{\mathfrak{A},A}\ ,\ [\![\phi]\!]^-_{\mathfrak{A},A}$$

to mean that ϕ is true (false) in the model $\mathfrak{A}$ under the variable assignment A. In particular, if P is a located n-ary relation symbol, then

$$[\![P(\ell,x_1,\ldots,x_n)]\!]^+_{\mathfrak{A},A} \quad \text{iff} \quad \langle A(\ell),\ [\![P]\!]_{\mathfrak{A}},\ A(x_1),\ldots,A(x_n),1\rangle \in \underline{in}_{\mathfrak{A}}.$$
$$[\![P(\ell,x_1,\ldots,x_n)]\!]^-_{\mathfrak{A},A} \quad \text{iff} \quad \langle A(\ell),\ [\![P]\!]_{\mathfrak{A}},\ A(x_1),\ldots,A(x_n),0\rangle \in \underline{in}_{\mathfrak{A}}.$$

We use $[\![P]\!]_{\mathfrak{A}}$ to denote the interpretaion of the relation symbol P in the model $\mathfrak{A}$, i.e. $[\![P]\!]_{\mathfrak{A}} \in R_n^{\ell}$ in $\mathfrak{A}$.

The assumption $\langle\vec{\alpha},1\rangle \in$ <u>in</u> implies $\langle\vec{\alpha},0\rangle \notin$ <u>in</u> guarantees that no atomic formula is both true and false. If we drop this requirement we get a notion of generalized model with an associated four-valued logic, which may have some interest in certain intensional contexts, but which is of great technical use in the present context.

The following set of axioms and rules give a <u>complete axiomatization</u> of the logic:

A1. $(\phi \vee \phi) \supset \phi$.

A2. $\phi \supset (\phi \vee \psi)$.

A3. $(\phi \vee \psi) \supset (\psi \vee \phi)$.

A4. $(\phi \supset \psi) \supset ((\chi \vee \phi) \supset (\chi \vee \psi))$.

A5, $(\phi \vee \psi) \supset \phi$.

A6. $(\phi \wedge \psi) \supset (\psi \wedge \phi)$.

A7. $\phi \supset (\psi \supset (\phi \wedge \psi))$.

A8. $\neg\phi \supset \neg(\phi \wedge \psi)$.

A9. $\neg(\phi \wedge \psi) \supset \neg(\psi \wedge \phi)$.

A10. $\neg(\phi \vee \psi) \supset \neg\phi$.

A11. $\neg(\phi \vee \psi) \supset \neg(\psi \vee \phi)$.

A12. $(\neg\phi \wedge \neg\psi) \supset \neg(\phi \vee \psi)$.

A13. $\neg(\phi \wedge \psi) \supset (\neg\phi \vee \neg\psi)$.

A14. $\phi \supset \neg\neg\phi$.

A15. $\neg\neg\phi \supset \phi$.

A16. $\phi \supset \neg\sim\phi$.

A17a. $\neg\sim\phi \supset \sim\sim\phi$.

A17b. $\neg\phi \supset \sim\phi$.

A18. t.

A19. $\sim\neg$t.

A20. $\neg$f.

A21. $\sim$ f.

A22. $\sim(\beta_1 < \beta_2) \equiv \neg(\beta_1 < \beta_2)$

A23. $\sim(\beta < \beta)$.

A24. $((\beta_1 < \beta_2) \wedge (\beta_3 < \beta_4)) \supset ((\beta_1 < \beta_4) \vee (\beta_3 < \beta_2))$.

In the following axioms α is an individual or location variable, β a constant or variable of the appropriate kind; we asume that α does not occur within the scope of any quantifier binding β in A25 and A26.

A25. $\forall\alpha\phi \supset \phi(\beta/\alpha)$.

A26. $\phi(\beta/\alpha) \supset \exists\alpha\phi$.

A27. $\neg\forall\alpha\phi \equiv \exists\alpha\neg\phi$.

A28. $\neg\exists\alpha\phi \equiv \forall\alpha\neg\phi$.

The logic has three rules of inference

R1. $\dfrac{\phi,\ \phi\supset\psi}{\psi}$

R2. $\dfrac{\chi\supset\phi}{\chi\supset\forall\alpha\,\phi}$

R3. $\dfrac{\phi\supset\chi}{\exists\alpha\phi\supset\chi}$

where α is not free in χ.

At this point we should add one further remark on how to evaluate a formula ϕ in a model $\mathfrak{A}$ under a given variable assignment A. We first of all make the stipulation

$$[\![\beta_1 < \beta_2]\!]^{+}_{\mathfrak{A},A} \text{ iff } \langle A(\beta_1), A(\beta_2)\rangle \in \underline{\text{precede}}$$

$$[\![\beta_1 < \beta_2]\!]^{-}_{\mathfrak{A},A} \text{ iff } \underline{\text{not}}\text{: } [\![\beta_1 < \beta_2]\!]^{+}_{\mathfrak{A},A}\text{,}$$

see axiom A22 above. The relation $(\beta_1 < \beta_2)$ thus obeys a completeness requirement in additon to the consistency requirements. The evaluation of non-atomic formulas proceeds in the standard inductive manner. Let us emphasize the difference between strong and weak negation:

$$[\![-\phi]\!]^{+}_{\mathfrak{A},A} \text{ iff } [\![\phi]\!]^{-}_{\mathfrak{A},A}.$$

$$[\![\sim\phi]\!]^{-}_{\mathfrak{A},A} \text{ iff } \underline{\text{not}}: [\![\phi]\!]^{+}_{\mathfrak{A},A}$$

The reader may verify that axioms A17b exactly differentiate between generalized and proper models. Dropping A17b gives a complete axiomatization of generalized models.

Persistence is a crucial notion in connection with partial models as in situaiton semantics. A formula ϕ is called 1-persistent if for any two models $\mathfrak{A} = \langle D,R,\Lambda,\underline{in}\rangle$ and $\mathfrak{A}' = \langle D,R,\Lambda,\underline{in}'\rangle$ with the same D,R,Λ and with $\underline{in} \subset \underline{in}'$: if ϕ is true in $\mathfrak{A}$, then ϕ is true in $\mathfrak{A}'$. A formula is 0-persistent if $\neg\phi$ is 1-persistent. ϕ is persistent if it is both 0- and 1-persistent. (Notice that the definitions do not mention precede; there is no partiality about location structures.)

We call a formula pure if it does not contain $\sim$; obviously every pure formula is persistent.

Theorem. A formula ϕ is 1-persistent iff there is a pure formula ψ such that $\phi \equiv \psi$ is provable.

The proof uses an interpolation theorem for our two-sorted language; see Langholm (1985). Our theorem gives us for any persistent ϕ two pure formulas ψ_1 and ψ_2 such that $\phi \equiv \psi_1$ and $\neg\phi \equiv \neg\psi_2$. Is it possible to choose $\psi_1 = \psi_2$? Langholm (1985) has shown that this is so if ϕ is quantifierfree. We conjecture that it is true in general.

We turn to the full modelstructure $\mathfrak{A} = \langle S,D,\Lambda,R,In\rangle$. We add to our previous language a modal operator $\Box$, and we refer to the old language as the restricted language. Formulas of the extended language are defined as follows:

- If ϕ is a restricted formula, then $\Box\phi$ is a formula of the extended language;
- if ϕ and ψ are extended formulas, so are $\phi \wedge \psi$, $\sim\phi$; $\forall x\phi$, $\forall \ell\phi$.

We introduce $\phi \supset \psi$ as an abbreviation for $\sim\phi \vee \psi$ and $\Diamond\phi$ as an abbreviation for $\sim\Box\sim\phi$. We have the following truth condition for $\Box\phi$

- $\Box\phi$ is true in $\mathfrak{A}$, $\mathfrak{A} \models \Box\phi$, iff ϕ is true in $\langle D,R,\Lambda,in_s\rangle$. for all $s \in S$.
- $\Box\phi$ is false iff $\Box\phi$ is not true.

We may introduce a strong implicaiton $\phi \Rightarrow \psi$ as $\Box(\phi \supset \psi)$, which implies that every situation s which supports the truth of ϕ also supports the truth of ψ.

We have the following axioms and rules for the extended system.

A'1. $\Box(\phi \supset \psi) \supset (\Box\phi \supset \Box\psi)$.

A'2. $\Diamond \neg\phi \supset \Box{\sim}\phi$, ϕ atomic

A'3. $\Diamond t$.

A'4. $\forall\alpha\phi \supset \phi(\beta/\alpha)$.

A'5. $\forall\alpha\Box\phi \supset \Box\forall\alpha\phi$.

R'1. If ϕ is a theorem of the restricted system, then $\Box\phi$ is a theorem.

R'2. If ϕ follows tautologically from ψ, then we can derive ϕ from ψ.

R'3. $\dfrac{\chi \supset \phi}{\chi \supset \forall\alpha\phi}$, α not free in χ.

In ESS we prove a completeness theorem for the system.

It is in the context of the extended language that the notion of <u>involvement</u> is of particular interest: We let $\rightharpoondown$ denote the relation of involvement and impose the requirement that

$$\phi \rightharpoondown \psi$$

is true in the extended system if every situation satisfying ϕ can be <u>extended</u> to a situation satisfying ψ. At present the notion of involvement is not well understood in the full system. Langholm (1985) has analyzed the situation in the quantifierfree case, obtaining a number of characterization theorems.

Some concluding remarks

In ESS we have a system which gives an algorithm for converting linguistic form to a format which we have called a situation schema. The schemata have a well-defined (algebraic) structure, suggestive of "logical form". This is a structure different from the standard model theoretic one. The latter is intended as a "slice" of the real world and may carry a rich "non-linguistic" structure, e.g. to correctly classify verbphrases describing spatio-temporal processes means to endow the location set Λ with a rich geometric structure to account for the "singularities" involved. Thus the model theory is an indispensable part of the enterprise.

The situation schemata structures can be put to other uses. They are not "semantically complete" in the sense that they can always be interpreted in a model structure $\mathfrak{A}$; they may e.g. leave questions of quantifier scope undetermined. But they do carry linguistic information in a structured form. And we claim that they are well suited to discuss such topics as tense and anaphora and other discourse phenomena. (For another approach using DRS theory see Hans Kamp's contribution to this volume.)

In ESS we have restricted ourselves to basic facts of linguistic form and semantic structure. Our contribution has been to show in precise technical details how to join the two. This we believe is necessary for further progress. And our method is extendible.

References

K. Ajdukiewicz (1936), Die Syntaktische Konnexität, Studia Philosophica 1.

J. Barwise, this volume.

J. Barwise, J. Perry (1983), Situations and Attitudes, MIT Press.

J. Bresnan (1982), The Mental Representation of Grammatical Relations, MIT Press.

N. Chomsky (1956), Syntactic Structures, Mouton, The Hague.

H.B. Curry (1961), Some Logical Aspect of Grammatical Structure, in R. Jakobson (ed.), The Structure of Language and its Mathematical Aspects, Am.Math.Soc, Providence, Rhode Island.

J.E. Fenstad, P.K. Halvorsen, T. Langholm, J. van Benthem (1985), Equations, Schemata and Situations, CSLI, Stanford.

P.K. Halvorsen (1983), Semantics for Lexical-Functional Grammar, Linguistic Inquiry 14.

H. Kamp (1981), A Theory of Truth and Semantic Representation, in Groenendijk, Janssen, Stokhof (eds.), Fomal Methods in the Study of Language, Amsterdam.

H. Kamp, this volume.

R. Kaplan, J. Bresnan (1982), Lexical-Functional Grammar: A Formal System for Grammatical Representation, in J. Bresnan (1982).

T. Langholm (1985), Studies on the Logical Theory of Situation Semantics.

R. Montague (1973), Universal Grammar, Theoria 36.

H. Reichenbach (1947), Elements of Symbolic Logic, Macmillan, New York.

L. Tesnière (1959), Eléments de Syntaxe Structurale, Paris.

LOGIC COLLOQUIUM '84
J.B. Paris, A.J. Wilkie, and G.M. Wilmers (Editors)
© Elsevier Science Publishers B.V. (North-Holland), 1986

EFFECTIVE CONSTRUCTION OF MODELS

Julia F. Knight[1]

University of Notre Dame and University of Illinois at Urbana-Champaign

The models considered here are all countable, with universe ω, and the languages are all recursive. For a given complete theory T, the Henkin construction yields a model $\mathfrak{A}$ such that the complete diagram $D^c(\mathfrak{A})$ and the open diagram $D(\mathfrak{A})$ are recursive in T. Of course, $Th(\mathfrak{A})$ is always recursive in $D^c(\mathfrak{A})$. However, $D(\mathfrak{A})$ may have much lower degree. For example, true arithmetic has degree $\underset{\sim}{0}^{(\omega)}$, and if $\mathfrak{N}$ is the standard model (presented in the standard way), then $D(\mathfrak{N})$ is recursive.

Harrington [1] showed that there is a non-standard model $\mathfrak{A}$ of P such that $\deg(Th(\mathfrak{A})) = \underset{\sim}{0}^{(\omega)}$ and $\deg(D(\mathfrak{A})) \leq \underset{\sim}{0}'$. Harrington's construction involves infinitely many workers. Worker n, for $n \geq 1$, has the job of producing the Σ_n-diagram, using an oracle for $0^{(n)}$. To assure coherence, Worker n must continually guess what Worker $n + 1$ has done. Marker [4] used a three-worker variant of Harrington's construction to show that if $\underset{\sim}{d} > \underset{\sim}{0}^{(n)}$ for all n, then there is a non-standard model $\mathfrak{A}$ of true arithmetic such that $\deg(D(\mathfrak{A})) = \underset{\sim}{d}'$.

In the present paper, Harrington's method is used to produce a model $\mathfrak{A}$, of an arbitrary theory, such that $\mathfrak{A}$ realizes a prescribed set of types of bounded complexity. Harrington's result follows as a corollary, but here the model $\mathfrak{A}$ that is obtained has the feature that $(\deg(D(\mathfrak{A})))' = \underset{\sim}{0}'$.

A <u>type</u> is a set of formulas all having precisely the same set of free variables. (The variables of the type may be determined by examining any one formula in the type.) The <u>type of $\vec{m}$</u> in M is the set of formulas $\varphi(\vec{x})$ true of $\vec{m}$ in M and having a fixed sequence $\vec{x}$ of variables to which $\vec{m}$ is assigned. The <u>C_n-type of $\vec{m}$</u> in M consists of the Σ_n and Π_n formulas in the type of $\vec{m}$. Let $C_n(M)$ be the set of all C_n-types realized in M, and let $C_\omega(M) = \bigcup_{n\in\omega} C_n(M)$.

If $\mathcal{A} \subseteq P(\omega)$, then an <u>enumeration</u> of $\mathcal{A}$ is a binary relation $R \subseteq \omega \times \omega$ such that $\mathcal{A}$ is the family of sets $R_i = \{k:(i,k) \in R\}$. If $A = R_i$, then i is said to be an <u>R-index</u> of A. Let M be a model and let $S \subseteq \omega$. Then $C_\omega(M)$ is said to be <u>accessible from S</u> if there are recursive functions f, g, and h such that for all $n \geq 0$, $\varphi^{S(n)}_{f(n)} = \chi_R n$ and $\varphi^{S(n)}_{g(n)} = \chi_P n$, and for $n \geq 1$,

$\varphi^{S^{(n)}}_{h(n)} = \chi_{Q^n}$, where R^n is an enumeration of $C_n(M)$, P^n is a ternary relation associating with each R^{n+1}-index i an enumeration of the $C_n(M)$-types consistent with R^{n+1}_i, and Q^n is a ternary relation associating with each pair (i,j) a set of R^n-indices of types consistent with R^{n+1}_i and $P^n_{i,j}$ (where $P^n_{i,j}$ is the jth C_n-type that P^n associates with R^{n+1}_i), such that for each C_{n-1}-type R^{n-1}_k consistent with $R^{n+1}_i \cup P^n_{i,j}$, there is an index of a completion of $P^n_{i,j} \cup R^{n-1}_k$ associated with (i,j). It will be shown that for the kind of model M considered here, $C_\omega(M)$ is always accessible from $D(M)$.

A model M is said to be <u>weakly</u> <u>homogeneous</u> if for all $\Gamma(\vec{u},x) \in C_\omega(M)$, all $\vec{m} \in M$, if $\Gamma(\vec{m},x) \cup D^c(M)$ is consistent, then some $a \in M$ realizes $\Gamma(\vec{m},x)$. Every model of P is weakly homogeneous. For a weakly homogeneous model M, $C_\omega(M)$ has the important feature that if it includes the C_{n+1}-type $\Sigma(x,y)$ and the C_n-type $\Gamma(x,z)$, where $\Sigma \cup \Gamma$ is consistent, then it also includes some extension of $\Sigma \cup \Gamma$. Here is the main theorem.

Theorem 1. Let M be weakly homogeneous, and let $S \subseteq \omega$. If $C_\omega(M)$ is accessible from S, then there is a weakly homogeneous model $\mathfrak{A}$ such that $(\deg(D(\mathfrak{A})))' \leq (\deg S)'$ and $C_\omega(\mathfrak{A}) = C_\omega(M)$.

Proof: As in Harrington [1], the complete diagram of the model $\mathfrak{A}$ is produced by infinitely many workers in infinitely many stages. Here Worker n, for $n \geq 0$, has the job of determining the C_n-diagram of $\mathfrak{A}$, using an oracle for $S^{(n)}$. Worker 0 also determines $(D(\mathfrak{A}))'$. At stage s, Worker n makes tentative assignments of C_n-types to certain finite sequences of Henkin constants, forming $p^s_{n\,i}$ for $i \leq s$. The constants are identified with natural numbers. To be precise, each $p^s_{n\,i}$ is an ordered pair $(k,\vec{c})$, where $\vec{c}$ is a sequence of constants and k is an R_n-index of the type assigned to the constants. Sometimes $p^s_{n\,i}$ will be used to denote $\Gamma(\vec{c})$, where $\Gamma(\vec{x}) = R^n_k$.

The hope is that the tentative assignments will stabilize. For each n and i, there should be some s_{0_∞} such that for all $s > s_0$, $p^s_{n\,i} = p^{s_0}_{n\,i}$. Then $p^{s_0}_{n\,i}$ may be denoted by $p^\infty_{n\,i}$. For $s \geq s_{0_\infty}$, $p^s_{n\,i}$ is said to be <u>stable.</u> It should be the case that if $i < j$, then $p^\infty_{n\,i} \subseteq p^\infty_{n\,j}$. Moreover, $\bigcup_{n,i} p^\infty_{n\,i}$ should be consistent, and should serve as the complete diagram of a model $\mathfrak{A}$ with the desired properties.

For all n and s, $p^s_{n\,0}$ assigns to the empty sequence of constants the first R^n-index of the C_n-theory, which is recognized as the type with no variables. In general, $p^s_{n\,i}$ will depend on some sub-collection of the following: $p^k_{n+1\,j}$ for $j \leq k < i$, $p^k_{n\,j}$ for $k < s$, $j \leq i,k$ or for $k = s$, $j < i$, and $p^k_{n-1\,j}$ for $j \leq i,k$. The set of $p^k_{m\,j}$ on which $p^s_{n\,i}$ actually depends will be called the <u>support</u> of $p^s_{n\,i}$. In particular, $p^s_{n\,0}$ has empty support.

Worker n has an oracle for $S^{(n)}$ and can enumerate $S^{(n+1)}$. If Worker n knows the strategy that Worker n + 1 is using, then Worker n can approximate Worker n +1's actions. The description below of the method for determining $p_{n\ i}^{\ s}$, for $n,s \in \omega$ and $i \leq s$, should be thought of as a recursive function σ such that if x is a strategy that works on the support of $p_{n\ i}^{\ s}$; i.e., if $p_{m\ j}^{\ k} = \varphi^{S^{(m)}}_{x}(m,k,j)$ for all $p_{m\ j}^{\ k}$ in the support, then $\sigma(x)$ works for $p_{n\ i}^{\ s}$, so that $p_{n\ i}^{\ s} = \varphi^{S^{(n)}}_{\sigma(x)}(n,i,s)$.

The Recursion Theorem gives a fixed point e for $\sigma(x)$. Then it is easy to show, by induction on i, n, and s, that $p_{n\ i}^{\ s} = \varphi^{S^{(n)}}_{e}(n,i,s)$ for all $n,s \in \omega$, all $i \leq s$. Note that if $p_{m\ j}^{\ k}$ is in the support of $p_{n\ i}^{\ s}$, then $j < i$, or $j = i$ and $m < n$, or $j = i$ and $m = n$ and $k < s$.

There is a recursive list of the Henkin tasks to be done in constructing the model $\mathfrak{A}$. These tasks have the following forms:

(a) Choose an R^n-index for the type of some constant b over whatever constants have been used so far. (If Worker n is about to attempt this task in forming $p_{n\ s}^{\ s}$, then the "constants used so far" are the ones in $p_{n\ s-1}^{\ s}$.) The constants $c < b$ will all be included, but there may also be constants d much greater than b. The task will be referred to simply as "choosing the C_n-type of b," with no attempt to list the other constants.

(b) Witness some C_n-type $\Gamma(\vec{c},x)$, if it is consistent with the C_{n+1}-type assigned to the parameters $\vec{c}$.

(c) Put some e into $(D(\mathfrak{A}))'$, if possible.

Choosing the C_n-type of b comes before choosing the C_{n+1}-type of b. If $b < c$, then choosing the C_n-type of b comes before choosing the C_n-type of c. Choosing the C_n-type of b is not one of the first b tasks on the list. Choosing the C_n-type of b is a task that Worker n should do. Witnessing a C_n-type must be done cooperatively by Workers $m \leq n$. Worker n is the first to suggest a witnessing constant, but if $n > 0$, then Worker $n - 1$ replaces the constant by a different one, and the other workers make further suggestions. The constant that is finally adopted is one suggested by Woker 0. Putting a number e into $(D(\mathfrak{A}))'$ is a job for Worker 0.

At stage 0, Worker n does nothing but determine $p_{n\ 0}^{\ 0}$. For many stages s, Worker n does nothing new. That is, $p_{n\ i}^{\ s} = p_{n\ \ i}^{\ s-1}$ for all $i \leq s - 1$ and $p_{n\ s}^{\ s} = p_{n\ \ s-1}^{\ s-1}$. Worker n never does more than one thing new at a given stage s, and the new work is always displayed in $p_{n\ s}^{\ s}$. It may be that for some j such that $0 < j < s$, $p_{n\ k}^{\ s} = p_{n\ \ k}^{\ s-1}$ for $k < j$, $p_{n\ k}^{\ s} = p_{n\ \ j-1}^{\ s-1}$ for $j \leq k < s$, and $p_{n\ s}^{\ s} \neq p_{n\ s-1}^{\ s}$. This happens when Worker n corrects a mistake or does some follow-through on a witnessing task. Or, it may be that $p_{n\ i}^{\ s} = p_{n\ \ i}^{\ s-1}$ for all $i < s$, and $p_{n\ s}^{\ s} \neq p_{n\ \ s-1}^{\ s-1}$. This happens when n attempts some new task in forming $p_{n\,s}^{s}$.

At stage $s > 0$, Worker n checks $p_{n\,j}^{s-1}$ for $j \leq i$ in deciding whether to make $p_{n\,i}^{s} = p_{n\,i}^{s-1}$. There is never any mistake in $p_{n\,0}^{s-1}$. Suppose that there is no mistake in $p_{n\,j}^{s-1}$ for $j < i$. If $p_{n\,i}^{s-1} \neq p_{n\,i-1}^{s-1}$, then $p_{n\,i}^{t}$ must have remained unchanged (as t varied) since state i. In checking $p_{n\,i}^{s-1}$, Worker n considers the support of $p_{n\,i}^{s-1}$ to be the support of $p_{n\,i}^{i}$. If n discovers a mistake, then $p_{n\,k}^{s} = p_{n\,i-1}^{s-1}$ for $i \leq k < s$, and the corrected version of $p_{n\,i}^{s-1}$ (if available at stage s) appears in $p_{n\,s}^{s}$. The upper part of the support for $p_{n\,s}^{s}$ (the part produced by Workers $n + 1$ and n) is the same as for $p_{n\,i}^{i}$, but the lower part of the support (the part produced by Worker $n - 1$, if $n > 0$) has grown. Worker n will look at $p_{n-1\,s}^{\infty}$, not just $p_{n-1\,i}^{\infty}$, in making the correction.

If there are no mistakes discovered at stage s, Worker n may do some follow-through on a witnessing task. Whatever follow-through Worker n does at stage s will make $p_{n\,s-1}^{s} = p_{n\,s-2}^{s-1}$, and will make $p_{n\,s}^{s}$ the same as $p_{n\,s-1}^{s-1}$ except for a change in witnessing constant. In choosing the new witnessing constant, Worker n looks down at $p_{n-1\,s}^{\infty}$ and does not look up at all. Worker n counts on Worker $n + 1$ to wait while this is going on.

If $\Gamma(\vec{u},x)$ is a C_n-type, then Worker n may initiate the task of witnessing $\Gamma(\vec{c},x)$ in forming $p_{n\,s}^{s}$ by choosing a special witnessing constant b and trying to find a type that amalgamates $\Gamma(\vec{u},x)$ with the type $\Sigma(\vec{u},v)$ from $p_{n\,s-1}^{s}$. Then, if $n > 0$, Worker n waits for Worker $n - 1$ to respond. The chosen constant b will alert Worker $n - 1$ to the fact that Worker n is trying to witness $\Gamma(\vec{c},x)$. Worker $n - 1$ will replace the constant b by a new constant b', and will choose a type for b' that is consistent with n's choice of type. Then Worker $n - 1$ joins Worker n in waiting for Worker $n - 2$ to respond. Finally, Worker 0 suggests a constant that everyone can accept.

Each constant carries the history of apparently correct work. If $\Gamma = R_{j}^{n}$, and Worker n first attempts to witness $\Gamma(\vec{c},x)$ at stage s, then the witnessing constant used in $p_{n\,s}^{s}$ will be $(n,j,\vec{c},(n,s))$. (Here tuples are identified with their Gödel numbers.) Suppose that at stage $t > s$, Worker n discovers a mistake in this attempt. The mistake will actually appear in $p_{n\,r}^{t-1}$ for some r such that $s \leq r \leq t - 1$, where $p_{n\,r}^{r}$ is just like $p_{n\,s}^{s}$ except for the constant. (The fact that the work on the task, the choice of amalgam type, appears in $p_{n\,r}^{t-1}$ instead of $p_{n\,s}^{t-1}$ results from the follow-through procedure.) If Worker n has a new amalgam type to try at stage t, then this appears in $p_{n\,t}^{t}$, with the witnessing constant $(n,j,\vec{c},(n,t))$.

If Γ is a C_r-type for some $r > n$, and Worker n first attempts at stage s to respond to Worker $n + 1$'s use of a witnessing constant a for $\Gamma(\vec{c},x)$, then n will use the constant $b = a\hat{}(n,s)$ in $p_{n\,s}^{s}$. If Worker n finds a mistake and attempts to do its part of this task again at stage t, then the constant used in $p_{n\,t}^{\,t}$ will be $\hat{a}(n,t)$.

Suppose that Worker n is doing the follow-through on the witnessing of $\Gamma(\vec{c},x)$, where Γ is either a C_n-type or a C_r-type for $r > n$. Suppose that in $p_{n\,s}^{\,s}$, Worker n either initiated the witnessing or made an initial response to something that Worker $n+1$ did on the task. Suppose that no mistake has been found up to stage t and follow-through is still called for. Let b be the constant appearing in $p_{n\;t-1}^{\,t-1}$, and suppose that Worker n is choosing the new constant to use, with the same type, in $p_{n\,t}^{\,t}$. There are two cases.

First, suppose that there is a "stable" response from Worker $n-1$. What this means is that $p_{n-1\,t}^{\,\infty}$ has a constant d with a history indicating that it and b are both descended from the constant introduced by n at stage s, and there have been responses by Worker k for all $k < n$. In this case, Worker n uses the constant d in $p_{n\,t}^{\,t}$. If no mistakes are found later, this will become $p_{n\,t}^{\,\infty}$. The follow-through process is complete. Second, suppose there is no stable response from $n-1$ ($p_{n-1\,t}^{\,\infty}$ does not have any witnessing constant descended from the one introduced by Worker n at stage s). Then in $p_{n\,t}^{\,t}$, Worker n uses the constant $b' = \hat{b}(n,t)$. At the next stage, $p_{n\,t}^{\,t}$ will be erased, and a new constant will be used in $p_{n\;t+1}^{\,t+1}$, so b' will not appear in $p_{n\,t}^{\,\infty}$.

If Worker n finds nothing to correct at stage s, and n does not anticipate any further response from Worker $n-1$, then n may attempt something new in forming $p_{n\,s}^{\,s}$. The method for choosing which task to attempt, if any, will be discussed below. The rules will assure that if Worker n attempts to do the task of choosing the C_n-type of b, then $b < s$. The method for choosing witnessing constants, discussed above, assures that if Worker n acts on a witnessing task in forming $p_{n\,s}^{\,s}$, then the witnessing constant will be some $b > s$. If $n > 0$, then there is at most one constant appearing in $p_{n\,s}^{\,s}$ and not in $p_{n\;s-1}^{\,s}$, and the constant tells what task was attempted in forming $p_{n\,s}^{\,s}$. Worker 0 may introduce more than one new constant at a time, in trying to control the jump of the open diagram.

If Worker n is to attempt some new task in forming $p_{n\,s}^{\,s}$, then the following conditions must be met.

(1) Worker $n+1$ must appear to have done, in $p_{n+1\;j}^{\,s-1}$ for $j \leq s-1$, all tasks of Type (a) that come before the task that Worker n is now considering

(2) Worker n must have done, in $p_{n\,j}^{\,s}$ for $j \leq s-1$ (in an apparently correct fashion), all possible tasks that come before the one being considered.

(3) If $n > 0$, then Worker $n-1$ must have done, in $p_{n-1\;j}^{\,\infty}$ for $j \leq s$, everything that could be done before Worker n does more.

(4) The task being considered must be among the first s tasks, so there must be some one of the first s tasks that still needs work.

(5) It must seem to Worker n that the task being considered can actually be done so as to maintain consistency with $p_{n+1\ j}^{\ \ s-1}$ for $j \leq s - 1$, and n must be able to see for sure that the proposed $p_{n\ s}^{\ s}$ is consistent with $p_{n-1\ s}^{\ \infty}$.

If Worker n never did anything new beyond what is in $p_{n\ s-1}^{\ s-1}$, then Worker n - 1 could only do finitely many tasks. There would be some t_0 such that for all $t > t_0$, $p_{n-1\ t}^{\ \infty} = p_{n-1\ t_0}^{\ \infty}$. When Worker n - 1 has done all that it can do, including follow-through on witnessing tasks, then Worker n - 1 must look for new constants appearing above. worker n can tell when Worker n - 1 is going to be forced in this way to look at $p_{n\ s-1}^{\ s-1}$. Worker n repeats certain things until Worker n - 1 responds.

Suppose that Worker r (far above n) initiates the witnessing of $\Gamma(\vec{c},x)$. Worker r - 1 will probably hand down some incorrect requests for help in witnessing, then a correct request, and then some imaginary requests. The imaginary requests don't contribute to $\bigcup_{t\in\omega} p_{r-1\ t}^{\ \infty}$, but they may be seen in $\bigcup_{t\in\omega} p_{r-2\ t}^{\ \infty}$. Could this prevent Worker n from ever getting to work on certain tasks? Suppose that there is some first Type (a) task, choosing the C_n-type of b, that is not done because old witnessing tasks (earlier on the master list) keep interfering. Suppose that the most complex of the old witnessing tasks involves a C_r-type. Worker r will eventually stop introducing new witnessing constants for the old tasks. Let $n < m < r$. If Worker m + 1 produces some $p_{m+1\ j}^{\ j}$ (that Worker m will look at when deciding what to do next) with no new witnessing constants for the old tasks, then Worker m will necessarily produce some $p_{m\ k}^{\ k}$ (that Worker m - 1 will look at) with no new witnessing constants for the old tasks. This means that eventually Worker n will be able to choose the C_n-type of b.

How does Worker n check the consistency of $\Gamma(\vec{x})$ and $\Sigma(\vec{x},\vec{y})$, where $\Gamma = R^{n+1}_{\ i}$ and $\Sigma = R^{n}_{\ j}$? At stage s, n uses a limited portion of P^n; namely, the first s types associated with the R^{n+1}-index i. If n finds that one of these matches $R^n_{\ j}$ up to s, then n assumes that the types are consistent. Otherwise, not. Similarly, if $\Gamma = R^n_{\ i}$ and $\Sigma = R^{n-1}_{\ j}$, then at stage s, n looks at the first s types associated by P^{n-1} with the R^n-index i. If one of these fully matches $R^{n-1}_{\ j}$, then n knows that the types are consistent. Otherwise, n assumes that they are not consistent.

How does Worker n attempt at stage s the task of choosing a C_n-type for b? First, consider n = 0. Worker 0 has guesses at $p_{1\ j}^{\ s-1}$ for $j < s$ and knows $p_{0\ s-1}^{\ s}$. Let the guess at $p_{1\ s-1}^{\ s-1}$ be $\Gamma(\vec{a})$, where $\Gamma(\vec{x}) = R^1_{\ i}$, and let $p_{0\ s-1}^{\ s} = \Sigma(\vec{a},\vec{c})$, where $\Sigma(\vec{x},\vec{y}) = R^0_{\ j}$. Worker 0 looks at the first s types associated by P^0 with the index i, searching for some $\Lambda(\vec{x},\vec{y},z)$ such that Λ extends Σ, up to s. Then 0 looks at the first s R^0-indices to see if there is some k such that $R^0_{\ k}$ matches Λ, up to s. If these searches are successful, then Worker 0 has in mind assigning the R^0-index k to $\vec{a}\hat{}\vec{c}\hat{}b$. Before doing this, however, Worker 0 makes a final check that this proposed $p_{0\,s}^{\,s}$ is consistent with $p_{1\,j}^{\,s-1}$ for $j < s$.

Worker 0 knows that if $\Gamma \cup \Sigma$ is consistent, then an appropriate type Λ and index k will eventually turn up. Therefore, if $\Gamma \cup \Sigma$ appears to be consistent, Worker 0 considers the task to be possible and refrains from going on to the next task (at stages $t > s$) until after this task appears to be done. Worker 0 may make errors, but there is no lower worker to worry about.

Next, consider $n > 0$. Worker n has guesses at $p_{n+1\,j}^{\,s-1}$ for $j < s$ and knows $p_{n\,s-1}^{\,s}$ and $p_{n-1\,s}^{\,\infty}$. Let the guess at $p_{n+1\,s-1}^{\,s-1}$ be $\Gamma(\vec{a})$, where $\Gamma(\vec{x}) = R^{n+1}_i$. Let $p_{n\,s-1}^{\,s} = \Sigma(\vec{a},\vec{c})$, where $\Sigma(\vec{x},\vec{y}) = R^n_j$, and let $p_{n-1\,s}^{\,\infty} = \Theta(\vec{a},\vec{c},b,\vec{d})$, where $\Theta(\vec{x},\vec{y},z,\vec{u}) = R^{n-1}_k$. Worker n checks the consistency of $\Gamma(\vec{x}) \cup \Sigma(\vec{x},\vec{y})$ and of $\Sigma(\vec{x},\vec{y}) \cup \Theta(\vec{x},\vec{y},z,\vec{u})$. In checking the consistency of $\Gamma \cup \Sigma$, Worker n looks at the first s types associated by P^n with the R^{n+1}-index i, hoping to find some $P^n_{i,j'}$ that matches R^n_j up to s. Then Worker n looks at the first s R^n-indices associated by Q^n with the pair (i,j'), hoping to find some $R^n_\ell = \Lambda(\vec{x},\vec{y},z)$ such that $\Lambda(\vec{x},\vec{y},z) \cup \Theta(\vec{x},\vec{y},z,\vec{u})$ is consistent.

Note that Worker n will not believe that $\Lambda \cup \Theta$ is consistent unless it actually is consistent. If the searches are successful, then Worker n has in mind assigning the R^n-index ℓ to $\vec{a}\hat{}\vec{c}\hat{}b$. Worker n makes a final check that this proposed $p_{n\,s}^{\,s}$ is consistent with $p_{n+1\,j}^{\,s-1}$ for $j < s$. If $\Gamma \cup \Sigma$ is consistent and Σ is "correct" (i.e., $p_{n\,s-1}^{\,s}$ is stable), then $\Sigma \cup \Theta$ should be consistent, and the desired Λ will eventually turn up. Therefore, if $\Gamma \cup \Sigma$ appears to be consistent, Worker n considers the task to be possible and does not go on to the next task (at later stages) until after this task has been done. If $p_{n+1\,j}^{\,s-1} \subseteq p_{n+1\,s-1}^{\,s-1}$ for $j < s - 1$ and $p_{n+1\,s-1}^{\,s-1} \cup p_{n\,s-1}^{\,s}$ is consistent, then once Worker n is correct about $p_{n+1\,s-1}^{\,s-1}$ and its consistency with $p_{n\,s-1}^{\,s}$, there will be no mistakes -- the task is done correctly or not at all. This is important because it means that Worker n - 1 cannot do too many tasks until Worker n has finished this one.

How does Worker n initiate at stage s the task of witnessing a C_n-type $\Gamma(\vec{c},u)$? Suppose that $p_{n+;\,s-1}^{\,s-1}$ appears to be $\Sigma(\vec{c},\vec{d})$, where $\Sigma(\vec{x},\vec{z}) = R^{n+1}_i$, and let $p_{n\,s-1}^{\,s} = \Lambda(\vec{c},\vec{d},\vec{a})$, where $\Lambda(\vec{x},\vec{z},\vec{y}) = R^n_j$. At stage s, Worker n looks at the first s types associated by P^n with the R^{n+1}-index i. If there is one that, up to s, appears to complete $\Lambda(\vec{x},\vec{z},\vec{y}) \cup \Gamma(\vec{x},u)$, then n considers using this type for $p_{n\,s}^{\,s}$. Worker n looks for an R^n-index $k < s$ for a type that matches this one up to s. If this search is successful, then n has in mind the R^n-index to use in $p_{n\,s}^{\,s}$. The choice of constant was described above.

Worker n makes a final check that the proposed $p_n{}^s{}_s$ is consistent with $p_{n+1}{}^{s-1}{}_j$ for $j < s$ and with $p_{n-1}{}^{\infty}{}_s$. If the search is unsuccessful, then Worker n assumes that the task doesn't need doing and goes on.

Suppose that a witnessing task was initiated by Worker r, for some $r > n$, and Worker n is trying to respond in $p_n{}^s{}_s$. Suppose that $p_{n+1}{}^{s-1}{}_{s-1}$ appears to be $\Sigma(\vec{a},\vec{b})$, where b is the witnessing constant calling for response, and $\Sigma(\vec{x},y) = R^{n+1}{}_i$. Let $p_n{}^s{}_{s-1} = \Lambda(\vec{a},\vec{c})$, where $\Lambda(\vec{x},\vec{z}) = R^n{}_j$. At stage s, Worker n looks at the first s C_n-types that P^n associates with the R^{n+1}-index i. What n hopes to find is a type $\Theta(\vec{x},y,\vec{z})$ that extends $\Lambda(\vec{x},\vec{z})$, at least up to s. If there is one, then n looks for an R^n-index $k < s$ for a type that matches Θ up to s.

If the searches are successful, then Worker n has in mind using the R^n-index k in $p_n{}^s{}_s$. The choice of constant was described above. Worker n checks that the proposed $p_n{}^s{}_s$ is consistent with $p_{n+1}{}^{s-1}{}_j$ for $j < s$ and with $p_{n-1}{}^{\infty}{}_s$. If $\Sigma \cup \Lambda$ is consistent, then Worker n knows that the searches for Θ and k should eventually succeed, so n does not go on to the next task until after this task is done. There may be mistakes, but if Worker n-1 goes on to later tasks, Worker n is not prevented from completing this one.

How does Worker 0 put the number e into $(D(\mathfrak{A}))'$? Suppose that $p_1{}^{s-1}{}_{s-1}$ appears to be $\Gamma(\vec{a})$, where $\Gamma(\vec{x}) = R^1{}_i$, and let $p_0{}^s{}_{s-1} = \Sigma(\vec{a},\vec{b})$, where $\Sigma(\vec{x},\vec{y}) = R^0{}_j$. Worker 0 looks at the first s types associated by P^0 with the R^1-index i, hopping to find a type $\Lambda(\vec{x},\vec{y},\vec{z})$ that extends $\Sigma(\vec{x},\vec{y})$, at least up to s, and such that for some choice of constants $\vec{c}$, having $\Lambda(\vec{a},\vec{b},\vec{c})$ as a subset of $D(\mathfrak{A})$ would cause $\varphi_e^{D(\mathfrak{A})}(e)$ to converge, with a computation less than s. If Worker 0 finds $\Lambda(\vec{a},\vec{b},\vec{c})$, with a convergent computation, then 0 looks at the first s R^0-indices to see if there is some $R^0{}_k$ that matches Λ up to s. If the searches are successful, then Worker 0 has in mind assigning k to $\vec{a}\hat{}\vec{b}\hat{}\vec{c}$ in $p_0{}^s{}_s$. Worker 0 checks the consistency of this with $p_1{}^{s-1}{}_j$ for $j < s$. If the searches are not successful, then Worker 0 assumes that the task is impossible and goes on.

If Worker n is checking at stage t the work done at stage s, which has seemed correct through state $t - 1$, then n looks at precisely the same upper support. Suppose, for example that in $p_n{}^s{}_s$, Worker n responded to the appearance of a new witnessing constant in $p_{n+1}{}^{s-1}{}_{s-1}$. The follow-through process will have pushed the type up from $p_n{}^s{}_s$ into $p_n{}^{t-1}{}_r$, for some r such that $s \leq r \leq t - 1$. Worker n will check the choice of type against $p_{n+1}{}^{s-1}{}_j$ for $j < s$. The guesses that Worker n makes about $p_{n+1}{}^{s-1}{}_j$ at stage t are more trustworthy than the ones at stage s. In addition, n looks at t types in carrying out various searches, and considers initial segments of types of length t.

Eventually, Worker n knows what $p_{n\,s}^{s}$ "should" have been; i.e., what it would have been with perfect information. If $n = 0$, and the task attempted in $p_{0\,s}^{s}$ was either choosing the C_0-type of some b or putting some e into $(D(\mathfrak{A}))'$, then a mistake is corrected at stage t by making $p_{0\,t}^{t}$ equal to what $p_{n\,s}^{s}$ should have been. For $n > 0$, if Worker n makes a correction at stage t, then the new $p_{n\,t}^{t}$ should be consistent with $p_{n-1\,t}^{\infty}$, not just with $p_{n-1\,s}^{\infty}$. Then what would have been correct for $p_{n\,s}^{s}$ may not be correct for $p_{n\,t}^{t}$.

Suppose that Worker n is correcting a mistake in witnessing, and at stage t, n has located the amalgam type that should have been used at stage s. The only adjustment needed is a change in the witnessing constant. Here $p_{n\,t}^{t}$ will be consistent with $p_{n-1\,t}^{\infty}$ if it is consistent with $p_{n-1\,s}^{\infty}$.

Now, suppose that Worker n is correcting the choice of the C_n-type for b. Suppose that $p_{n+1\,j}^{s-1} \subseteq p_{n+1\,s-1}^{s-1}$ for $j < s - 1$, and suppose that $p_{n+1\,s-1}^{s-1} \cup p_{n\,s-1}^{s}$ is consistent. Once n knows $p_{n+1\,s-1}^{s-1}$ and is correct about the consistency of $p_{n+1\,s-1}^{s-1} \cup p_{n\,s-1}^{s}$, n will not make further mistakes of the kind that put something new into $p_{n\,t}^{t}$.

What remains in the task is a pair of searches. First, n must look for evidence that $p_{n\,s-1}^{s} \cup p_{n-1\,t}^{\infty}$ is consistent, and then n must search for a C_n-type, in the appropriate variables, that is consistent with $p_{n+1\,s-1}^{s-1} \cup p_{n\,s-1}^{s}$ and that has among the C_{n-1}-types consistent with it a match for the type in $p_{n-1\,t}^{\infty}$. When the searches terminate, Worker n has the correct C_n-type to put into $p_{n\,t}^{t}$.

The searches would certainly end if $p_{n-1\,t}^{\infty}$ stayed fixed as t varied. However, if $p_{n-1\,t}^{\infty}$ kept growing with t, then they might not. There is some first Type (a) task, say fixing the C_n-type of c, that is not attempted by Worker n before the start of the first search and cannot then be attempted until after the correct C_n-type for b is obtained. Worker $n - 1$ is limited to a fixed finite set of tasks in $p_{n-1\,t}^{\infty}$, so long as the constant c does not appear in $p_{n\,t-1}^{t-1}$. With $p_{n-1\,t}^{\infty}$ staying constant (while Worker $n - 1$ waits for c to appear above), Worker n can complete the searches and find the correct C_n-type for b.

Now, the method for determining all $p_{n\,i}^{s}$ has been described. It should be clear that $p_{n\,i}^{\infty} \subseteq p_{n\,j}^{\infty}$ if $i < j$. It must be shown that $\bigcup_{n,i} p_{n\,i}^{\infty}$ is consistent.

<u>Claim</u>: If $p_{n\,s}^{s}$ is stable, then $p_{n\,s}^{s}$ is consistent with $p_{n+1\,j}^{j}$ for any stable $p_{n+1\,j}^{j}$ such that $j < s$.

The claim is clearly true if $p_{n\,s}^{s}$ attempts something new (i.e., is not a correction). Suppose that $p_{n\,s}^{s}$ is a correction of $p_{n\,j}^{j} = p_{n\,j}^{s-1}$. Then $p_{n\,s}^{s}$ is consistent with $p_{n+1\,k}^{j-1}$ for $k < j$, and n does not look up at $p_{n+1\,k}^{\ell}$

for $k \le \ell$, $j \le \ell < s$. Now, Worker n + 1 cannot attempt anything new in forming these $p_{n+1\,\ell}^{\ell}$ unless n has done all that it can do. For $j \le \ell < s$, $p_{n+1\,\ell}^{\infty} = p_{n+1\,j-1}^{\infty}$. Similarly, if $p_{n\,s}^{s}$ is the final version of a witnessing task, Worker n + 1 will wait for n to do something new.

Each $p_{n\,i}^{\infty}$ is equal to $p_{n\,j}^{j}$ for some stable $p_{n\,j}^{j}$, $j \le i$. For $i = 0$, $p_{n\,0}^{0}$ is stable, and $\bigcup_n p_{n\,0}^{0}$ is consistent. Suppose that $\bigcup_n p_{n\,i}^{\infty}$ is consistent. If $\bigcup_n p_{n\,i+1}^{\infty}$ is inconsistent, then there is some first m such that $\bigcup_n p_{n\,i}^{\infty} \cup \bigcup_{n \le m} p_{n\,i+1}^{\infty}$ is inconsistent. Note that the constants of $p_{n+1\,k}^{\infty}$ are among those of $p_{n\,k}^{\infty}$. (Permanent choices of witnesses come up from below, and if the C_{n+1}-type of b is being chosen when b has not been used for witnessing, then the C_n-type will have been chosen first.) Then $p_{m\,i+1}^{\infty} \cup p_{m+1\,i}^{\infty}$ is inconsistent. It must be the case that $p_{m\,i+1}^{\infty} \neq p_{m\,i}^{\infty}$. Then $p_{m\,i+1}^{i+1}$ is stable. Also, $p_{m+1\,i}^{\infty} = p_{m+1\,j}^{j}$ for some $j \le i$, and $p_{m+1\,j}^{j}$ is stable. By the Claim, $p_{m\,i+1}^{i+1} \cup p_{m+1\,j}^{j}$ must be consistent, a contradiction. Then $\bigcup_n p_{n\,i+1}^{\infty}$ is consistent, and by induction, $\bigcup_n p_{n\,j}^{\infty}$ is consistent for all j.

Let $\mathfrak{A}$ be the model such that $D^c(\mathfrak{A}) = \bigcup_{n,i} p_{n\,i}^{\infty}$. It is not difficult to see that $\mathfrak{A}$ is weakly homogeneous, with $C_\omega(\mathfrak{A}) = C_\omega(M)$. For each $n \ge 0$, the assignment of C_n-types to sequences of constants is recursive in $S^{(n+1)}$. It must be shown that $(D(\mathfrak{A}))'$ is recursive in S'. To decide whether $e \in (D(\mathfrak{A}))'$, first locate s such that the task of putting e into $(D(\mathfrak{A}))'$ is not attempted in $p_{0\,t}^{t}$ for any $t > s$. Then check to see if the task is done in some *stable* $p_{0\,j}^{j}$ for $j \le s$.

This completes the proof of Theorem 1. Note that if $(D(\mathfrak{A}))'$ is recursive in S', then the Σ_1-diagram of $\mathfrak{A}$ is recursive in S', and then the Σ_n-diagram is recursive in $S^{(n)}$ for all $n \ge 1$.

In Harrington [1], the model and its theory are produced simultaneously. Theorem 1 above produces models for arbitrary theories. Applying Theorem 1 to a particular theory yields a sharper version of Harrington's result. This will be done in Corollary 2, but a little background is needed first.

A *Scott set* is a set $\mathcal{A} \subseteq P(\omega)$ such that (1) if $A \in \mathcal{A}$ and B is recursive in A, then $B \in \mathcal{A}$, (2) if $A, B \in \mathcal{A}$, then so is $\{2n : n \in A\} \cup \{2n+1 : n \in B\}$, (3) if $\mathcal{J} \subseteq 2^{<\omega}$ is a tree such that $\mathcal{J} \in \mathcal{A}$, then $\mathcal{J}$ has a *path* (branch of maximum length) in $\mathcal{A}$. If T extends P, then a set $S \subseteq \omega$ is *representable* with respect to T if there is a formula $\varphi(x)$ such that $T \vdash \varphi(n)$ for $n \in S$ and $T \vdash \sim\varphi(n)$ for $n \notin S$. The family of sets representable with respect to T is denoted by Rep(T). Scott [5] showed that $\mathcal{S}$ is a countable Scott set iff $\mathcal{S} = \mathrm{Rep}(T)$ for some completion T of P. If $\mathfrak{A}$ is a model of P, then the family of standard parts of sets definable with parameters in $\mathfrak{A}$ is a Scott set, denoted by $\mathcal{SS}(\mathfrak{A})$. If T is a completion of P, then the countable Scott sets that can serve as $\mathcal{SS}(\mathfrak{A})$ for $\mathfrak{A}$ a model of T are precisely the ones that include Rep(T).

Corollary 2. There is a non-standard model $\mathfrak{A}$ of P such that $(\deg(D(\mathfrak{A})))' \leq \underset{\sim}{0}'$ and $\deg(Th(\mathfrak{A})) = \underset{\sim}{0}^{(\omega)}$.

Proof: Let T^* be a completion of P such that $(\deg(T^*))' = \underset{\sim}{0}'$. (Jockusch and Soare [2] proved the existence of such a T^*.) Let $\mathcal{A} = Rep(T^*)$. Then $\mathcal{A}$ is a countable Scott set with an effective enumeration recursive in T^*. This enumeration will be denoted by R^*. Actually, R^* consists of a binary relation enumerating $\mathcal{A}$ and three functions, picking out indices of paths through trees, etc., but R^*_i will be used to denote the set with index i in the binary relation.

There is a recursive sequence of sentences $(\sigma_n)_{n\in\omega}$ such that σ_n is Π_{n+1} and for any C_n-theory K such that $P \cup K$ is consistent, σ_n and $\sim\sigma_n$ are both consistent with $P \cup K$. When the theory $T = Th(\mathfrak{A})$ is determined, the sequence $(\sigma_n)_{n\in\omega}$ will code a set of degree $\underset{\sim}{0}^{(\omega)}$. Let A be a set of degree $\underset{\sim}{0}^{(\omega)}$ such that for some recursive function α, $\varphi^{\underset{\sim}{0}^{(n)}}_{\alpha(n)} = \chi_{A\cap(n+2)}$ for all $n \in \omega$.

Let $T_0 = R^*_{e_0}$ be the complete C_0-theory, consistent with P, that is picked out by the effective enumeration R^*. Given $T_n = R^*_{e_n}$, a complete C_n-theory consistent with P, let $T_{n+1} = R^*_{e_{n+1}}$ be the complete C_{n+1}-theory picked out by R^*, such that T_{n+1} is consistent with $P \cup T_n$ and $\sigma_n \in T_{n+1}$ iff $n \in A$. Then let $T = \bigcup_{n\in\omega} T_n$. Note that for some a, $\varphi^{R^{*(n)}}_a(n) = e_{n+2}$, and $T_{n+2} = R^*_{e_{n+2}}$ for all n. There is a non-standard model M of T such that $\mathcal{SS}(M) = \mathcal{A}$. To apply Theorem 1, what is needed is to show that $C_\omega(M)$ is accessible from R^*.

For all n and i, let $R^n_i = R^*_i$ if this is a C_n-type consistent with T_{n+1}. Otherwise, there is some first k such that $R^*_i \cap k$ looks like an initial segment of a consistent C_n-type, and $R^*_i \cap (k+1)$ does not. Let R^n_i be the C_n-completion of $R^*_i \cap k$ obtained by the natural recursive procedure, using T_{n+1}. Given $R^n_i \cap m$, where m is a formula in the set of variables used so far, put m into $R^n_i \cap (m+1)$ if this is consistent with T_{n+1}. Note that for some b $\varphi^{R^*}_b(n,e_{n+1},i)$ gives an R^*-index for R^n_i. Then there is a recursive function f such that $\varphi^{R^{*(n)}}_{f(n)} = \chi_{R^n}$.

For any i, it is possible to enumerate the C_n-types consistent with R^{n+i}_i in the same way as the C_n-types consistent with T_{n+1}. There is some c such that $\varphi^{R^*}_c(n,e_{n+2},i,j)$ gives an R^*-index for the jth C_n-type consistent with $R^{n+1}_i = \varphi^{R^*}_b(n+1,e_{n+2},i)$. It follows that there is a recursive function g such that $\varphi^{R^{*(n)}}_{g(n)} = \chi_{P^n}$.

Finally, let Q^n consist of the triples $(i,j,q(k))$ such that if i' is the R^*-index of R^{n+1}_i, j' the R^*-index for the jth C_n-type consistent with R^{n+1}_i, and k' the R^*-index for R^{n-1}_k, then $q(k)$ is the R^*-index of the C_n-type picked out by R^* as a completion of $R^*_{j'} \cup R^*_{k'}$ that is consistent with $R^*_{i'}$. This $q(k)$ is also an R^n-index of the type. Now, Theorem 1 yields a model $\mathfrak{A}$ of T such that $(D(\mathfrak{A}))'$ is recursive in $R^{*}{}'$, where R^* is recursive in R^* and $(\deg(T^*))' = \underset{\sim}{0}'$.

The condition of accessibility in Theorem 1 can be relaxed slightly. Let M be a model, and let $S \subseteq \omega$. Then $C_\omega(M)$ is said to be <u>almost accessible</u> from S if there are recursive functions f, g, and h such that for all $n \geq 0$ and for all sufficiently large s, $\varphi^{S(n)}_{f(n,s)} = \chi_{R^n}$ and $\varphi^{S(n)}_{g(n,s)} = \chi_{P^n}$, and for all $n \geq 0$ and all sufficiently large s, $\varphi^{S^{(n)}}_{h(n,s)} = \chi_{Q^n}$, where R^n, P^n, and Q^n are all the same as in the original definition of accessibility.

The proof of Theorem 1, with a very minor modification, yields the following.

<u>Theorem 3</u>. Let M be a weakly homogeneous model, and let $S \subseteq \omega$. If $C_\omega(M)$ is almost accessible from S, then there is a weakly homogeneous model $\mathfrak{A}$ such that $C_\omega(\mathfrak{A}) = C_\omega(M)$ and $(\deg(D(\mathfrak{A})))' \leq (\deg S)'$.

This version of the main theorem is useful in constructing models whose bounded complexity types come from a Scott set with a given enumeration. For any theory T, let T_n denote the C_n portion of T. Let $\mathcal{A}$ be a Scott set such that $T_n \in \mathcal{A}$ for all $n \in \omega$. Let $C_n(\mathcal{A},T)$ be the set of C_n-types Γ such that $\Gamma \cup T$ is consistent and $\Gamma \in \mathcal{A}$. Let $C_\omega(\mathcal{A},T) = \bigcup_{n\in\omega} C_n(\mathcal{A},T)$. A model M of T is said to <u>represent</u> $\mathcal{A}$ if $C_\omega(M) = C_\omega(\mathcal{A},T)$. If M is a non-standard model or P, then M represents $\mathcal{A}$ iff $\mathcal{A} = \mathcal{S}\mathcal{A}(M)$.

<u>Lemma 4</u>. Let $\mathcal{A}$ be a countable Scott set, and let T be a theory such that $T_n \in \mathcal{A}$ for all n. Then T has a weakly homogeneous model M such that M represents $\mathcal{A}$.

Proof: Note the following facts:

(1) If $\Sigma(\vec{u}) \in C_\omega(\mathcal{A},T)$ and $\Sigma(\vec{u}) \cup \{\varphi(\vec{u},x)\} \cup T$ is consistent, then there is some $\Gamma(\vec{u},x) \supseteq \Sigma(\vec{u}) \cup \{\varphi(\vec{u},x)\}$ such that $\Gamma(\vec{u},x) \in C_\omega(\mathcal{A},T)$.

(2) If $\Sigma(\vec{u},\vec{v})$, $\Gamma(\vec{u},x) \in C_\omega(\mathcal{A},T)$, and $\Sigma(\vec{u},\vec{v}) \cup \Gamma(\vec{u},x) \cup T$ is consistent, then there is some $\Lambda(\vec{u},\vec{v},x) \supseteq \Sigma(\vec{u},\vec{v}) \cup \Gamma(\vec{u},x)$ such that $\Lambda(\vec{u},\vec{v},x) \in C_\omega(\mathcal{A},T)$.

With these facts in mind, it is not difficult to carry out a Henkin construction of a weakly homogeneous model M such that $C_\omega(M) = C_\omega(\mathcal{A},T)$.

The following result, which is an application of Theorem 3, says that certain information about a theory is sufficient to construct a model of the theory. The result has the virtue that it deals with arbitrary theories, but it certainly ought to be improved upon.

Theorem 5. Let $\mathcal{A}$ be a countable Scott set, and let T be a theory such that $T_n \in \mathcal{A}$ for all $n \in \omega$. Let $S \subseteq \omega$, and suppose that (a) $\mathcal{A}$ has an enumeration R that is recursive in S, and (b) there is a recursive function α such that $\varphi^{S^{(n)}}_{\alpha(n)} = \chi_{T_{n+2}}$ for all $n \in \omega$. Then there is a weakly homogeneous model $\mathfrak{A}$ of T such that $\mathfrak{A}$ represents $\mathcal{A}$ and $(\deg(D(\mathfrak{A})))' \leq (\deg S)'$.

Proof: By Lemma 4, T has a weakly homogeneous model M such that $C_\omega(M) = C_\omega(\mathcal{A}, T)$. By results of Macintyre and Marker [3], there is an effective enumeration R^* of $\mathcal{A}$ such that R^* is recursive in S. (Macintyre and Marker obtained a general result on degrees of recursively saturated models, and Marker applied this result in a clever way to show that if R is an enumeration of a Scott set $\mathcal{A}$, then there is an effective enumeration R^* of $\mathcal{A}$ such that R^* is recursive in R.)

The function α in Condition (b) above can be used to get a recursive function β such that for all n and for all sufficiently large s, $\varphi^{S^{(n)}}_{\beta(n)}(s)$ is the first R^*-index of T_{n+2}. Now, by an argument like that in the proof of Corollary 2, $C_\omega(M)$ is almost accessible from S. Therefore, by Theorem 3, there is a weakly homogeneous model $\mathfrak{A}$ such that $C_\omega(\mathfrak{A}) = C_\omega(M)$ and $(\deg(D(\mathfrak{A})))' \leq (\deg S)'$.

Solovay, in his paper in this volume [6], showed that the degrees of non-standard models of true arithmetic are precisely the degrees of enumerations of Scott sets containing the arithmetic sets, and the degrees of models representing a particular Scott set $\mathcal{A}$ are the degrees of enumerations of $\mathcal{A}$. Solovay's proof makes use of the fact that if R is an enumeration of a Scott set containing the arithmetic sets, then the theory of true arithmetic is recursive in R''. Theorem 5 has as a corollary the following weak version of Solovay's result.

Corollary 6. Let $\mathcal{A}$ be a countable Scott set including the arithmetic sets. Let R be an enumeration of $\mathcal{A}$. Then there is a non-standard model of true arithmetic such that $\mathcal{A}\mathcal{A}(\mathfrak{A}) = \mathcal{A}$ and $(D(\mathfrak{A}))'$ is recursive in R'.

Proof: There is a recursive procedure for deciding what is in $0^{(n+1)}$, given R' and an R-index for $0^{(n)}$. It is possible to guess, eventually correctly, the first R-index for $0^{(n)}$. It follows that if T is the theory of true arithmetic, then there is a recursive function α such that $\varphi^{R^{(n)}}_{\alpha(n)} = \chi_{T_{n+2}}$ for all $n \in \omega$. (A great deal of the power of $R^{(n)}$ is wasted here; R'' would serve for all $n \geq 2$.) Theorem 5 yields a model $\mathfrak{A}$ of

T such that $C_\omega(\mathfrak{A}) = C_\omega(\mathfrak{B},T)$ and $(D(\mathfrak{A}))'$ is recursive in R'. Since $C_\omega(\mathfrak{A})$ contains types for infinite numbers, $\mathfrak{A}$ is non-standard, and $\mathfrak{S}\mathfrak{B}(\mathfrak{A}) = \mathfrak{B}$.

It can be shown that for a weakly homogeneous model $\mathfrak{A}$, $C_\omega(\mathfrak{A})$ is accessible from $D(\mathfrak{A})$ in a trivial way.

Theorem 7. For any weakly homogeneous model $\mathfrak{A}$, $C_\omega(\mathfrak{A})$ is accessible from $D(\mathfrak{A})$.

Proof: The idea is simple: think of $\vec{a}$ as an index for the C_n-type realized by $\vec{a}$ in $\mathfrak{A}$. This won't quite do, because there are different possible sequences of variables that could be used in the type of $\vec{a}$. The only constraint is that the same variable should not be assigned to two distinct elements of $\vec{a}$. Fix a recursive enumeration of the pairs $(\vec{a}_i,\vec{x}_i)$ such that $\vec{a}_i$ is a finite sequence of elements of $\mathfrak{A}$, $\vec{x}_i$ is a sequence of variables, the lengths of $\vec{a}_i$ and $\vec{x}_i$ are the same, and if two entries in $\vec{x}_i$ match, then the corresponding two entries in $\vec{a}_i$ match.

Let R^n be the enumeration of C_n-types such that R^n_i is the C_n-type having variables $\vec{x}_i$ and realized by $\vec{a}_i$ in $\mathfrak{A}$. For each pair (i,j), let $J(i,j) = j$ if matching entries in $\vec{x}_i{}^\frown\vec{x}_j$ correspond to matching entries in $\vec{a}_i{}^\frown\vec{a}_j$, and if this is not the case, let $J(i,j)$ be the first ℓ such that $\vec{a}_\ell = \vec{a}_j$ and matching entries in $\vec{x}_i{}^\frown\vec{x}_\ell$ correspond to matching entries in $\vec{a}_i{}^\frown\vec{a}_j$. Let $P^n_{i,j} = R^n_{J(i,j)}$. Let Q^n associate with each pair (i,j) the set of indices k such that matching entries in $\vec{x}_i{}^\frown\vec{x}_{J(i,j)}$ and $\vec{x}_k$ correspond to matching entries in $\vec{a}_i{}^\frown\vec{a}_j$ and $\vec{a}_k$.

There is a recursive function δ such that $\varphi^{D(\mathfrak{A})(n)}_{\delta(n)}$ is the characteristic function of the C_n-diagram of $\mathfrak{A}$. Hence, there are recursive functions f and g such that $\varphi^{D(\mathfrak{A})(n)}_{f(n)} = \chi_{R^n}$ and $\varphi^{D(\mathfrak{A})(n)}_{g(n)} = \chi_{P^n}$ for all $n \geq 0$. The relation Q^n, which is the same for all $n \geq 1$, is recursive.

It is natural to ask whether Theorems 1 and 3 could be improved to give a model $\mathfrak{A}$ such that $D(\mathfrak{A})$ is recursive in the set S. If the answer is positive, then Theorem 5 could be improved, and Solovay's result would follow as Corollary 6.

FOOTNOTE:

[1]This work was partially supported by the National Science Foundation.

REFERENCES

[1] Harrington, Leo, Building nonstandard models of Peano arithmetic, handwritten notes, 1979.

[2] Jockusch, Carl G., and Robert I. Soare, Π^0_1-classes and degrees of theories, Trans. Amer. Math. Soc. 173(1972), 33-56.

[3] Macintyre, Angus, and David Marker, Degrees coding recursively saturated models, Trans. Amer. Math. Soc., 282(1984), 539-554.

[4] Marker, David, Degrees of models of true arithmetic, Proc. of the Herbrand Symposium: Logic Colloquium 1981, ed. by J. Stern, North-Holland, Amsterdam, 1982.

[5] Scott, Dana, Algebras of sets binumerable in complete extensions of arithmetic, Recursive Function Theory: Proc. of Symp. in Pure Math., 5 Amer. Math. Soc., Providence, R. I., 1967, pp. 117-121.

[6] Solovay, Robert M., Degrees of true arithmetic, this volume.

LOGIC COLLOQUIUM '84
J.B. Paris, A.J. Wilkie, and G.M. Wilmers (Editors)

TWENTY YEARS OF P-ADIC MODEL THEORY

Angus Macintyre*

Yale University
and
Oxford University

§0. INTRODUCTION:

I have friends who ridicule the above title, perhaps construing it as mildly pretentious. (Actually I can't remember whether it was I or one of the organizers who suggested it.) At any rate, it makes good sense for me. When I gave my survey lectures in Manchester p-adic model theory was exactly twenty years old, and I had thought about it through most of that time. About half way through those twenty years I had the good luck to make a discovery which is currently regarded as basic (and yet, in the three years following publication of my result, only Kreisel, van den Dries and Denef showed any understanding of it). The phrase "good luck" above is important, not because it confirms my modesty, but because it is meant to prepare the reader for reflections on the somewhat odd and instructive development of p-adic model theory. I will be happy if I succeed in providing an elaborate example for Kreisel's "shifts of emphasis" in logic.

§1. THE CLASSICAL IMPORTANCE OF THE P-ADICS:

1.1. Hensel [Hasse 1980] invented the p-adic numbers. The most obvious use of these objects is to allow free use of rational commutative algebra in the systematic study of congruences. Thus, if $f \in Z[\vec{x}]$ then f has a zero in the p-adic integers if and only if for each k f has a solution modulo p^k. The study of an individual Z/p^k must take account of nilpotent elements, but if one is interested only in the question of solving diophantine equations in all the Z/p^k, then one need only consider solvability in the characteristic 0 <u>domain</u> Z_p (the ring of p-adic integers). The complications due to nilpotents disappear. Moreover, it is entirely natural to replace Z_p by its quotient field Q_p, the field of p-adic numbers.

Hasse [Hasse 1980] points out that Steinitz undertook his fundamental (and model-theoretically influential) treatise on field theory [Steinitz 1930] largely because of these new fields discovered by Hensel.

1.2. As long as one considers only one prime p at a time, one expects to use Z_p (or Q_p) in connection only with a <u>necessary</u> condition for solvability (in Z or Q) of a diophantine equation. But even this can be very powerful. Skolem ([Skolem 1938] or [Borevich-Shafarevich 1966]) used the idea to obtain profound finiteness theorems for certain norm equations. The technique rests on p-adic analytic function theory. Skolem converted the norm equations to p-adic exponential equations, and then employed analytic function theory. It is noteworthy that the method is highly ineffective.

---

*Supported by N.S.F. Grants
The author takes this chance to record his gratitude to the National Science Foundation of the U.S.A. for very generous support over twelve years at Yale.

Skolem's method is however of less theoretical importance than the method of local to global transfer. In the latter the ideal is to pass from the knowledge that an equation is solvable in <u>every</u> Q_p to the knowledge that it is solvable in Q. In fact this deduction is rarely justified, but if one adds the information that the equation is solvable in R, then there are very important situations in which we get a <u>sufficient</u> (and obviously necessary) condition for solvability in Q of the equation. The most famous case is that of quadratic forms [Cassels 1978], and there are various subtle generalizations culminating in Hasse's Theorem [Tate 1967] which is more naturally expressed cohomologically. A main theme of contemporary number theory is to <u>measure</u> the failure of such local to global principles, and here the cohomological formulation is essential. (For a lucid account of the conjectured finiteness theorems in this area, i.e. the conjectures of Shafarevich-Tate and Birch-Swinnerton-Dyer, one should read [Manin 1971].)

1.3. It was predictable that against the background of these facts logicians would formulate decision problems for the various p-adic fields, maybe in connection with the compelling 10th Problem of Hilbert. The actual history is less tidy. Tarski (quoted in [J. Robinson 1965]) made the very rash conjecture that C, R and the finite fields are the only decidable fields -- I say "rash" because there seems to have been no positive evidence for it, but of course it did go unrefuted until 1964. To my knowledge, Tarski never mentioned the p-adics in print prior to 1964. (Nor, I believe, did Abraham Robinson.) The first published remarks of substance by logicians on the subject of Q_p seem to be by Nerode ([Nerode 1963]) and Julia Robinson ([J. Robinson 1965]). Both are of considerable importance and will be analyzed later (in 8 and 1.8 respectively). Methodologically, Julia Robinson's is of greater importance, for it stresses a resemblance between Q_p and R, and clearly justified the conjecture that decidability and definability in Q_p would be mastered as in R.

1.4. There are several ways to look at the construction of Q_p:

(1) Form $Z_p = \lim\limits_{\overleftarrow{n}} Z/p^n$, and let Q_p be the field of fractions of Z_p;

(2) Complete Q under a metric defined in terms of divisibility by powers of p.

Using (1), Z_p is a compact domain, the projective limit of the finite rings Z/p^n. There are many ways to exhibit a metric for the topological ring Z_p. For perhaps the most sophisticated see [Weil 1967]. It turns out that Z_p is complete under an <u>absolute value</u>

$$|\cdot| : Z_p \to R .$$

The axioms for an absolute value $|\cdot|$ are:

AV0 : $|\cdot|$ is a map to R;
AV1 : $|x| \geq 0 \wedge (|x| = 0 \leftrightarrow x=0)$;
AV2 : $|xy| = |x|\cdot|y|$;
AV3 : $|x+y| \leq |x| + |y|$.

In this particular case one has even the stronger

AV4 : $|x+y| \leq \max(|x|,|y|)$.

The $|\cdot|$ on Z_p is unique if we demand $|p| = p^{-1}$.

It is trivial that $|\cdot|$ extends to Q_p to satisfy AV0 - AV4. Q_p is locally compact in the resulting topology.

If one goes via (2), one fixes a real c with $0 < c < 1$, and defines, for $m,n \in Z$ and $n \neq 0$, $|m/n| = c^{v(m)-v(n)}$, where $v(x)$ is the exponent to which p divides x. Then, defining $d(x,y) = |x-y|$, d is a metric on Q. If we complete Q with respect to this metric we get the same topological field as in procedure (1). $|\cdot|$ is an absolute value, and the choice of c as 1/p will give $|p| = p^{-1}$, as in the discussion of (1).

The first point of this is that Q has very few metric topologies given by absolute values. Aside from the discrete topology ($|x| = 1$ for all $x \neq 0$), the only examples are

(i) the order topology
and (ii) the p-adic topologies, for p a prime.

For this, see [Artin 1967].

The corresponding completions are R and the Q_p. R is connected, and the Q_p are totally disconnected (see [Serre 1968]). So R is not homeomorphic to any Q_p. But in fact all the Q_p are homeomorphic, being separable, noncompact, totally disconnected spaces without isolated points. However, if $p_1 \neq p_2$ then Q_{p_1} and Q_{p_2} are not isomorphic as rings. I do not claim this is obvious, and it will be discussed later (in 3.1).

At any rate, here is one analogy between R and the Q_p. They arise as the completions of Q under nontrivial absolute values, and we have the all-important family of dense embeddings

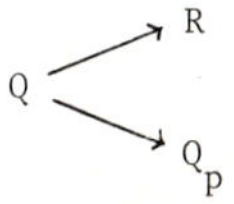

(p prime) .

Less obviously, from this perspective there are privileged absolute values on the completions. Consider the following normalizations:

(i) on R, $|x| = x$ if $x > 0$;
(ii) on Q_p, $|p| = 1/p$.

Then let T be the set of metric topologies on Q given by nontrivial absolute values. For $t \in T$, let $|\cdot|_t$ be the above normalized absolute value. Then one has the basic

Product Formula. $\prod_t |x|_t = 1$ if $x \in Q$, $x \neq 0$.

This is readily proved by computation ([Cassels 1967]), but admits a measure-theoretic interpretation ([Cassels 1967] or [Tate 1967]).

The formula is highly constrained. Artin and Whaples ([Artin 1967]) showed that the only way to get such a formula on Q is to fix some real $\alpha \neq 0$ and replace each $|\cdot|_t$ by $|\cdot|_t^{\alpha}$.

1.5. Generalizations to number fields. One replaces Q by an arbitrary number field K (i.e. a finite extension of Q). As usual $\mathcal{O}_K$ is the ring of integers of K, i.e. the ring of elements integral over Z. From a prime P on $\mathcal{O}_K$ one easily constructs a P-adic absolute value, and a completion K . Using the completion process (2) as in the p-adic case, it turns out that the natural constant c to use is NP, the cardinality of $\mathcal{O}_K/P$.

Other absolute values come from the (finitely many) field embeddings $\sigma : K \to C$, by restriction of the standard absolute value on C. Now the completions are homeomorphic to either R or C.

Write $|\cdot|_p$ (resp. $|\cdot|_\sigma$) for the absolute value defined from P (resp. σ). The $|\cdot|_P$ and the $|\cdot|_\sigma$ give the only nontrivial absolute value topologies on K. It turns out (but again there is a measure-theoretic explanation) that there is a normalization (and essentially only one) making the Product Formula true for K. Cassels (Cassels [1967]) gives a concise account of the functoriality involved in these Product Formulas.

1.6. The analogy with function fields. The observation that the product formula resembles the function-theoretic principle that the sum of the residues is zero is the source of some of the most powerful ideas in modern number theory.

The analogy is with the function field of a curve, but let us consider just the case $K = L(x)$, where L is any field and x is transcendental over L. There are many equivalent basic geometrical notions (places, valuations, absolute values) but here we consider absolute values $|\cdot|$ on K which are trivial on L. From a valuation $v : K \to Z \cup \{\infty\}$ one gets an absolute value by fixing a real c and putting $|x|_v = c^{v(x)}$. There are two ways to get nontrivial v. One way is to take an irreducible $f \in L[x]$, define v on the unique factorization domain $L[x]$ by $v(g)$ = the exponent to which f divides g, and extend v to $L(x)$. The other way is to define v on $L[x]$ by $v(f)$ = the degree of f in x, and extend v to $L(x)$. It turns out tha t these are the only nontrivial v which are trivial on L (see [Artin 1967]), and if we choose one fixed c to define the corresponding absolute values the Product Formula holds.

The completions of K with respect to these absolute values are readily identified. Case 1. $|\cdot|$ defined via an irreducible f. Let L_1 be the field $L[x]/f$, a finite extension of L. Then the completion is $L_1((t))$, the field of formal Laurent series in t over L_1.

Case 2. $|\cdot|$ defined via degree. Then the completion is the field of formal Laurent series in $1/x$ over L.

1.7. There is an important topological difference between the number field case and the general function field case. In the number field case all the completions are locally compact fields, in the list consisting of C, R and finite extensions of the Q_p's. In the function field case, no completion is locally compact unless L is finite, and then they all are. Finally there is the satisfying theorem of Pontrjagin (see [Weil 1967]) that the only locally compact fields are C, R, the finite extensions of the Q_p, and the Laurent series over finite fields.

1.8. There are no general methods in logic for exploiting the compactness of a model. A little can be done if the model is profinite (and this is the basis of Nerode's 1963 work on solving equations in Z_p).

In the same way little can be got logically from the assumption of completeness of a model. For example, in the case of R all first-order consequences of completeness (or local compactness) are consequences of a formal Completeness Scheme. If a polynomial $f(x)$ changes sign on an interval then f has a zero in that interval.

One of the main successes of logic prior to 1964 had been Tarski's discovery that the theory of ordered fields satisfying the Completeness Scheme is complete. Indeed, Tarski had given an effective elimination of quantifiers, deduced decidability, and made some informative remarks about the structure of definable subsets of R (though not of R^n for $n > 1$). Beyond this, Abraham Robinson stressed the important algebraic information (especially Hilbert's 17th Problem) encapsulated in the model-completeness of the theory of real closed fields (that is, the theory of ordered fields satisfying the Completeness Scheme). A translation, frequently more efficient, is obtained by the methods of ultraproducts and saturated models, notably in [Kochen 1961].

It was well-known among algebraists that fields complete under any valuation satisfy Hensel's Lemma. This will be explained later. For now we need know only that it is a "Formal Completeness Scheme", closely related to Newton's computational method in classical real and complex algebra. Like the Completeness Scheme for R, Hensel's Lemma gives a sufficient condition for a one variable polynomial to have a zero. The scheme is a variant of the Implicit Function Theorem, and has a higher dimensional version which has no immediate analogue in R. (See for example [Birch and McCann 1967] or [Cassels 1966]).

Now, Julia Robinson's 1963 paper seems to predict that the logic of Q_p will be understood by focussing on Hensel's Lemma, just as the logic of R follows from the Completeness Scheme above. This has been amply borne out. This paper seeks to map the progress that has been made.

§2. THE ANALOGY BETWEEN Q_p AND $F_p((t))$:

Fix a prime p throughout.

2.1. It is well-known ([Jacobson 1964]) that there are several equivalent notions on which one can base the analysis of Q_p.

2.1.1 Valued fields. Here one has a field K, and an ordered abelian group Γ extended by an element ∞ with $\infty > \gamma$ and $\gamma+\infty = \infty+\gamma = \infty$ for all $\gamma \in \Gamma$. The key notion is that of a valuation $v : K \to \Gamma \cup \{\infty\}$, satisfying

$v(0) = \infty$;

$v(x) \in \Gamma$ if $x \in K^*$ $(= K\setminus\{0\})$;

$v(xy) = v(x) + v(y)$;

$v(x+y) \geq \min (v(x), v(y))$.

From the last axiom one easily deduces:

if $v(x) \neq v(y)$ then $v(x+y) = \min (v(x), v(y))$.

2.1.2. Places. Here one has a field K, a field L, and a ring morphism $\pi : V \to L$, where V is a subring of K with fraction field K, and if $t \notin V$ then $t^{-1} \in V$ and $\pi(t^{-1}) = 0$. π is called a place on K with values in L.

2.1.3. Local fields. Here K is the field of fractions of a domain V which is a local ring (i.e. has a unique maximal ideal).

2.1.4. Fields with absolute value into an ordered field. Here we have $|\cdot| : K \to R$, where R is an ordered field, and the axioms AV1 - AV4 of 1.4 are satisfied with R replacing R.

The equivalence of 2.1.1, 2.1.2 and 2.1.3 is standard. 2.1.4 is thrown in for amusement -- that is, the amusement will come from showing the equivalence of 2.1.4 with the other formulations.

It will be worthwhile to recall the passage from 2.1.1 to 2.1.2 and 2.1.3.

From 2.1.1 to 2.1.2. Let V be the set of all x with $v(x) \geq 0$. Then V is a local domain, with maximal ideal $I = \{x : v(x) > 0\}$. Let L be the field V/I, and let π be the natural $V \to V/I$. Since $v(x^{-1}) = -v(x)$, it is clear that K is the field of fractions of V.

From 2.1.1 to 2.1.3. Similar.

The L described above is the residue class field, and V the valuation ring.

Finally, I outline how to get from 2.1.4 to 2.1.3. Just let $V = \{x : |x| \leq 1\}$.

Each of the notions 2.1.1 - 2.1.4 suggest a formal language for the study of valued fields. All but 2.1.3 (the local field formulation) naturally use a many sorted language. The classes corresponding to 2.1.1, 2.1.2 and 2.1.3 are mutually bi-interpretable in first-order logic. It will be a further amusement to show that 2.1.4 is not interpretable in any of the others. However, the others are interpretable in 2.1.4.

I should stress that the elementary classes isolated above involve no reference to completeness. At this moment we have not imposed any "Formal Completeness Scheme".

2.2. The above discussion is quite general. In the special case of Q_p, the basic notions specialize as follows. The absolute value $|\cdot|$ is as described in 1.4. The valuation v is as in 1.4. The value group Γ is Z with the usual order. The residue class field L is F_p, the field with p elements, because Z/pZ is isomorphic to F_p. Finally, $v(p) = 1$.

The analogy with $F_p((t))$ is clear. In this case too the residue field is F_p and the value group Z $v(t) = 1$. v is explicitly defined by:

$$v(\sum_{j\geq n} a_j t^j) = n \quad \text{if} \quad a_n \neq 0 ,$$

where each $a_j \in F_p$.

2.2.1. The elements of $F_p((t))$ are given formally as expansions with coefficients in F_p. F_p is a subfield of $F_p((t))$, so the latter has characteristic p.

On the other hand, the elements of Q have p-adic expansions $\sum_{j=n}^{m} a_j \cdot p^j$, where the $a_j \in Z$ and $0 \leq a_j < p$. (So the a_j represent elements of F_p, though the a_j are not in F_p.) By the density of the embedding $Q \to Q_p$ one readily deduces that the elements of Q_p have infinite expansions $\sum_{j\geq n} a_j \cdot p^j$ (which are interpreted via limits, and are unique).

2.2.2. It seems as well to point out here the further analogy between the valuation rings of Q_p and $F_p((t))$. For Q_p, the valuation ring is Z_p, whose elements are of the form $\sum_{j=0}^{\infty} a_j \cdot p^j$. For $F_p((t))$, the valuation ring is $F_p[t]$, whose elements are of the form $\sum_{j=0}^{\infty} a_j \cdot t^j$. Each of these rings is compact, in fact profinite. Thus

$$Z_p \cong \varprojlim_{n} Z/(p^n) ,$$

and

$F_p[t] \cong \varprojlim_n F_p[t]/(t^n)$.

The main difference is that Z_p has characteristic 0, and $F_p[t]$ has characteristic p. This is of course an essential difference. The first major progress came from considering the resemblance, for "nonstandard primes".

2.2.3. A flaw in the analogy. In the above we think of a correspondence between p and t. This induces correspondence between

(i) $\{p^n : n \in Z\}$ and $\{t^n : n \in Z\}$

and

(ii) the maps $n \to p^n$

and $n \to t^n$.

The point of (i) is that v restricted to either of these sets is a bijection with the value group. The point of (ii) is that these maps ("cross-sections") give splittings of v.

A flaw is that p is a definable element of the field Q_p, whereas t is not definable in $F_p((t))$. Indeed, Q_p has no nontrivial automorphisms, whereas the automorphism group of the field $F_p((t))$ is uncountable.

At present we know essentially nothing about definability or decidability in the individual $F_p((t))$, and this is certainly related to the above-mentioned automorphisms.

§3. THE ANALOGY BETWEEN Q_p AND R:

3.1. Analogy 1. This concerns the interaction between topology and algebra in those fields.

In R one may define a basis of open sets in terms of <, i.e. the sets $\{x : |x-\alpha| < \epsilon\}$.

In Q, one may define a basis in terms of v or $|\cdot|$, i.e. the sets $\{x : |x-\alpha| < \epsilon\}$ or $\{x : v(x-\alpha) > \delta\}$.

Now, in the field R, < is algebraically definable, thus:

$x \geq 0 \Leftrightarrow (\exists y)(y^2 = x)$.

Moreover, R has a unique structure of ordered field.

In Q_p, Z_p (or the relation $v(x) \geq 0$) is algebraically definable thus:

(i) For $p \neq 2$, $v(x) \geq 0 \Leftrightarrow (\exists y)(y^2 = 1 + px^2)$;

(ii) For $p = 2$, $v(x) \geq 0 \Leftrightarrow (\exists y)(y^2 = 1 + p^3 x^2)$.

The justification for these depends on Hensel's Lemma, to be discussed below.

These equivalences imply that Q_p has a unique structure of valued field subject only to the conditions that $v(p) > 0$ and $v(p)$ is not divisible by 2 in the value group. This is easily checked. Indeed, using variants of the above equivalences, one sees that Q_p has a unique structure of valued field subject only to the condition that $v(p)$ is not infinitely divisible. A more elaborate argument using Henselizations and uniqueness of Henselian structure (see [Nagata 1962]) will show that Q_p has a unique structure of valued field if we demand only that $v(p)$ is not infinitely divisible <u>or</u> that the residue class field is not algebraically closed. <u>But</u>, perhaps unfortunately, it is possible to find a structure of valued field on Q_p with $v(p) > 0$, $v(p)$ infinitely divisible, and the residue class field algebraically closed. I don't know if this has been pointed out before. Anyway, the example is so uninteresting that I omit details.

Similar considerations apply to $F_p((t))$, provided one uses t in the definitions. But in fact the valuation is algebraically definable without using t, and this is an instance of an interesting general theorem. C. U. Jensen, (unpublished?) completing earlier work of Ax and Frey, showed that if K is a not algebraically closed valued field satisfying Hensel's Lemma then the valuation is algebraically definable.

From this we can deduce that for all locally compact fields except C one has an algebraically definable basis for the topology. It is a worthwhile exercise to show that for C one cannot define a basis.

3.2. The preceding implies that we can interpret, for each locally compact field except R and C, Presburger arithmetic, i.e. the theory of the ordered group Z. It is quite easy to see that Presburger is not interpretable in C, but rather less obvious that it cannot be interpreted in R. The proof of the latter seems to need cylindric decomposition for R (see [Collins 1975]).

3.3. <u>Analogy 2</u>. This concerns the topology on the multiplicative group K^*, where K is respectively R or Q_p.

<u>$K = R$</u>. The group of squares $(K^*)^2$ is of index 2 in K^*, and the two cosets disconnect K^*. The cosets (which are evidently open in K) are represented by 1 and -1. Each coset has in K a single boundary point 0.

<u>$K = Q_p$</u> Fix any $n \geq 2$. $(K^*)^n$ is of finite index in K^*. The cosets (which are open in K) are represented by elements of Z. Each coset has in K a single boundary point 0.

3.4. <u>Analogy 3</u>. In both cases $(K^*)^n$ is <u>effectively open</u> in K^*. The notion is explained by the following two remarks.

3.4.1. $K = R$. If $x \in (K^*)^n$ and $|x-y| < |x|$ then $y \in (K^*)^n$.

3.4.2. $K = Q_p$. Take the usual normalization of $|\cdot|$, i.e. $|p| = p^{-1}$. Then if $x \in (K^*)^n$ and $|x-y|$ $p^{-(1+2v(n))} \cdot |x|$ then $y \in (K^*)^n$.

3.5. <u>Analogy 4</u>. <u>Formal Completeness Schemata</u>. I first give an unconventional version which follows from 3.4.

3.5.1. If $f \in K[x]$ and arbitrarily close to α f takes values in two cosets of $(K^*)^n$ then $f(\alpha) = 0$.

3.5.2. The usual schemata are:

<u>$K = R$</u>. <u>Sign Change Scheme</u>: If $f(x)$ changes sign on $[a,b]$ then f has a root in $[a,b]$.

<u>$K = Q_p$</u>. <u>Hensel Scheme</u>: If $f \in Z_p[x]$, f monic, and f has a simple root β modulo p, then f has a unique root α congruent to β modulo p.

<u>Remarks</u> (1). [Cherlin 1976] has a suggestive discussion of the above analogy. (2) There are many useful variants of the Hensel Scheme. See for example [Ribenboim 1968].

3.6. Analogy 5. Finite Extensions. The only algebraic extensions of R are R and C, and the absolute Galois group G_R is $Z/(2)$.

By Krasner's Lemma ([Lang 1964]) one can show that for each n Q_p has only finitely many extensions of degree n. Not all extensions of Q_p are normal, but each normal extension is solvable ([Serre 1965]). The profinite group G_{Q_p} is topologically generated by finitely many elements, and generators and relations are known ([Serre 1965]).

This knowledge is highly relevant to a logician aiming to eliminate quantifiers in some natural language. The key problem of this elimination will be:

When does $a_0 + a_1x + \ldots + a_nx^n$ have a root?

This can be suggestively rephrased as:

Which of the finitely many extensions of dimension $\leq n!$ provides the roots of $a_0 + a_1x + \ldots + a_nx^n$?

One expects to need auxiliary predicates corresponding to each of the finitely many types of extension of dimension $\leq n!$

For example, for R, the extensions of dimension ≤ 2 correspond to the cosets of the squares in R^*. The remark about solvability of G_{Q_p} should suggest that the nth powers are relevant for Q_p, and of course 3.1 confirms this.

§4. METHODS FOR UNDERSTANDING Th(R):

4.1. Quantifier elimination. This was the original method of Tarski. It was later eclipsed by the methods of model-completeness and saturated models, but from the 1970's on, with the emphasis on complexity of computation, it was again prominent.

The natural language for the elimination is that of ordered rings. Without order one cannot eliminate quantifiers, as the example $(\exists y)(y^2 = x)$ shows. The general elimination problem easily reduces to that for

$$(\exists y)[p(y,\vec{x}) = 0 \wedge \bigwedge_j q_j(y,\vec{x}) > 0]$$

where p and the q_j are in $Z[y,\vec{x}]$. Another inevitable problem is to calculate, as the function of $\vec{x}$, the number of zeros of $p(y,\vec{x})$ in R. Except for the $\vec{x}$ such that all y coefficients of $p(y,\vec{x})$ vanish, this number is bounded by the formal y-degree of p. Once the exact number of roots is known, the roots can be distinguished by their position in the order <. That done, the general problem with the p and q_j comes down to asking the sign, at the kth root of p, of each q_j. In the case when p is missing, or has no roots, one is asking a more complex question. This can be reduced to questions of earlier type by a systematic use of the formal Euclidean algorithm. I will make a detailed account of the method available in [Macintyre 1986?]. The classical texts are [Tarski 1951] and [Cohen 1969].

For the problem of counting the number of zeros of p (maybe satisfying various inequalities) Tarski used the result of Sturm ([Jacobson 1964]) giving this number as the number of sign changes in a so-called Sturm sequence. Sturm's method also depends on the formal Euclidean algorithm. Many years later [Ben-Or, Kozen and Rief 1983] added new parallel algorithms for the elimination theory to get the fastest known quantifier-elimination for R.

Cohen's 1969 method is more compressed, and makes no direct use of Sturm sequences. Cohen appeals directly to the Sign Change Scheme, and stresses (in his formalism of effective functions) the elimination theory of such conditions as: $q(y,\vec{x})$ is positive at the kth root of $p(y,\vec{x})$.

4.2. Consequences.

4.2.1. Decidability. The elimination is constructive, and leads easily to a primitive recursive decision procedure for the first-order theory of R. But even the elementary recursive procedures discovered in the 1970's have remained without application. By now it is clear that as far as R is concerned the most fertile ideas in Tarski's work concern the structure of the definable sets.

4.2.2. Definability. In his paper Tarski gave applications to the structure of definable sets in R. Such sets are exactly the finite unions of all kinds of intervals. There seem to be no convincing direct applications of this result, whereas a thorough study of the definable subsets of $R^n (n \geq 1)$ has proved fertile and profitable. The special case of definable functions has had some applications to differential operators ([Hörmander 1969]), with the most useful cases being functions "distance to A", where A is a definable set.

The R-definable sets are nowadays blandly called semi-algebraic ([Brumfiel 1979]). I would prefer real-constructible, by analogy with the constructible subsets over an algebraically closed field. This would then suggest a suitable name for the p-adic analogue.

The finer structure of definable sets was achieved nearly forty years after Tarski's work. Attention was focussed on the topological and analytic structure of the decompositions of R^n induced by polynomials $f(x_1,\ldots,x_n)$. To f one associates its zero set, and its sets of positivity and negativity. A fundamental result of [Whitney 1957] says that each of these sets has only finitely many connected components, and later work of [Collins 1975] exhibits an effective "cylindric decomposition" of R^n. This, with a constructive method for sampling from connected components, gives a fast quantifier elimination quite different from that of Ben-Or, Kozen and Rief.

From the standpoint of direct effective definability theory, one may fairly say that Tarski left out the richest part of the theory. To be fair, however, one must point out that [van den Dries 1984A] showed that even the cylindric decomposition (at least in a slightly weakened version) follows just from the assumption that the definable sets are unions of intervals. Actually, van den Dries proved such a result for theories and later [Knight, Pillay and Steinhorn 1985] did it for single models.

4.3. Model theory of real closed fields. Tarski was of course aware that his result applied not only to R but to any real closed ordered field. The latter class can be defined in many different ways, but I prefer to define it as the class of models of the Sign Change Scheme. The essential point is that one need not use full completeness of R, but rather only a very restricted formal completeness. Because of the quantifier-elimination one then has the Transfer Principle that any first-order property holding in R holds in all real-closed fields. There is a little stock of examples in this area, notably concerning division algebras (see [Jacobson 1964]).

From the quantifier-elimination follows the model-completeness of the class of real-closed fields. Tarski did not go in this direction, but Abraham Robinson found much of interest there. It was down this path that one first spotted the p-adic model theory.

4.4. Model-completeness. For an overview of this subject, see [Macintyre 1977].

In Robinson's approach to the case of R, one considers two elementary classes, namely ordered fields and real-closed fields, and maps from the former to the latter.

The Artin-Schreier theory (based also on Sturm sequences) gives existence and uniqueness of real closures (that is, prime model extensions from ordered fields to real closed fields). Using this, the methodology of Robinson's Test gives model-completeness of the theory of real-closed fields, and, indeed, quantifier-elimination, albeit much less effectively than by Tarski's method. One should point out that by [van den Dries 1984B] the existence and uniqueness of real-closures follows from the quantifier-elimination.

The main difference of method between Tarski and Robinson is this. Tarski works over a given structure, but uniformly for all structures. Robinson uses essentially maps between structures, and the mathematical theory of the category of models. The difference seems minor, and ultimately it is, but for the art of invention or discovery the difference has proved significant.

4.5. Saturated models. This was a powerful method in the hands of Kochen and Keisler around 1960. The essential point seems to be to have a universal domain in which the global theory of 4.4 is essentially encapsulated. Again, there can be no doubt of the value of the method in discovery. Of course much of the paraphernalia of the isomorphisms theorems (e.g. assumption of CH) has been dropped in favour of restricted notions of saturation, e.g. ω_1-saturation or recursive saturation.

4.6. Some other sources of analogy. I now collect a few loosely connected structural items about real-closed fields, and will soon turn to p-adic analogues.

4.6.1. If K is real-closed, and L is a subfield of K, then the relative algebraic closure of L in K is real-closed.

4.6.2. If K is ordered, $[a,b] \subseteq K$, and $f \in K[x]$, then the number of zeros of f in $[a,b]$ in any real-closure $K \to L$ is independent of L.

4.6.3. If $f \in K[x]$ as above, the roots of f in L can be isolated by their position in the natural order.

4.6.4. The real-closure of K is algebraic over K, unique up to isomorphism over K, and rigid over K.

4.6.5. The theory of real-closed fields has definable Skolem functions. See [van den Dries 1984B] for this and the connection with 4.6.4.

4.6.6. The theory of real-closed fields has definable selectors from definable equivalence relations. That is, every definable equivalence relation on n-space is equivalent to one of the form

$$\bar{x} \equiv \bar{y} \Leftrightarrow F(\bar{x}) = F(\bar{y})$$

where F is a definable function. This was pointed out by van den Dries, and relates to Poizat's elimination of imaginaries ([Poizat 1983]).

4.6.7. Though unstable, the theory of real-closed fields does not have the independence property. See [Pillay 1983]. The proof uses the stability-theoretic notion of coheir of a type, and because of quantifier-elimination this is readily identified for real-closed fields.

4.6.8. [Robinson 1959] proved the completeness of the theory of pairs (K,L) where K and L are distinct real-closed fields and L is a dense subfield of K. This was perhaps the first case where the saturated models approach ([Macintyre 1968]) is really easier than model-completeness (Robinson) or bare-hands quantifier-elimination (P. J. Cohen, unpublished).

4.6.9. [Robinson 1965] give a very memorable presentation of the solution to Hilbert's 17th Problem, using model-completeness. It should be stressed that one of the main ideas from the original proof by Artin-Schreier remains intact, namely the connection between sums of squares and orderings. Robinson's contribution was to collapse the rest of the proof to a few lines of model-completeness.

4.9.10. Despite some suggestive early ideas of [Robinson 1965] on metamathematical theory of ideals, it took until the 1970's and the work of the Costes ([Coste-Roy 1982]) before one had a sound theory of real spectra for commutative rings. Essentially because of the rigidity of real-closures, the real spectrum is spatial (in contrast to the Zariski spectrum).

§5. THE ADVENT OF P-ADIC MODEL THEORY:

5.1. I turn now to the events of 1964-5, the work of Ax-Kochen and Ersov. Nerode's work of 1963 remains important, but it is not model theory. It will be discussed later in Section 8.

5.2. The method was based on global considerations (in contrast to the local setting of Tarski's elimination). Ax-Kochen used saturated models, and Ersov model-completeness.

Let us repeat once more the ingredients in the real case:

$(1)_R$ R;

$(2)_R$ the class of ordered fields;

$(3)_R$ the class of real-closed fields;

$(4)_R$ the global theory (Artin-Schreier) linking $(2)_R$ and $(3)_R$.

Write $(k)_{Q_p}$ for the conjectural analogue of $(k)_R$. Of course $(1)_{Q_p}$ is Q_p. An unequivocal answer was obtained for $(3)_{Q_p}$. For (2), and therefore (4), there are several answers depending on the logical problem to hand.

Of course some help on ((2) is given by looking at subfields of Q_p. (Notice that these are all dense in Q_p.) Such fields have residue class field and value group Z. v(p) = 1, the least positive element of Z.

So one possibility for $(2)_{Q_p}$ is the class of all characteristic 0 valued fields with residue-class field F_p, valued in a Z-group and with v(p) = 1. Call this the <u>pure alternative</u>.

The other possibility, the <u>impure alternative</u>, comes from considering suitably saturated elementary extensions of Q_p. We still have F_p as residue field, the value group a model of Presburger arithmetic, and $v(p) = 1$. But if we pass to subfields, the value group need be only a subgroup of a Z-group, containing 1. To understand this restriction we need a structure theory of Artin-Schreier type for Presburger. That is readily available. The analogue of (2) is the class of discretely ordered abelian groups with least element 1, and then (3) is the subclass of Z-groups. In (4) we consider embeddings of ordered groups. (Notice in particular that a pure subgroup of a Z-group is a Z-group.) All this leads to the impure alternative, namely we relax the assumption on the value group, assuming only that it is discretely ordered with least element 1.

For (3), we want the class to consist of valued fields elementarily equivalent to Q_p. Such fields certainly have Z-groups as value groups. By analogy with R one seeks to axiomatize the class by a Formal Completeness Scheme. We shall see below that Hensel's Lemma is an appropriate scheme. But it has to be justified by a p-adic "Artin-Schreier" theory. Anyway, on either alternative, $(3)_{Q_p}$ will be the class of Henselian fields with residue-field F_p, valued in discretely ordered groups with $v(p) = 1$.

On either alternative $(4)_{Q_p}$ will be based on valued-field embeddings connecting the elements of (2) to the elements of (3). Beyond this one considers <u>immediate</u> embeddings between elements of (2), i.e. embeddings extending neither the residue-class field nor the value group. The importance of these embeddings in the structure theory of complete valued fields was known from [Kaplansky 1942] and [Kaplansky 1945]. There is however nothing in those papers directly applicable to either version of $(2)_{Q_p}$, and the main effort came in extending Kaplansky's analysis to the $(2)_{Q_p}$.

<u>Analogy 6</u>. The real closed fields are the ordered fields which have no proper ordered algebraic extensions. Under suitable hypotheses, of Kaplansky or Ax-Kochen-Ersov, Henselian valued fields are exactly those with no proper immediate algebraic extensions.

<u>Analogy 7</u>. Every ordered field has an algebraic real-closed extension. Every valued field has an algebraic immediate extension which is maximal with this property.

However, there is a difference at this level of generality (all valued fields). Whereas the real closure is unique, the maximal algebraic immediate extension is not, unless stringent hypotheses are made on the residue field and value group. For a systematic account of the successful hypotheses, see [Delon 1981]. The most important special cases are:

(0) characteristic 0 residue class field;

(B) the value group discrete, the base field of characteristic 0, residue field of characteristic p, and $v(p) \in Z$ (the subgroup of the value group generated by 1).

Under either hypotehsis, Hensel's Lemma axiomatizes the property of having no proper immediate algebraic extensions. Moreover, under the same hypotheses, valued fields have (up to isomorphism over them) unique immediate algebraic extensions to Henselian fields, and these fields are in fact the Henselizations (in the sense of [Nagata 1962]) of the original fields. For this too see [Delon 1981]. The Henselization of K is rigid over K.

The case (B) covers any element of $(2)_{Q_p}$ for either alternative.

A very important property of the p-adic case, not always given due prominence, is that any relatively algebraically closed subfield of an element of $(3)_{Q_p}$ is an element of $(3)_{Q_p}$. The computational crux here is to show that the value group of such a field is a Z-group, i.e. that the Euclidean algorithm holds in the value group. This is achieved by Hensel's Lemma (or the related Hensel-Rychlik Lemma) applied to various polynomials X^n-a. This kind of computation has occurred in most of the influential papers. I first saw it in [Ax-Kochen 1965A].

Let us now call the elements of $(3)_{Q_p}$ p-adically closed (valued fields). So we now have the analogue of 4.6.1 for p-adically closed fields.

It is now a short step to a proof, in Robinson's style, of model-completeness of the theory of p-adically closed fields. This was first done by Ersov. We need one more analogy relating to the valuation-theoretic counterpart of cut.

If $K \to L$ is an embedding of ordered fields, there is a clear sense to the notion of a basic open set of L defined over K. Namely, we consider sets

$\{x \in L : x-a > 0\}$,

where $a \in K$. If K and L are instead valued fields, we consider sets $\{x : v(x-a) > 0\}$ instead.

Then we have:

Analogy 8. If K is real-closed and x is transcendental over K, then any order on $K(x)$ extending the K-order (but actually this is automatic) is determined by the set of basic open sets of K to which x belongs.

If K is Henselian, and (0) or (B) holds, and the value group of $K(x)$ is a pure extension of that of K, then any valuation on $K(x)$ which is an immediate extension of that on K is determined by the set of basic open sets of K to which x belongs. (If K is not algebraically closed, any valuation on $K(x)$ extends that on K.)

The valuation-theoretic part of this has traditionally been based on Ostrowski's pseudo-convergence ([Kaplansky 1942]). With luck, van den Dries will publish a presentation of the entire subject on a more elegant basis (but at present one has only his handwritten lecture notes from Stanford 1984).

The real case of model-completeness goes thus:

(i) Use 4.6.1 and Robinson's Test to reduce to showing that if $K \to L$, K,L real-closed, and transcendence degree of L over K is 1, then $K \prec_1 L$;

(ii) Use the real part of analogy 8, the uniqueness of real closure, and the compactness theorem to reduce the $\prec_1$ problem to the special (and trivial) case of showing that if U is a finite intersection of basic open sets over K, and $U \cap L \neq \phi$ then $U \cap K \neq \phi$.

The classic reference is [Robinson 1965].

No great changes are required for p-adically closed fields. The analogue of (i) goes through, and if $K \to L$ is immediate the whole thing goes through, using Analogy 8. So we are reduced to the case $K \to L$ not immediate, and of transcendence degree 1. Let Γ_K, Γ_L be the respective value groups. The residue-fields of K and L are the same, so by standard valuation theory Γ_L has Q-rank 1 over Γ_K. Suppose $K \not\prec_1 L$, and choose a finite tuple $\vec{x}$ witnessing this. Again by pure valuation theory one shows that there exists $t \in L$ with $v(t) \notin \Gamma_K$, such that $K(\vec{x},t)$ is immediate over $K(t)$. In particular, $\vec{x}$ is in the Henselization of $K(t)$. By Analogy 8, $K(t)$, qua valued field, is determined by the cut $v(t)$ makes in Γ_K, and the Henselization of $K(t)$ is so determined. This Henselization is present in any Henselian L containing t. We now finish by the standard Robinson compactness argument and reduction to the case of a basic open set problem.

This is a short, sweet proof which rapidly yields the traditionally important metamathematical classifications for Q_p (and for a huge variety of other fields). The completeness of the theory of p-adically closed fields comes easily from the prime model test (the algebraic p-adic numbers being the prime model). Decidability is immediate from the recursive axiomatization.

There is however a real sense in which the proof is too slick. Notice that it makes essential use of "uniqueness of closures" not only for p-adic fields, but also for fields with value groups which are not Z-groups. In fact the result is an instance of a very general result on transfer of model-completeness, worked out by [Ziegler 1972] and [Delon 1981]. There is an analogous transfer of completeness theorem. The form of such transfer is:

If the theories of the residue field and the value group each have property P, then so does the theory of the Henselian field.

The success of these general arguments seems to have diverted attention from the specifics of Q_p. As outlined in the above proof, the analogy between R and Q_p is perfect only if we restrict to immediate maps. There is nothing in the real case corresponding to the pure/impure distinction. As we shall see, a great deal was left unsaid about this matter.

5.3. Saturated models. Ax-Kochen used mainly the method of ultraproducts, or equivalently saturated models. There were various elements now obsolete, such as the assumption of GCH and absoluteness considerations, but many of the details of their work remain important.

Firstly, in common with Ersov, they had to develop the purely algebraic theory surrounding Hypothesis B.

Next, they had to isolate the essential properties of κ-saturated Henselian fields (under hypotheses like (O) or (B)). The model for this was the identification of the η_α property as the essential property of ω_α-saturated real closed fields (see [Kochen 1961]). They produced a natural candidate, ω_α-pseudo-completeness, an infinitary version of a key notion from the Ostrowski-Kaplansky theory.

Analogy 9. Suppose κ uncountable. Any κ-saturated real-closed field contains a (non-canonical) copy of R. Any κ-saturated p-adically closed field contains a (non-canonical) copy of Q_p.

There is a Hypothesis O analogue of this, which the reader may care to formulate.

The crux, in Ax-Kochen, was to find isomorphism theorems justifying the above putative identifications of the saturated models. Nowadays we ask less, that is we give criteria for extending partial maps.

The spin-off from their method is that one identifies obstructions, not expressible in the natural language of valued fields, to extension of isomorphisms.

They then construct isomorphisms respecting the extra structure, and thereby get stronger completeness theorems. The obstructions in question relate directly to the pure/impure distinction.

The basic problem is to find a criterion for being able to fill in the following diagram:

$$\begin{array}{ccc} K & & L \\ \uparrow & & \uparrow \\ K_1 & \cdots > \cdots & \cdot \\ \uparrow & & \uparrow \\ K_0 & \xrightarrow{\quad f \quad} & L_0 \end{array}$$

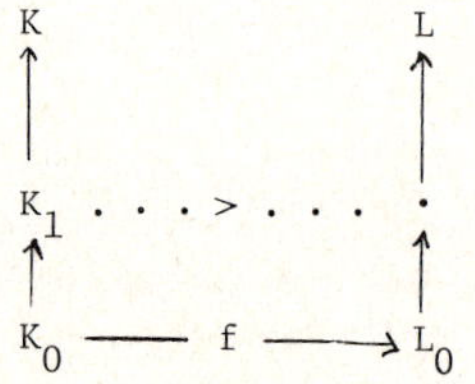

where K, L are p-adically closed satisfying some saturation conditions, K_0, K_1, L_0 are valued subfields as indicated and f is an isomorphism of valued fields. Typically K_0 is countable, K_1 is a simple extension of K_0, and K, L are ω_1-saturated.

One of the most regrettable features of this situation is that it was not until my 1976 paper that it was pointed out that f is not always extendable. It seems likely that the obstruction was known to Ax-Kochen, or Ersov, or Cohen.

Suppose for example $K_0 = Q(u)$, where $v(u) > Z$, and $L_0 = Q(w)$, where $v(w) > Z$. Then $K_0 \cong L_0$ via an f which sends u to w. f is an isomorphism of valued fields. If, as can perfectly well happen, $2 | v(u)$ but $2 \nmid v(w)$ there is is no chance of f extending to an isomorphism of K and L. At this level, we are just exploiting the fact that Presburger does not admit quantifier-elimination in the language of valued groups. The obstruction goes deeper, however. Even if there is an isomorphism of Z-groups sending $v(u)$ to $v(w)$ it may not be possible to extend f to an isomorphism of K and L. That is, even if we add predicates $n | v(x)$ inspired by the known quantifier elimination for Presburger using divisibility and order, we don't attain quantifier-elimination in the valued field set-up. The difficulty is simply that $v(u)$ and $v(w)$ may have the same Presburger type without u and w having the same valued field type. We can represent the obstruction easily under Hypothesis 0. Imagine u as a (perhaps generalized) power series $\alpha \cdot t^{v(u)}$ + higher terms, and w as $\alpha \cdot t^{v(w)}$ + higher terms. Even if $v(u)$ and $v(w)$ have the same type, the "leading coefficients" α and β may not. (As we shall see below, a similar representation exists for p-adically closed fields.) This suggests adding a primitive for "leading coefficient", but one should notice how non-canonical it is (cf. 2.2.3.).

Ax and Kochen added an equivalent primitive, a cross-section. Getting α from u is clearly equivalent to getting $t^{v(u)}$ from u. Write $\pi(u)$ for $t^{v(u)}$. The crucial property of π is that π is a homomorphism from the value group to the multiplicative group, and $v \circ \pi$ is the identity. This notion, in contrast to the series representation above, has clear sense in any valued field. For p-adically closed fields, one would naturally impose the normalization $\pi(1) = p$.

Note that π is highly noncanonical. Clearly Q_p has a normalized cross-section, as does any $K((t))$. Ax and Kochen showed that saturation gives a cross-section, and [Cherlin 1976] gave the decisive result that an ω_1-saturated valued field has a cross-section (which can be chosen normalized if $v(p) = 1$).

Returning to the above diagram, one now assumes K and L have normalized cross-sections, and f preserves those. It is clearly necessary from the earlier discussion to assume that f respects the predicates $n|v(x)$. By the theory of closures one can assume K_0 and L_0 are Henselian. To avail oneself best of Analogy 8 one needs the value groups of K_0 and L_0 to be Z-groups. Essentially this is a matter of passing to pure closures of value groups. Unfortunately this may destroy preservation of π unless one assumes originally that f preserves the (partial) functions $\pi(v(x)/n)$. The latter is respected by passage to pure closures. Finally, with these new primitives in place, one can show, by a complex and vital argument using radical extensions, that f extends. The details are not easily summarized, and the reader should consult Ax-Kochen.

The outcome is a <u>quantifier elimination</u> for Q_p, in an elaborate language involving π, $n|v(x)$, and $\pi(v(x)/n)$. No consequences were drawn from this in the original investigations, and in fact it is only now (1984-5) that important details are emerging.

By their analysis Ax and Kochen mastered the structure of the saturated model obtained another very striking analogy. For this one needs a generalization of the classical power series construction. Let K be a field, Γ an ordered abelian group, and κ an uncountable cardinal. $K((t^{\Gamma}))_{\kappa}$ is the field of formal power series with coefficients in K, exponents in Γ, and support well-ordered of cardinality $< \kappa$. (See [Kochen 1974]). Then:

<u>Analogy 10</u>. (Assume $2^{\kappa} = \kappa^{+}$). Let D be a saturated divisible ordered abelian group of power κ^{+}. Then $R(t^{D})_{+}$ is the saturated real closed field of power κ^{+}, and $Q_p(t^{D})_{\kappa^{+}}$ is the saturated p-adically closed field of power κ^{+}.

Neither representation is canonical. Notice that $Q_p(t^{D})_{\kappa^{+}}$ satisfies Hypothesis B canonically, and Hypothesis 0 non-canonically.

5.4. <u>Hypothesis 0</u>. A major restriction under Hypothesis B is that the proof in 5.3 makes strong use of the finiteness of the residue class field. See 5.7 below for the very little that is known in the general case.

The finiteness of the residue field eliminates the need to consider residue field extensions. This might well induce pessimism for the case of Hypothesis 0. Miraculously this is not so.

Assume K is Henselian and satisfies (0). Then the residue-class field is represented by a subfield (non-canonical) of K. Call any such K_{res}. For the analysis of saturation via extendibility, analogous to 5.3, one consider maps f respecting K_{Res}. Indeed, the main problems reduce to consideration of

$$\begin{array}{ccc} K & & L \\ \uparrow & & \uparrow \\ K_1 & & \\ \uparrow & & \\ K_0 & \overset{f}{\cong} & L_0 \\ \uparrow & & \uparrow \\ K_{res} & \overset{f}{\cong} & L_{res} \end{array},$$

and then the exact analogue of 5.3 goes through. The bonus is that one now has completeness and decidability not only for the original language, and for the cross-section formalism, but even for the natural formalism taking account both of cross-section and residue field representative. The complications needed to handle model completeness and a suitable quantifier elimination can be found in the works of Ziegler and Delon cited earlier, and in [Weispfenning 1976].

In light of 5.3 the following consequence is especially noteworthy: (GCH) Suppose K is Henselian, satisfies (0), has residue field L, and value group Γ. If K is saturated of power κ^+ then $K \cong L((t^\Gamma))_{<\kappa}$.

Eliminating the set theory this implies:

If K_i $(i = 1,2)$ are Henselian and satisfy (0), with residue fields L_i and value groups Γ_i, then

$$K_1 \equiv K_2 \Leftrightarrow L_1 \equiv L_2 \text{ and } \Gamma_1 \equiv \Gamma_2.$$

This gives striking precision to the analogy of Section 2.

5.5. Comparing the generic Q_p and the generic $F_p((t))$. Suppose D is a non-principal ultrafilter on the set P of primes. Compare

$$\prod_D Q_p/D$$

and $\prod_D F_p((t))/D$.

By Łoś Theorem each has residue field $\prod_D F_p/D$, and value group Z^P/D. Thus each satisfies Hensel's Lemma, and Hypothesis (0), so they are elementarily equivalent. Notice how little one knew about $\prod_D F_p/D$ in 1964! The mere fact that it has characteristic 0 has impressive consequences by the foregoing. For let D vary. There follows:

The $Q_p/F_p((t))$ Analogy (Ax-Kochen): For any first-order sentence Φ of the language of valued fields (even with cross-section) there is a prime $p_0(\Phi)$ such that if $p \geq p_0(\Phi)$ then

$$Q_p \models \Phi \Leftrightarrow F_p((t)) \models \Phi.$$

This offers the prospect of transfer from Q_p to $F_p((t))$ or vice-versa. Basic algebraic computations are easier in $F_p((t))$, whereas analysis or algebraic geometry at the level of desingularization are easier in Q_p. Till now, the method has had one great success, going from $F_p((t))$ to Q_p, in connection with a conjecture of Artin.

Recall that a field K is said to be C_i if every homogeneous form $f(x_1,\ldots,x_n)$ of degree d, with $n > d^i$, has a nontrivial solution in K. Algebraically closed fields are C_0, and finite fields are C_1. By a major result of [Lang 1952] $F_p((t))$ is C_2. Neither $F_p((t))$ nor Q_p is C_1. (It is a good exercise to give a counterexample with $n=3$, $d=2$, using the definability of the valuation ring as in 3.1). Artin conjectured that Q_p is C_2.

By the Ax-Kochen analogy this is approximately true. That is, given $n > d^2$, if p is large enough then any f of degree d in n variables has a nonzero solution. Of course n is irrelevant (put some variables equal to zero), so in fact we get a function $F(d)$ so that if $n > d^2$ and $p \geq F(d)$ then any f of degree d in n variables has a nonzero solution in Q_p.

The naturality of the method has been confirmed by a series of examples showing that $F(d)$ cannot be chosen to be 0. One should consult the recent [Lewis-Montgomery 1983] for references, and for some new information of a quantitative kind. The method of Ax-Kochen does not guarantee a primitive recursive F, though it does give an ϵ_0- recursive F. For what improvements logic can give see Section 10. It is now known that Q_p is not C_i for any i. See Lewis-Montgomery for an intriguing conjecture about the counterexamples for fixed p. The structure of these counterexamples is certainly a natural topic in the advanced definability theory of Q_p.

5.6. Defects in the analogy. There has been almost no progress on understanding the elementary theory of $F_p((t))$. Further, one has the ominous result that $F_p((t))$ with cross-section is undecidable (despite the fact that cross-section is allowed in the Ax-Kochen analogy). Apparently Ax first proved this, and it was rediscovered by Jacob. The details inspired Cherlin ([Cherlin 1984]) to show that for valued fields of characteristic p there is no possibility of Ax-Kochen isomorphism theorems preserving cross-section and residue field. In fact, almost always the last "and" can be replaced by an "or".

Ax's main observation is that in the field $F_p((t))$ one can define the elements of residue 0, i.e. the series $\Sigma a_n t^n$ where $a_{-1} = 0$. (The underlying computation is well-known in local-class-field theory.) Recall that t is not definable.

5.7. Algebraic extensions of Q_p. Since every finite extension of Q_p is of the form $Q_p(\alpha)$ where α is algebraic over Q (an easy consequence of Krasner's Lemma), every finite extension of Q_p is interpretable in Q_p and so decidable. However, these methods are a little too crude for informative definability results.

The issue was not immediately considered. However in 1974 Kochen published an elegant paper outlining the modifications necessary for the analogues of the original Ax-Kochen isomorphism theorems. For unramified extensions no modifications are needed, but in the general case one has to distinguish an α as above. In particular, to capture the theory of $Q_p(\alpha)$ one needs more than just the residue field and the degree of ramification. There are no surprises, and adding cross-section makes no extra complication. Note that Hypothesis B applies.

I do not know if anyone wrote out the Ax-Kochen analogy in this setting, but there are no difficulties in finding a natural formulation. Because of lack of space I leave the exercise to the reader. (The version in Kochen is not quite what I have in mind.) Recently Belair, van den Dries and I noted that even if $K \equiv Q_p(\alpha)$, K need not be of the form $L(\beta)$ where $L \equiv Q_p$.

Infinite Extensions. As pointed out in 5.4 the Ax-Kochen approach to the mixed characteristic case (valued field of characteristic 0, residue field characteristic p) works best for finite residue field. The reason is that in other cases there will be no obvious way to keep the residue field fixed from the beginning of the construction of an isomorphism, with consequent complications in the study of non-immediate extensions. Recall that there is no problem under Hypothesis 0 (cf. 5.4).

Ax and Kochen found a brilliant trick for reducing the mixed case under Hypothesis B to Hypothesis 0. This involves an elaboration of the observation that saturated p-adically closed fields are (generalized) power series fields over Q_p.

[Kochen 1974] exploited the observation to prove decidability of the maximal unramified extension of Q_p. Here we have a Henselian field K where the residue class field is algebraically closed, the value group is Z, and $v(p) = 1$. The idea is to look at a saturated extension L with value group Γ (necessarily saturated). Z is a convex subgroup of Γ, and so Γ/Z has a natural order. The map

$$L^* \to \Gamma \to \Gamma/Z$$

gives a new valuation $v_1 : L^* \to \Gamma/Z$. L^* is Henselian with respect to v_1, and readily seen to be saturated.

Γ/Z is divisible and saturated. The main point is that the residue field F for v_1 has characteristic 0. F can be identified thus. Let V be the valuation ring of v. Consider the projective system of rings V/p^n $(n \geq 1)$, and its projective limit R. R can be identified with $V/\bigcap_n p^n V$, which is a domain of characteristic 0. F is naturally isomorphic to the field of fractions of R.

It follows then from the Hypothesis 0 analysis that L will be identified, as a field with valuation v_1, once F is identified as a field. Since $v(p) = 1$, the argument of shows that v is algebraically definable, so to identify (L,v) we need only identify F.

We may suppose $F \subseteq L$. The restriction of v to F takes values in Z, and F is complete under v. The residue field of v on F is (canonically) the residue field of v on L, and so is saturated. As Kochen points out, F is then uniquely determined as a Witt-Teichmuller construction over the residue field.

It follows easily that the theory of K is axiomatized by the theory of its residue field, plus the information that the value group is a Z-group and $v(p) = 1$. The decidability of the maximal unramified extension of Q_p follows. (Note that Kochen slips up by identifying the above field with the maximal cyclotomic extension of Q_p.)

By the above technique one may handle all Hypothesis B fields, as far as complete axiomatizations are concerned. This includes all unramified extensions of Q_p, and there is no trouble in handling finite ramification. But, as far as I know, there has been no systematic study of model-completeness, let alone quantifier-elimination.

There are two interesting infinite extensions of Q_p which elude the above methods. The first is the maximal abelian extension, with algebraically closed residue field and value group $\{m/p^k : m,k \in Z\}$. The other is the totally ramified extension got gy adjoining the p^j th roots of unity, for all j. This has residue field F_p and value group as above.

A clue to their analysis comes from some work of van den Dries (unpublished, early 1980's). He looked at the general mixed characteristic Henselian case, with no restrictions on value group Γ. Instead of the convex subgroup Z of Γ one now considers Γ_p, the convex subgroup generated by $v(p)$. The role of Γ/Z is taken by Γ/Γ_p. With this modification, v_1 is constructed as before. The residue field admits the description given before, via $R = \varprojlim V/p^n V$. By a typically clever short proof, van den Dries (unpublished) obtained the following completeness theorem:

The theory of the mixed characteristic Henselian field K, with residue field of characteristic p, valuation ring V and value group Γ, as determined by the theories of

$(\Gamma, v(p))$ and V/p^nV, $n \geq 1$.

I noticed in writing this paper the fact that for $n > 1$ V/p^nV is a Witt construction ([Jacobson 1964]) over V/pV, so in van den Dries' result one needs only $(\Gamma, v(p))$ and V/pV.

These neat observations have not yet produced new applications. Even for the infinite extensions listed above V/pV is from a logical point of view quite complex (a local ring of characteristic p with lots of nilpotent elements).

5.8. Cohen's Elimination. Cohen's work was published in 1969, but available in 1966-7. The most obvious advances over Ax-Kochen-Ersov are

(i) a primitive recursive decision procedure for Q_p;

(ii) primitive recursive bounds in the Ax-Kochen analogy.

However, the following elements of his proof have proved of greater interest and utility:

(1) His elimination takes place inside a fixed Henselian field, using only Hensel's Lemma, and in particular, no global Ostrowski-Kaplansky theory is needed;

(2) He gives a procedure for "isolating" the roots of polynomials, and simultaneously reducing conditions Φ (the kth root of f) to simple conditions on the coefficients of f.

Cohen's proof is rather demanding. The reader is lured into it by a memorable real analogue (a quantifier elimination for R using only the Sign Change Property and a similar isolation/elimination technology).

It is not at all easy to say exactly what Cohen proved. Anyone with aspirations to research in this area must make a patient study of the paper (and will be rewarded). Roughly, effective functions (like k^{th} root) can be eliminated in favour of a stock of functions on the value group and on the various residue rings, together with cross-section and a constant for an element of valuation 1. Cohen works under the assumption of characteristic 0 field and discrete valuation into a Z-group.

How then can he obtain information about $F_p((t))$? His general method produces, for a batch of $f(x)$ of degree $\leq n$, what he calls an effective graph for f. That is, various functions associated with the geometry of f are eliminable in terms of functions at the level of the value group or the residue rings. For given n, the reduction will be valid even in characteristic p, provided p is bigger than a certain primitive recursive function of n. (Typically the reductions fail because of singularities of f the residue field.) Then one shows easily that if Φ is given Z/p^n and $F_p[t]/t^n$ agree on Φ for n large enough, thus yielding the Ax-Kochen analogy.

This approach strongly suggests that our understanding of $F_p((t))$ will reach that of Q_p if we acquire systematic desingularization techniques in characteristic p. Observations on this can be found in [Roquette 1977], [McKenna 1980], and various talks by Denef. In particular one should be looking for restricted decision/elimination methods, say for f as above with some limitations on its complexity.

5.9. The analogue of Hilbert's 17th Problem. In 4.6.8 I mentioned Robinson's approach to Hilbert's 17th Problem. This subtracts nothing from the part of the original proof connecting order and sums of squares, but trivializes the specialization argument.

[Kochen 1969] established a very satisfying p-adic analogue. The order-theoretic "positive definite" is replaced by the valuation-theoretic "integral definite" -- essentially $x > 0$ is replaced by $v(x) \geq 0$. (Note that this is not entirely reasonable -- by our earlier discussion $v(x) \geq 0$ corresponds to $|x| \leq 1$.)

The main difficulty is to find the analogue of the sums of squares/order link. Kochen found such an analogue. One notable difference in the two cases is that on R one can bound, in terms of the number of variables the complexity of the sums of squares representation. No such bound is known in Kochen's analogue.

§6. TEN YEARS LATER:

As indicated above, some important work of consolidation was done by Ziegler and Weispfenning. Beyond that, the scene quietened very rapidly.

The next important advance came from a reconsideration of some peripheral work I had done on the R/Q_p analogy in 1966.

Robinson 1959 had proved that the theory of pairs (L,K), of real-closed fields with K a proper dense subfield of L, is complete. His proof is not attractive. He proved model-completeness in a language which certain predicates for algebraic dependence over K. I noticed in 1966
completeness theorem has a memorable proof by saturated models. About the same time Cohen sketched a constructive elimination theorem for this problem.

I went on (in 1966) to establish

Analogy 11. The theory of pairs (L,K) where K,L are real-closed (resp. p-adically closed) and K is a proper dense subfield of L is complete.

In both cases the proof uses the theory of closures and some lemmas about dense transcendence bases.

In 1974 I had many discussions with Peter Winkler in connection with his thesis begun under Robinson and finished under my direction. We catalogued the familiar theories in terms of algebraic boundedness and Vaughtian pairs (see Macintyre 1975]). Algebraic boundedness for R comes from the quantifier-elimination, and the lack of Vaughtian pairs for Th(R) comes from Tarski's observation that infinite definable sets have interior, together with the lemma that an interval in an ordered field has the cardinality of the field. The latter is used in Analogy 11. This set me thinking first about algebraic boundedness and Vaughtian pairs for Q_p, and then about an abstract (topology-free) formulation of the theorem about dense embeddings. Straightaway I came up against the problem whether an infinite definable subset of Q_p has interior.

The answer is no, if one allows cross-section as primitive, and yes if one does not. The former is easy ([Macintyre 1976]), but the latter is not. One has to understand the definable sets, and that tends to mean that one must eliminate quantifiers in terms of some natural primitives. I solved the problem by focussing on elements which had occurred in Ax-Kochen, but were used there only as a book-keeping device (to reduce to the case of pure value-group).

The relevant primitives are the sets P_n of nth powers.

Analogy 12. (a) The field R has quantifier-elimination in terms of the field language and P_2.

(b) The field Q_p has quantifier-elimination in terms of the field language and all P_n.

It can be shown that for Q_p all P_n are needed.

From 12 there follow quickly:

Analogy 12. ($K = R$ or Q_p). Infinite definable sets have interiors.

Analogy 13. ($K = R$ or Q_p). Th(K) is algebraically bounded and has no Vaughtian pairs.

Analogy 14. ($K = R$ or Q_p. See [Pillay 1983]). Th(K) is unstable but does not have the independence property.

My proof of quantifier-elimination in [Macintyre 1976] used quite a lot of the Ax-Kochen paraphernalia. Much later [Weispfenning 1984] gave a primitive recursive quantifier-elimination in my formalism.

In a very useful text published in 1984, Prestel and Roquette extended the method to finite extensions of Q_p. As well as the P_n one must distinguish a generator α as in 5.7.

I do not know if the case of infinite (even unramified) extensions has been checked out.

Tung raised in 1984 the possibility that α is irrelevant for quantifier-elimination. [Yasumuto 1985] refuted this for a ramified extension using ideas of [E. Robinson 1985B]. However, van den Dries

§7. DEVELOPMENTS IN THE P_n FORMALISM:

7.1. Model-theoretically the next step should have been to identify the universal theory of Q_p in the P_n-formalism. In fact this has been done only recently, by [Robinson 1983] and [Belair 1985], whose work I discuss later.

For real-closed fields we have quantifier-elimination using either $<$ or P_2. The $<$-substructures are exactly the ordered fields. The P_2-substructures are exactly the fields with a subset P_2 such that if we define $x > 0$ by $x \neq 0 \wedge P_2(x)$ we get a structure of ordered field. Note that P_2 on the substructure is not in general the set of squares.

For p-adically closed fields $v(x) \geq 0$ is definable using P_2 (and no extra quantification). So in seeking the analogue of ordered field we look for fields with sets P_n and suitable universal axioms.

Let $T_{\forall,p}$ be the universal theory of p-adically closed fields in the above formalism. I propose to call the models of $T_{\forall,p}$ potentially p-adic fields.

Without identifying $T_{\forall,p}$ [van den Dries 1984B] proved:

Analogy 15. (cf. 5.2.). Each potentially p-adic field K has an algebraic extension (for the P-formalism) to a p-adically closed field. This is a prime extension which is rigid over K.

This is not a trivial consequence of the preceding, and depends essentially on the fact that the group of roots of unity in a p-adically closed field is reduced (i.e. has trivial divisible part).

In consequence (see [van den Dries 1984B]).

Analogy 16. The theory of p-adically closed fields has definable Skolem functions.

There now comes an interesting flaw in the analogy. Consider the equivalence relation $v(x) = v(y)$ on Q_p. One can show that this is not equivalent to any relation $F(x) = F(y)$ where F is a definable function in the field formalism. Such an F would necessarily be of the form $\pi(v(x))$ for a cross-section π! By [Macintyre 1976] no such F exists. Recall however 4.6.6.

This provokes the

Conjecture. In the language with cross-section $Th(Q_p)$ eliminates imaginaries.

All the evidence supports this, but as of September 1985 no proof has been circulated. Such a proof would tend to rehabilitate the cross-section formalism.

7.2. Identifying $T_{\forall,p}$. This was done from 1983 on by E. Robinson and L. Belair independently. Common to their axiomatizations are the obvious universal conditions about the $P_n \cap K^*$, i.e. that these are groups with certain prescribed integers as coset representatives. One wants also the universal description of the valuation in terms of P_2.

These axioms leave out a more subtle constraint. Robinson and Belair identified this in different ways. Belair's extra axiom scheme is perhaps ad hoc, giving the "skeleton" of the projective system of P_n -- in particular it codifies the behaviour of roots of unity. For details see [Belair 1985]. Robinson's scheme is much more memorable. It is simply a transcription of 3.4 of this paper, saying precisely how $P_n \cap K^*$ is effectively open.

Robinson's axiomatization adds yet another analogy.

Analogy 17. ([Macintyre, McKenna, van den Dries 1983]).

(1) Any ordered field with quantifier elimination is real closed.

(2) Any "Belair-Robinson" field with quantifier-elimination is p-adically closed.

The point is that in the joint paper just cited we made apparently essential use of openness of the $P_n \cap K^*$.

7.3. Definable functions. I do not know to whom to credit the next analogy. It was surely evident to anyone who knew the quantifier-elimination results and Analogy 12.

Analogy 18. ($K = R$ or Q_p). Let f be a definable partial function on K^n. Then there are definable subsets $X_1,\ldots,X_k$ of K^n such that $\bigcup_i X_i$ is the domain of f, and on each X_i f is algebraic.

7.4. Defining open sets. In [van den Dries 1982] there is a pretty little proof that a definable open subset of R^n can be defined using $\wedge,\vee$ and strict polynomial inequalities $f(x_1,\ldots,x_n) > 0$.

Now let us modify the P_n formalism slightly, using rather D_n, where $D_n(x) \leftrightarrow x \neq 0 \wedge P_n(x)$. The above result can be expressed in terms of the D_n, just replacing $f(x_1,\ldots,x_n) > 0$ by $D_2(f(x_1,\ldots,x_n))$.

In [Robinson 1983] the following analogy was obtained:

Analogy 19. ($K = R$ or Q_p). Any definable open subset of K^n can be defined using $\wedge, \vee$ and the D_n.

7.5. Spectra. The work of Belair and Robinson, though perfectly natural in the context of the classical model theory of fields, was actually a foundational component of the construction of the p-adic spectrum of a commutative ring. Here one was pursuing the analogy with the intensively studied real spectrum of a commutative ring (see Coste-Roy 1982].

There is a wealth of material on the topic, for example in [Robinson 1983] in which one will find many sophisticated ideas from categorical logic. There had been regrettably little trade between categorical logic and the more conventional applied model theory, until Robinson's work. He uses the work of Cohen and me, but gives back fine information on coherent axiomatizations as well as preservation theorems for various classes of local rings with p-adic structure. Both he and Belair relate spatially of the p-adic spectrum to rigidity of p-adic closure.

Lack of space prohibits further discussion of this exciting work (indispensible for any future p-adic algebraic geometry). I recommend that the reader make a serious study of the theses [Belair 1985] and [Robinson 1983], and the papers [Robinson 1985A, B] and [Belair 1985A].

7.6. More on definable sets in K^n. Analogy 18 was really very weak, unless we could better visualize the definable sets in K^n. This has been done in [van den Dries-Scowcroft 1985]. I will cite just two more analogies:

Analogy 20. ($K = R$ or Q_p). Suppose $S \subseteq K^m$ has nonempty interior, and $S = \bigcup_{i=1}^{n} S_i$, each S_i definable. Then some S_i has nonempty interior.

Think of this as a "formal Baire theorem".

Analogy 21. ($K = R$ or Q_p). Let S be a definable subset of K^m. Then S is equal to a union $\bigcup_{i=1}^{n} S_i$, where each S_i is definable, and for each i either S_i is open or has no interior and is homeomorphic by a bianalytic projection along certain coordinate axes to an open subset of some K^r where $r < m$.

7.7. Dimension. If we make a slight reformulation of Analogy 21 we reach a plausible definition of real and p-adic dimension. Namely, with K fixed, define the dimension of S as the largest d such that S is a union $\bigcup_{i=1}^{n} S_i$, each S_i definable, and for some i S_i is homeomorphic, by a bianalytic projection along certain coordinate axes, to an open subset of K^d. (For $S = \phi$ put dimension $(S) = -\infty$).

This was one of the proposals of [van den Dries-Scowcroft 1985]. It leads in the p-adic case to an attractive dimension theory (at least partially anticipated by [Robinson 1983]).

Writing dim for dimension one then has the following remarkable analogy:

Analogy 22. ($K = R$ or Q_p).

(i) $\dim K^m = m$;

(ii) If X and Y are definable subsets of K^m $\dim(X \cup Y) = \max(\dim(X), \dim(Y))$, and $\dim(X) \leq \dim(Y)$ if $X \subseteq Y$;

(iii) If $X \subseteq K^{m+n}$ is definable and $X_{\bar{a}} = \{\bar{b} : (\bar{a},\bar{b}) \in X\}$ then for every $d \in N$ $X(d) = \{\bar{a} : \dim(X_{\bar{a}}) = d\}$ is definable;

(iv) Definable maps do not increase dimension;

(v) If X is definable, $\dim(X)$ is equal to the algebraic-geometric dimension of the Zariski closure X.

(iii) is of course remeniscent of stratification formulas in stability theory, and I would expect it to be worthwhile to attempt an axiomatic dimension theory subsuming the cases of C, R and Q_p. There is already some unpublished work on groups (by van den Dries and Scowcroft) which confirms this.

It appears also that it will be necessary to have a finer dimension theory for the cross-section formalism. Scowcroft has work in progress on this.

§8. DENEF'S APPLICATION:

8.1. Since Z_p is the projective limit of the finite rings Z/p^n, one knows that for $f \in Z_p[x_1,\ldots,x_n]$ f has a zero in Z_p if and only if f has a solution in each Z/p^n.

Before passing to the finer analysis of the above, one should observe that the remark above already implies that the set of $f \in Z[x_1,\ldots,x_n]$ unsolvable in Z_p is recursively enumerable. For if f is unsolvable, this will be revealed by enumerating the finite rings Z/p^n and verifying unsolvability of f in one of those.

In 1963 Nerode managed to show that the set of $f \in Z[x_1,\ldots,x_n]$ solvable in Z_p is recursively enumerable, and thereby recursive. Nerode's method depended on two observations:

(a) if f has a solution, it has a solution algebraic over Q;

(b) the set of algebraic elements of Z_p is a recursive ring.

An alternative would have been to prove (the true result) that there is a recursive function $\beta(f)$ such that f is solvable in Z_p if and only if f is solvable in Z/p^n for $n = \beta(f)$. A result of this kind, for varieties V rather than a single f was given in [Birch-McCann 1967].

It is Hensel's Lemma (suitably generalized) which provides such a bound. For example, for $p \neq 2$, and $v(a) \geq 0$, x^2-a is solvable in Z_p if and only if a is a square modulo $p^{v(a)+1}$.

Life would be much simpler if to test solvability of f one needed only to test solvability mod p. Even the above example shows that this doesn't work. Hensel's Lemma adds suitable hypotheses of nonsingularity mod p.

This suggests classifying Z_p solutions in terms of their singularities mod p^n, each n. The classification is codified by various Poincare' series, as follows. Fix $f_1, \ldots, f_r \in Z_p[x_1, \ldots, x_m]$. Let $\tilde{N}_n$ be the number of solutions of $f_1 = \ldots = f_r = 0 \mod p^n$, and let N_n be the number of such solutions coming from p-adic solutions. Define the formal series

$$\tilde{P}(T) = \sum_{n=0}^{\infty} \tilde{N}_n \cdot T^n$$

and $P(T) = \sum_{n=0}^{\infty} N_n \cdot T^n$.

Borevic and Shafarevic (page 63) conjectured that $\tilde{P}(T)$ is a rational function of T, and this was proved by Igusa and Meuser using resolution of singularities.

There are various related series in several variables. For example, let $N_{n,j}$ be the number of solutions p^n which lift to solutions mod p^{n+j}. Consider then

$$P(T,U) = \Sigma N_{n,j} T^n U^j.$$

[Denef 1984] showed this is rational, and related the rationality to the linear growth of the following function γ. $\gamma(n)$ is defined as the least $\gamma \geq n$ such that for any solution $\bar{y}$ mod p^γ there is a p-adic solution congruent to $\bar{y}$ mod p^n. See also [Greenberg 1966]. Notice that γ is definable, not in the pure language of field-theory, but in a natural many-sorted language having in particular a sort for the value-group. As Weispfenning showed, there is a good quantifier-elimination for this formalism, and Denef observed that one could deduce from this alone that there is a finite partition of N into congruence classes such that on each class, and for sufficiently large arguments, γ is linear.

Denef's main achievement was to show that both $\tilde{P}$ and P are rational, without using resolution of singularities, relying only on quantifier-elimination and some refined work on definability.

By standard formal manipulations the rationality in either case can be reduced to:

Theorem (Denef). Let S be a definable subset of Q_p^m, contained in a compact subset. Let h be a definable function from Q_p^m to Q_p such that $|h(x)|$ is bounded on S. Let $e \in N$, $e \geq 1$. Suppose that $v(h(x)) \in eZ \cup \{\infty\}$ for $x \in S$. Then

$$\int_S |h(x)|^{s/e} \cdot |dx| \qquad (s \in R,\ s > 0)$$

is a rational function of p^{-s}.

The analysis is relative to normalized Haar measure on Q_p. Note that it even requires proof that $|h(x)|^{s/e}$ is measurable.

It is easily seen that one need only consider S given by a condition $P_n(f(x))$. Also, without loss of generality h is simply a rational function.

Denef's main idea is a process of separation of variables, which depends on a rather detailed analysis, of Cohen's type, of the graph, vis a vis the P_n, of a generic polynomial g. This enables him to express the above integral as the product of a similar one in dimension m-1 and a "Poincare' series of Presburger type" $J(s) = \sum_{(k_1,\ldots,k_m) \in \bar{L}} p^{-\sum_{i=1}^{m} k_i A_i(s)}$ where L is a subset of Z^m definable in Presburger, and the A_i are linear polynomials with integer coefficients. By techniques of Meuser, J is a rational function of p^{-s}.

Denef's paper is not easy to summarize. Aside from the main theorem, there are important contributions to effective Skolemization and effective graphing. At the end there are intriguing remarks about the possibilities for analogous results with cross-section. Apparently rationality can be obtained, but less information about the multiplicity of poles.

Methodologically it is worth pointing out that Denef's proof needs not only quantifier-elimination (over-view of definable sets) but also the existence of definable Skolem functions. This has some bearing on the uniformity in p of Denef's result.

§9. ALL p SIMULTANEOUSLY:

The order may be confusing. The proof of the Ax-Kochen analogy involved $\prod_D Q_p$, $\prod_D F_p((t))$, and $\prod_D F_p$. Since the residue fields are uniformly interpretable in the valued fields, one certainly cannot understand the theory of all (or almost all) Q_p without understanding the theory of all (or almost all) F_p. On the other hand it was fairly clear from the Hypothesis 0 case that one would understand these global theories of Q_p once one understood the global theories for the F_p. The latter were mastered by [Ax 1968], and the consequences for decidability were stressed.

Less attention was paid at that time to definability. In fact Ax showed quantifier-elimination for the above global theories, in a language with extra predicates Sol_n interpreted by

$$Sol_n(x_1,\ldots,x_n) \leftrightarrow (\exists y)(y^n + x_1 y^{n-1} + \ldots + x_n = 0).$$

The elimination is uniform for all theories of pseudofinite fields. See [Cherlin-Macintyre-van den Dries 1981].

We want to make uniform (in p) the elimination theory of §6. First note ([Belair 1985A]) that there is a bound, independent of p, for the index of (P_n^*) in Q_p^*. This suggests adding to the language of field theory, for each n, constants for coset representatives of P_n^*. To combine this with the elimination theorem of the previous paragraph, one can appeal to the comprehensive study made by [Delon 1981], and obtain a uniform elimination based on the P_n, the above constants, and predicates $\overline{Sol_n}$ which hold of $x_1,\ldots,x_n$ iff the x_i are in the valuation ring and $Sol_n(x_1,\ldots,x_n)$ holds modulo the valuation ring. The exact result is spelled out in [Belair 1985A].

The implications of this have not been worked out systematically. One amusing consequence is that any set $\{p : Q_p \models \Phi\}$, where Φ is first-order, is also of the form $\{p : F_p \models \psi\}$, where Ψ is first-order.

A desirable application would be to uniformity of Denef's result. This is prima facie blocked because there is no uniform definability of Skolem functions for the Q_p. See [Belair 1985A]. I still believe, however, that further efforts in this direction are worthwhile.

§10. COMPLEXITY THEORY:

10.1. From the early 1970's significant progress was made in understanding the complexity (usually in terms of computation time, but sometimes in terms of space) of the classical decidable theories. It is notable that some of the new algorithms involved refinements of both quantifier-elimination and effective Skolemization.

Let us for example consider the ideas in the work of Collins and Monk-Solovay on Th(R).

(1) A prenex sentence $(Q_1x_1) \dots (Q_nx_n)\ A(x_1,\dots,x_n)$ has various constituent polynomials $f(x_1,\dots,x_n) \in Z[x_1,\dots,x_n]$. The truth of Φ is to be determined by how the zero positivity and negativity sets of the f decompose R^k into finitely many "regions of equivalence".

(2) One has to compute a finite set of algebraic needs so that the truth of Φ is equivalent to the truth of Φ relativized to the algebraic sample set.

(3) One should calculate rapidly the truth-value of the relativization.

(2) and (3) are obviously related to effective elimination and Skolemization.

Roughly, the quantitative result obtained is that there are constants C_1, C_2 such that with Φ as above and m = length of A, the sample set can be chosen of size (see below) $2^{2^{C_1(2m+n)}}$, and Φ can be treated in time $2^{2^{C_2n}}$. The size of a real algebraic number is the least length of an integral polynomial of which it is a root.

10.2. The analogue of the above ought to be done for Q_p. Since Q_p interprets Presburger arithmetic, and the latter cannot be decided in time $2^{2^{Cn}}$ ([Ferrante-Rackoff 1979]), it is reasonable to conjecture that Q_p and Presburger have the same time complexity for a decision procedure. This has certainly not been established, and in fact no proof is available that Q_p has an elementary recursive decision-procedure.

10.3. Oddly enough, at the level of the Ax-Kochen analogy, elementary recursive bounds have been obtained. By methods analogous to those of Collins, Monk and Solovay, [Brown 1978] proved:

If Φ is a sentence of the language of valued fields, m = length Φ, and

$$P > 2^{2^{2^{2^{2^{11m}}}}}$$

then Q_p and $F_p((t))$ agree about Φ. It is noteworthy too that methods of this kind are used in the important papers [Kiehne 1979A, B] on constructive model-completeness.

10.4. Concluding Remarks. Very little of a systematic nature has been published on effective definability theory for Q_p (or even R!). There is a wealth of topics -- Newton's method, fast factorization of polynomials, diophantine approximation and efficient sample sets. It seems to me likely that this will not be the last survey needed on the logic of the p-adics. I hope the next one will report shifts of emphasis to and fro from definability to decidability.

REFERENCES:

[Artin 1967]. E. Artin, Algebraic Numbers and Algebraic Functions, Gordon and Breach, 1967.

[Ax 1968]. J. Ax, The elementary theory of finite fields, Annals of Math. 88 (1968), 239-271.

[Ax-Kochen 1965A]. J. Ax and S. Kochen, Diophantine problems over local fields I, Amer. Jour. of Math. 87 (1965), 605-630.

[Ax-Kochen 1965B]. ________________, Diophantine problems over local fields II, Amer. Jour. of Math. 87 (1965), 631-648.

[Ax-Kochen 1966]. ________________, Diophantine problems over local fields III, Annals of Math. 83 (1966), 437-456.

[Belair 1985A]. L. Belair, Topics in the model theory of p-adic fields, Ph.D. Thesis, Yale, May 1985.

[Belair 1985B]. ________, Model theory of p-adic fields, Preprint 1985.

[Birch-McCann 1967]. B. J. Birch and K. McCann, A criterion for the p-adic solubility of diophantine equations, Quant. J. Math. 18 (1967), 59-63.

[Borevich-Shafarevich 1966]. Z. I. Borevich and I. R. Shafarevich, Number Theory, Academic Press 1966.

[Brown 1978]. S. Brown, Bounds on Transfer Principles ..., Memoirs of A.M.S. 204, 1978.

[Brumfiel 1979]. G. Brumfiel, Partially Ordered Rings and Semi-algebraic Geometry, Cambridge 1979.

[Cassels 1966]. J. W. S. Cassels, Diophantine equations with special reference to elliptic curves, J. London Math. Soc. 41 (1966), 193-291.

[Cassels 1967]. ______________ and A. Fröhlich, Algebraic Number Theory, Academic Press 1967.

[Cassels 1978]. ______________, Rational Quadratic Forms, Academic Press 1978.

[Cherlin 1976]. G. Cherlin, Model Theoretic Algebra, Selectic Topics, Lecture Notes in Mathematics 521, Springer 1976.

[Cherlin 1984]. _________, Paper in Logic Colloquium 1982 (ed. G. Lolli et. al.), North Holland 1984.

[Cherlin, Macintyre, van den Dries 1981]. _________, A. Macintyre, L. van den Dries, The elementary theory of regularly closed fields, Preprint 1981, to appear in J. reine angew. Math.

[Cohen 1969]. P. J. Cohen, Decision procedures for real and p-adic fields, Comm. on Pure and Applied Math. XXII (1969), 131-151.

[Collins 1975]. G. E. Collins, Quantifier elimination for real closed fields by cylindrical algebraic decomposition, pages 134-183 in Lecture Notes in Computer Science 33, (ed. H. Brakhage), Springer 1975.

[Coste-Roy 1982]. M. Coste and M. F. Roy, La topologie du spectre réel, in Ordered Fields and Real Algebraic Geometry, (D. W. Dubois and T. Recio, eds.), Contemporary Mathematics Vol. 8, A.M.S. 1982.

[Delon 1981]. F. Delon, Quelques propriétés des corps values en théorie des modèles, Thesis, Univ. Paris VII, 1981.

[Denef 1984]. J. Denef, The rationality of the Poincare series associated to the p-adic points on a variety, Inventiones Math. 77 (1984), 1-23.

[Ersov 1965]. Ju. L. Ersov, On elementary theories of local fields, Algebra i Logika 4 (1965), 5-30.

[Ferrante-Rackoff 1979]. J. Ferrante and C. Rackoff, The Computational Complexity of Logical Theories, Lecture Notes in Mathematics 718, Springer 1979.

[Greenberg 1966]. M. Greenberg, Rational points in henselian discrete valuation rings, Publ. Math. IHES 31 (1966), 59-64.

[Hasse 1980]. H. Hasse, Number Theory, Springer 1980.

[Hörmander 1969]. L. Hörmander, Linear Partial Differential Operators, Springer 1969.

[Jacobson 1964]. N. Jacobson, Lectures in Abstract Algebra, Vol. III, Van Nostrand, 1964.

[Kaplansky 1942]. I. Kaplansky, Maximal fields with valuations, Duke Math. J. 9 (1942), 313-321.

[Kaplansky 1945]. ___________, Maximal fields with valuations II, Duke Math. J. 12 (1945), 243-248.

[Kiehne 1979A]. U. Kiehne, Bounded products, J. reine angew. Math. 305 (1979), 9-36.

[Kiehne 1979B]. _________, Zeros of polynomials over valued fields, J. reine angew. Math. 305 (1979), 37-59.

[Kochen 1961]. S. Kochen, Ultraproducts in the theory of models, Annals of Math. 74 (1961), 221-261.

[Kochen 1969]. _________, Integer valued rational functions over the p-adic numbers. A p-adic analogue of the theory of real fields, in A.M.S. Proc. Symp. Pure Math. XII (1969), 57-73.

[Kochen 1975]. _________, Model theory of local fields, in Logic Conference, Kiel 1974. (Ed. G. Muller et. al.), Lecture Notes in Mathematics 499, Springer 1975.

[Knight, Pillay, Steinhorn 1985]. J. Knight, A. Pillay, C. Steinhorn, preprint, Notre Dame 1985.

[Lang 1952]. S. Lang, On quasi algebraic closure, Annals of Math. 55 (1952), 373-390.

[Lang 1964]. _______, Algebraic Numbers, Addison-Wesley 1964.

[Lewis-Montgomery 1983]. D. J. Lewis and H. L. Montgomery, On zeros of p-adic forms, Michigan Math. Journal 30 (1983), 83-87.

[Macintyre 1968]. A. Macintyre, Classifying pairs of real closed fields, Ph.D. Thesis, Stanford 1968.

[Macintyre 1975]. Dense embeddings I: A theorem of Robinson in a general setting, in Lecture Notes in Mathematics 498, Springer 1975.

[Macintyre 1976]. ___________, On definable subsets of p-adic fields, J.S.L. 41, No. 3, 1976, 605-610.

[Macintyre 1977]. ___________, Model-completeness, in Handbook of Mathematical Logic (ed. J. Barwise), North Holland 1977.

[Macintyre 1986?]. Lectures on Real Exponentiation, book in preparation.

[Macintyre, McKenna, van den Dries 1983]. ___________, K. McKenna, L. van den Dries, Elimination of quantifiers in algebraic structures, Advances in Mathematics 47 (1983), 74-87.

[Manin 1971]. Ju. I. Manin, Cyclotomic fields and modular curves, Russian Mathematical Surveys 26 (1971), 7-78.

[McKenna 1980]. K. McKenna, Some diophantine Nullstellensätze, pages 228-247 in Springer Lecture Notes in Mathematics 834. Springer 1980.

[Monk 1975]. L. Monk, Elementary-recursive decision procedures, Ph.D. Thesis, Berkeley 1975.

[Nagata 1962]. M. Nagata 1962]. M. Nagata, Local Rings, Wiley Interscience, 1962.

[Nerode 1963]. A. Nerode, A decision method for p-adic integral zeros of diophantine equations, Bull. A.M.S. 69 (1963), 513-517.

[Pillay 1983]. A. Pillay, An Introduction to Stability Theory, Oxford University Press 1983.

[Poizat 1983]. B. Poizat, Une theorie de Galois imaginaire, J. S. L. 48 (1983), 1151-1170.

[Prestel-Roquette 1984]. A. Prestel and P. Roquette, Formally P-adic Fields, Lecture Notes in Mathematics 1050, Springer 1984.

[Ribenboim 1968]. P. Ribenboim, Theorie des Valuations, 2e. edition, Les Presses de l'Universite de Montreal, 1968.

[A. Robinson 1959]. A. Robinson, Solution of a problem of Tarski, Fund. Math. 47 (1959), 179-204.

[A. Robinson 1965]. __________, Model Theory, North Holland, 1965.

[E. Robinson 1983]. E. Robinson, Affine Schemes and P-adic Geometry, Ph.D. Thesis, Cambridge 1983.

[E. Robinson 1985A]. __________, The p-adic spectrum, Preprint, Peterhouse, Cambridge 1985.

[E. Robinson 1985B]. __________, The geometric theory of p-adic fields, Preprint, Peterhouse, Cambridge, 1985.

[J. Robinson 1965]. J. Robinson, The decision problem for fields, in Theory of Models (ed. Addison et. al.), North Holland 1965, 299-311.

[Roquette 1977]. P. Roquette, A criterion for rational places over local fields, J. reine angew. Math 292 (1977), 90-108.

[Scowcroft, van den Dries 1985]. P. Scowcroft and L. van den Dries, On the structure of semi-algebraic sets over p-adic fields, preprint 1985 (available from Stanford).

[Serre 1965]. J-P. Serre, Cohomologie Galoisienne, Lecture Notes in Mathematics 5, Springer 1965.

[Serre 1968]. _________, Corps Locaux, Hermann 1968.

[Skolem 1938]. T. Skolem, Diophantische Gleichungen, Springer 1938.

[Steinitz 1930]. E. Steinitz, Algebraische Theorie der Körper, Berlin 1930.

[Tarski 1951]. A. Tarski, A Decision Method for Elementary Algebra and Geometry, 2nd ed., revised, Berkeley and Los Angeles 1951.

[Tate 1967]. J. Tate, Global class field theory, in [Cassels 1967].

[Van den Dries 1982]. L. van den Dries, Some applications of a model-theoretic fact to (semi-) algebraic geometry, Indagat. Math. 44 (1982), 397-401.

[Van den Dries 1984A]. ______________, Remarks on Tarski's problem concerning (R, +, •, exp), pages 97-121 in Logic Colloquium 1982, (ed. G. Lolli et. al.), North Holland 1984.

[Van den Dries 1984B]. ______________, Algebraic theories with definable Skolem functions, J.S.L. 49 (1984), 625-629.

[Weil 1967]. A. Weil, Basic Number Theory, Springer 1967.

[Weispfenning 1976]. V. Weispfenning, On the elementary theory of Hensel fields, Ann. Math. Logic 10 (1976), 59-93.

[Weispfenning 1984]. _____________, Quantifier elimination and decision procedures for valued fields, pages 419-472 in Lecture Notes in Mathematics 1103 (eds. G. Muller and M. Richter), Springer 1984.

[Whitney 1957]. H. Whitney, Elementary structure of real algebraic varieties, Annals of Math. 66 (1957), 545-556.

[Yasumoto 1985]. M. Yasumoto, Personal communication.

[Ziegler 1972]. M. Ziegler, Die elementare Theorie der henselschen Körper, Dissertation, Köln 1972.

LOGIC COLLOQUIUM '84
J.B. Paris, A.J. Wilkie, and G.M. Wilmers (Editors)

MALAISE ET GUERISON

Bruno Poizat
Université Pierre & Marie Curie
Paris

Le premier jet de cette conférence, dont quelques privilégiés, à Paris et à Wittenberg, ont eu la primeur, était intitulé "Malaise dans les modèles récursivement saturés". J'y expliquais combien je trouvais malsaines toutes ces notions de récursivité, décidabilité, et autres fariboles, qu'on voit encore trop souvent traîner dans des exposés de Théorie des Modèles, et j'avouais mon désarroi devant des travaux récents à propos de modèles resplendissants - une notion à laquelle j'attachais une signification structurelle - de théories très stables, qui donnaient l'impression que des arguments récursivistes allaient intervenir de façon essentielle dans des théorèmes de structure.

Pour me guérir de mon malaise, comme je ne pouvais éviter les modèles récursivement saturés, il me fallait les neutraliser, et je risquai la conjecture suivante :

Conjecture de Poizat : Si T est une théorie complète, totalement transcendante, dans un langage L fini, tout modèle resplendissant de T est saturé, tout modèle récursivement saturé de T est oméga-saturé.

Julia Knight a eu la bonté de m'envoyer un contre-exemple à cette conjecture avant la tenue de cette réunion.

Craignant une rechute grave, j'ai trituré dans tous les sens ce contre-exemple, et j'ai enfin compris que mon malaise ne venait que d'un aveuglement passager dû à l'idéologie ambiante : contrairement à ce qu'on pourrait penser, la notion de modèle resplendissant n'a pas un caractère exclusivement structurel, elle ne peut se réduire à quelque chose de modèle-théoriquement sain ; on peut montrer cela au moyen d'un argument mathématique à la fois simple et décisif.

C'est ainsi que j'ai trouvé la voie de la guérison ; à l'intention de ceux qui pensent que ces bonnes nouvelles de ma santé mentale n'ont pas un intérêt général, je précise que ce malaise nous donnera l'occasion de rencontrer quelques beaux théorèmes de Théorie des Modèles. Ces résultats ne sont pas de moi, et les éminents savants dont j'expose les travaux ne portent aucune responsabilité dans les positions sectaires qui sont exprimées ici.

1 - LA RECURSIVITE EN THEORIE DES MODELES

On rencontre encore aujourd'hui des gens qui vous présentent la décidabilité de la théorie de je ne sais quelle structure, par exemple un corps, un groupe ... comme quelque chose de digne d'intérêt. Cette manie a son origine dans la perversité de nos pères, qui, rompant avec une tradition millénaire, ont introduit en mathématique des objets indéterminés, tout cela pour s'offrir le plaisir d'un doute ontologique à leur sujet : pour être bien sûr de l'existence de quelque chose, il fallait en donner un algorithme de construction.

Il était naturel à l'époque de chercher une "théorie des ensembles" qui fût non seulement vraie et complète, mais aussi décidable, le caractère algorithmique de la logique sous-jacente étant une garantie de la solidité de la construction. Mais on ne voit pas quel enjeu se cache derrière la décidabilité d'un corps de séries formelles, et on serait bien en peine de citer un exemple où un tel résultat, en soi, serait utile à quelque chose pour la Théorie des Modèles, en dehors du domaine spécifique de l'étude des modèles de l'Arithmétique.

Ce placage artificiel de notions récursivistes sur la Théorie des Modèles n'apporte rien non plus à l'étude de la récursivité, ou de la complexité des algorithmes, qui n'y intervient que par des techniques routinières. On montre généralement l'indécidabilité de la théorie d'une structure en y interprétant quelque chose comme l'Arithmétique ; et pourquoi donc, quand on a interprété l'Arithmétique du 25° ordre dans M, en tirer le corollaire triomphant que M a une théorie indécidable ?

Quant aux arguments de décidabilité, ils reposent uniformément sur une généralisation de la Thèse de Church, à savoir que toute fonction de N dans N est récursive primitive.

En fait, quand un théoricien des modèles proclame qu'une théorie est décidable, il a en tête tout autre chose ; il veut dire que cette théorie est <u>simple</u>, qu'il sait la maîtriser, qu'il sait en décrire les types, peut-être même en classer les modèles. La décidabilité d'une théorie mesure, d'une certaine façon, la complexité d'un système d'axiomes ; on sait qu'elle est sensible à toute sorte de manipulations artificielles sur la présentation du langage ; elle ne concerne pas le théoricien des modèles, dont le souci est de décrire une classe de structures. Plutôt que de faire appel à des notions déplacées, il devrait être capable d'exprimer ses résultats dans un langage autonome et adéquat.

Cela dit, il n'y a pas là grand mystère, ni de quoi provoquer un malaise : tout au plus une légère irritation. Notre tranquilité d'esprit n'aura été que légèrement perturbée par l'apparition des "modèles récursivement saturés", une sorte de hochet pour récursivistes ; mais le malaise s'installe quand apparait le Théorème de Ressayre qui montre, dans le cas dénombrable, l'équivalence de la saturation récursive avec ce qu'on appelle, depuis Barwise et Schlipf, la resplendance, une notion qui semble si naturelle du point de vue modèle-théorique !

Et le malaise s'est aggravé récemment, lorsque sont parus des travaux de Buechler, Knight, Pillay, et d'autres, portant sur les modèles resplendissants de théories superstables, ou même totalement transcendantes (donc les plus éloignées de l'Arithmétique), où il semblait que des techniques de récursivité étaient indispensables pour mettre en évidence des propriétés <u>structurelles</u> de ces modèles.

2 - QUELQUES DEFINITIONS

Avant d'entrer dans le vif du sujet, je rappelle les principaux thèmes de la saturation récursive et de la resplendance.

On considère une théorie complète T, <u>dans un langage fini</u> L. Un modèle

M de T est dit récursivement saturé si pour tout uple $\bar{a}$ d'éléments de M, et tout ensemble p, récursif et consistant, de formules $f(x,\bar{a})$, il existe un élément de M qui les satisfait toutes.

Comme L et $\bar{a}$ sont finis, la notion d'"ensemble récursif de formules" ne pose pas de problème, soit qu'on définisse les ensembles récursifs directement comme ensembles de mots dans un alphabet fini, soit qu'on les définisse comme ensembles de nombres entiers, et qu'on représente les formules au moyen d'un codage non exotique. Ce serait une notion beaucoup moins signifiante si le langage L était dénombrable, car elle serait sensible à la façon dont ce langage serait présenté.

Dès les premiers mots on voit qu'il y a un ver dans le fruit ; en effet, dans la définition ci-dessus, on peut remplacer "récursif" par "récursivement énumérable" ; on le voit par une méthode classique de pléonasme, le pléonasme étant indissociable de la logique du premier ordre ; en effet, si p est un ensemble récursivement énumérable, l'appartenance de f à p se traduit par la satisfaction d'une formule arithmétique (E u) a(f,u) , où a est à quantifications bornées ; considérons l'ensemble q des formules g, obtenues en mettant u doubles négations devant une formule f, le couple (f,u) satisfaisant a ; pour un élément de M, satisfaire p ou q, c'est pareil, et q est récursif, puisque l'appartenance de g à q s'exprime au moyen d'une formule à quantifications bornées en fonction de la complexité de g. A l'inverse, on verrait par la même méthode, en dilatant artificiellement la taille des formules, qu'on peut remplacer "récursif" par P ou NP ; en conséquence, les modèles récursivement saturés ne font pas des distinctions qui sont pourtant fondamentales dans l'étude des procédés algorithmiques, et il est difficile de les justifier en prétendant qu'ils pourraient avoir des applications importantes à l'Informatique (Théorique), comme c'est la mode aujourd'hui.

Il faut aussi remarquer que les modèles récursivement saturés forment une classe pseudo-élémentaire ; vous pouvez le montrer de différentes façons, suivant votre degré de sophistication en "soft model theory", mais le mieux est de procéder de la manière la plus directe ; pour cela, on se remémore un vieux (1958) théorème de Craig-Vaught, qui affirme qu'une théorie récursivement axiomatisable, dans un langage fini, a une expansion finiment axiomatisable. On ajoute alors au langage L de T celui de l'Arithmétique et de la combinatoire des parties finies de M, et aussi un prédicat de satisfaction Sat(f,u) destiné à noter la satisfaction d'une formule f du langage L par un uple u d'éléments de M, plus une fonction t(n,u). Nous pouvons rassembler tous les symboles rajoutés à L en un seul, R, et considérer l'énoncé rs(R) qui assure un minimum d'arithmétique, qui dit que tout uple d'éléments de M est représenté, que le prédicat de satisfaction a les propriétés qu'on pense, et que pour tout uple u, et tous entiers n et m, s'il existe un x satisfaisant les m premières formules du n° ensemble p_n défini par une formule arithmétique à quantifications bornées, une fois les paramètres substitués par u, alors t(n,u) est un tel x.

Je dis que M est récursivement saturé si et seulement si on peut y interpréter R de manière à satisfaire rs(R). Si M est récursivement saturé, interpréter R de façon standard ; et si M se transforme en un modèle de rs(R), l'interprétation de l'Arithmétique et de la Combinatoire sera peut-être non-standard, introduisant des notions (entiers, uples, formules) sans signification réelle, mais ce que nous avons mis dans rs(R) l'empêche de mentir à propos des notions standards dont nous avons besoin (et en particulier l'empêche de déclarer faussement qu'un entier standard satisfait une formule arithmétique standard à quantifications bornées).

Passons maintenant à la deuxième notion ; M est dit resplendissant si pour tout énoncé $f(\bar{a},R)$, dans un langage faisant intervenir, outre L, un uple $\bar{a}$ d'éléments de M et un nouveau symbole relationnel R, et qui soit consistant avec la théorie de M (je veux dire le type de $\bar{a}$), alors on peut interpréter R de manière à transformer M en un modèle de $f(\bar{a},R)$. Par exemple un modèle resplendissant est oméga-fortement-homogène, car si $\bar{a}$ et $\bar{b}$ ont même type, il est consistant de supposer qu'il existe un automorphisme de M (notion exprimable par un énoncé, puisque le langage est fini) qui les échange.

Nous voyons que l'énoncé rs(R) ci-avant est toujours consistant avec M, ne serait-ce que parce que M a une extension oméga-saturée, si bien qu'un modèle resplendissant est récursivement saturé ; on peut même préciser que si $f(\bar{a},R)$ est consistant avec M, M se transforme en un modèle (M,R) récursivement saturé de cet énoncé : si on peut mettre "resplendissant" au lieu de "récursivement saturé" dans cette phrase, on dit que M est chroniquement resplendissant ; on ne sait rien de bien général sur la splendeur chronique.

Le Théorème de Ressayre affirme la réciproque dans le cas dénombrable : si M est dénombrable (le langage étant fini) et récursivement saturé, il est resplendissant (et même chroniquement !) ; on le montre en faisant à l'intérieur de M une construction de Henkin pour $f(\bar{a},R)$: comme la notion de conséquence est récursivement énumérable, et M récursivement saturé, on peut toujours interpréter les témoins par des éléments de M.

Si T est oméga-catégorique, ses modèles resplendissants forment une classe pseudo-élémentaire : la consistance de $f(\bar{a},R)$ ne dépendant que du type de $\bar{a}$, qui est isolé, introduire un symbole $R(\bar{x},\bar{y})$ et déclarer que $f(\bar{a},R(\bar{a},\bar{y}))$ est vrai quand $\bar{a}$ a le type convenable. Cela n'est pas vrai en général : si T est la théorie d'une relation d'équivalence E qui, pour chaque entier n, a exactement une classe avec n éléments, les modèles resplendissants de T sont saturés (il est consistant de supposer que chaque classe infinie est en bijection avec le modèle, et qu'il existe un ensemble A formé d'éléments deux-à-deux non équivalents et en bijection avec le modèle), et la notion de resplendance ne se conserve pas par ultrapuissance.

On voit facilement, en itérant oméga fois le Lemme de consistance disjointe, que tout modèle a une extension élémentaire resplendissante de même cardinal ; il est également facile de voir qu'un modèle saturé (i.e. kappa-saturé de cardinal kappa) est resplendissant (et même kappa-resplendissant : voir ci-après). Par contre, il est souvent délicat de déterminer si un modèle donné est resplendissant ou non.

Il y a un lemme de dilatation de Schmerl qui est extrêmement utile pour cela : si M est un modèle récursivement saturé dénombrable, il y a en tout cardinal lambda un modèle resplendissant de ce cardinal qui réalise les mêmes types que M ; comme il s'agit de structures oméga-homogènes, ces deux modèles sont élémentairement équivalents dans $L_{\infty\omega}$.

Ce lemme est le résultat d'une commande passée à Schmerl par Buechler, et il est publié dans (BUECHLER 1984) ; il repose sur un maniement assez délicat d'indiscernables.

3 - UN PEU DE STABILITE

Nous commençons par généraliser à la fois la notion de resplendance et celle de kappa-saturation ; nous dirons que M est kappa-resplendissant si toute théorie* dans un langage faisant intervenir strictement moins de kappa nouveaux symboles, et utilisant strictement moins de kappa éléments de M, est réalisable sur M. J'introduis également, si le langage de M est fini, une notion intermédiaire entre la resplendance et la oméga-resplendance (que certains appellent relation-universalité) ; étant donné un réel r, je dis que M est (récursivement en r)-resplendissant si on a ça pour les théories, dans un langage fini, qui sont récursives en r ; on voit aisément, en ajoutant un peu d'Arithmétique, que récursivement-resplendissant = resplendissant.

Avec une hypothèse de stabilité, la splendeur a tendance à impliquer carrément la saturation ; la chose semble avoir été remarquée pour la première fois par Buechler dans son PhD.

Montrons que, si T est oméga-stable, un modèle M oméga-resplendissant de T est saturé. Soit A une partie de M, de cardinal strictement inférieur à celui de M, et soit p un 1-type sur A ; soit p' un fils de p sur M, isolé des types de même rang de Morley par une formule $f(x,\bar{a})$; on sait que p' est l'unique fils non-déviant de sa restriction q à $\bar{a}$; on ajoute au langage un prédicat unaire I, un symbole de fonction s, on déclare que s est une bijection entre M et I, et que I est une copie de la suite de Morley de q, c'est-à-dire que tout n-uple extrait de I est formé de réalisations de q indépendantes ; comme c'est consistant, ça existe ; et on sait que tous les éléments de I, sauf un nombre fini si A est fini, sauf card(A) si A est infini, réalisent sur A l'unique fils non-déviant de q, c'est-à-dire p.

Une démonstration semblable montre que si T est stable un modèle de T qui est $\mathrm{card}(T)^+$-resplendissant, ou même seulement $\mathrm{card}(T)^+$-saturé et oméga$^+$-resplendissant est saturé ; de même, pour une théorie superstable, un modèle oméga$^+$-resplendissant est saturé.

Ce type de propriétés caractérise la stabilité : pour une théorie instable, aucune condition de resplendance ne peut impliquer la saturation. En effet, si T est instable, on peut fabriquer de manière très souple des modèles qui sont non-saturés non pas parce qu'ils omettent des types, mais parce qu'ils en réalisent ! Plus précisément, si on part d'un modèle quelconque M_0, puis qu'on réalise dans M_1 tous les lambda-types sur M_0, puis dans M_2 tous les lambda-types sur M_1, et qu'on répète lambda fois, le modèle M_λ obtenu à la fin ne peut pas être lambda$^+$-saturé (pour les détails, le mieux sera de consulter (POIZAT 198?) quand cet excellent ouvrage aura trouvé un éditeur ; en attendant, voir (POIZAT 1983)) ; il ne sera pas non plus lambda$^+$-homogène si par exemple M_0 est lambda$^+$-saturé. Comme dans cette construction la seule chose qu'on demande aux M_α est de réaliser des types, on peut mettre en sandwich une autre construction qui garantit que M_λ soit kappa-resplendissant, pour un kappa raisonable par rapport à lambda ; et ensuite, si le cardinal de M_λ vous semble trop gros, le faire maigrir par un théorème de Löwenheim.

Pour une théorie stable non-superstable, la même construction peut se faire pour lambda = oméga (elle fonctionne dans le cas stable si lambda est strictement inférieur à kappa(T)), si bien qu'en tout cardinal supérieur ou égal à 2^ω une théorie non-superstable a des modèles oméga-resplendissants qui ne sont pas oméga$^+$-saturés, ni même oméga$^+$-homogènes.

* consistante avec celle de M !

NOTE Si on y efface les oméga et les oméga$^+$, ce dernier résultat a été montré dans (KNIGHT 1982), par un argument où interviennent des définitions récursives de types, qui a été pour beaucoup dans mon malaise ; Julia Knight montre que si pour un lambda non-dénombrable tous les modèles resplendissants de T sont homogènes, alors cela se produit pour tout lambda non-dénombrable ; T est alors nécessairement superstable. Il y a des théories superstables pour lesquelles c'est le cas (par exemple toutes les théories oméga-stables, comme nous allons le voir), d'autres non ; j'aimerais qu'on me donnât une caractérisation modèle-théoriquement signifiante du cas où ça se produit, mais je ne sais que penser ; une telle caractérisation n'existe peut-être pas !

4 - THEORIES OMEGA-STABLES (DANS UN LANGAGE FINI)

Dans une théorie T oméga-stable, chaque type (complet) p est déterminé par la formule $f(x,\bar{a})$ qui l'isole des types de même rang de Morley, et le théoricien des modèles sait d'expérience qu'on peut alors, à partir de considérations sur des formules, montrer des théorèmes qui demandent des conditions beaucoup plus globales dans le cas où T est seulement stable. Cela explique la conjecture de l'introduction ; elle serait aisément falsifiable si le langage était infini, car la resplendance ne fait intervenir le langage que fragment fini par fragment fini.

Si nous cherchons à prouver la conjecture en adaptant la démonstration de la section 3, nous pouvons, par un seul énoncé, imposer à chaque élément de I de satisfaire la formule $f(x,\bar{a})$ qui détermine q, et à l'ensemble I d'être indiscernable par un ensemble récursif de conditions, qu'on transforme en un seul énoncé en accroissant le langage ; mais comment imposer de cette manière à la suite indiscernable d'être bien la bonne, d'être bien la suite de Morley de q ?

Il suffit, en fait, de pouvoir imposer le type sur $\emptyset$ de cette suite de Morley ; en effet, on sait que la classe d'un type dans l'ordre fondamental ne dépend que du type sur $\emptyset$ de sa suite de Morley (voir (POIZAT 1983a)), et que l'extension non déviante de q est l'unique type dans la classe de q à satisfaire $f(x,\bar{a})$. Réciproquement, à n'importe quelle suite indiscernable est associée la classe de son type limite, si bien que l'ordre fondamental est en bijection avec le fermé de $S_\omega(\emptyset)$ formé des types de suites indiscernables.

Le problème est donc d'identifier, finiment ou récursivement, une suite indiscernable parmi d'autres ; il se résoud de lui-même si l'ordre fondamental est fini (ou même s'il n'y a qu'un nombre fini de classes de types non-réalisés). D'ailleurs, on pourrait utiliser directement l'ordre fondamental en ajoutant des prédicats qui forcent l'existence d'une longue suite croissante M_i de restrictions élémentaires de M, avec a_i dans $M_{i+1} - M_i$, satisfaisant $f(x,\bar{a})$, et représentant sur M_i les formules de la borne de q.

En conséquence, la conjecture est vraie si T est oméga-stable et oméga-catégorique ; mais ce n'est pas une trivialité de voir que l'ordre fondamental d'une telle théorie est fini ; c'est un corollaire de l'analyse de Cherlin-Harrington-Lachlan, qui montrent que chaque suite indiscernable y est déterminée par ses deux premiers éléments !

Je sais aussi la démontrer pour tous les modules oméga-stables (dans un langage fini, par exemple si l'anneau est finiment engendré), et aussi pour tous les exemples qui me viennent à l'esprit, par exemple pour les corps différentiellement clos de caractéristique nulle.

Anand Pillay, par un argument à la fois simple et ingénieux, a montré qu'elle était vraie pour une théorie finidimensionelle. Il suffit alors d'avoir de longues suites de Morley pour des représentants $p_1, \ldots p_n$ des classes minimales de l'ordre de Rudin-Keisler, et Pillay remarque qu'on peut les choisir de manière qu'à p_i soit associées des formules $f_i(x)$, $g_{i,1}(x), \ldots g_{i,k}(x)$ avec paramètres dans le modèle premier, telles que dans une paire de modèles où $g_{i,1}, \ldots g_{i,k}$ n'augmentent pas, tous les nouveaux éléments satisfaisant f_i réalisent p_i. Cela lui permet, en introduisant des chaînes de modèles comme ci-dessus, à forcer par un seul énoncé la suite de Morley à être celle de p_i.

Pour le cas général, comme je l'ai dit, la conjecture est fausse ; on peut toutefois remarquer que l'ordre fondamental d'une théorie oméga-stable est dénombrable (car il y a un modèle saturé dénombrable : considérer les types sur ce modèle !), si bien qu'on peut coder en un seul réel r tous les types sur Ø de oméga-suites indiscernables. Par conséquent, si T est oméga-stable, dans un langage fini, il existe un réel r tel que tout modèle de T (récursivement en r)-resplendissant soit saturé. Ce n'est pas un théorème bien sérieux, mais il explique pourquoi toute théorie satisfait la conjecture, à moins d'être spécialement fabriquée pour être un contre-exemple : c'est une conséquence de la Thèse de Church généralisée !

Le résultat vrai le plus proche de la conjecture est de Buechler, qui affirme que tout modèle M resplendissant d'une théorie oméga-stable, dans un langage fini, est homogène ; on sait qu'un modèle homogène est caractérisé par son cardinal et par les types (de n-uples) qu'il réalise ; on retrouve donc le résultat pour le cas oméga-catégorique (car homogène = saturé), et on voit aussi que dès que M est resplendissant et oméga-saturé, il est saturé. Le réel r' qui code tous les types sur Ø de n-uples suffit lui aussi à saturer le modèle ; je ne vois pas trop quel est son rapport avec le r précédent.

L'argument de Buechler repose sur un lemme délicat, qui, comme beaucoup de résultats profonds, donne l'impression première de devoir être faux, mais se trouve finalement être juste : si I est une suite infinie, indiscernable sur $\bar{a}$, dont tous les éléments satisfont $f(x,\bar{a})$, et dont le type limite est $f(x,\bar{a})$-régulier, alors, pour que I soit indépendante au-dessus de $\bar{a}$, il suffit que ses éléments le soient deux-à-deux ; en d'autres termes, sous ces hypothèses, on est sûr qu'une suite indiscernable est la suite de Morley qu'on croit dès qu'on en connait les deux premiers éléments ; en cherchant des contre-exemples dans une théorie comme celle des corps différentiellement clos, on constatera que l'hypothèse de régularité forte du type limite est nécessaire ; il ne suffit pas qu'il soit de poids un.

Ce résultat de Buechler trivialise la classification des modèles resplendissants d'une théorie oméga-stable ; par un argument de Löwenheim, un modèle resplendissant de cardinal lambda a une restriction élémentaire dénombrable resplendissante qui réalise les mêmes types ; et pour aller de oméga à lambda, on utilise le lemme d'étirement de Schmerl ; comme un modèle homogène est caractérisé par les types qu'il réalise, on voit qu'il y a le même nombre de modèles resplendissants en tout cardinal. Pour une théorie oméga-stable, un modèle resplendissant, c'est un modèle homogène et récursivement saturé ; il y a un exemple, instable, de modèle homogène et récursivement saturé qui n'est pas resplendissant ; il est aussi de Julia Knight.

Je croyais que ce nombre de modèles resplendissants était un ; Julia Knight m'a envoyé une liste de contre-exemples où il y a un, deux, ... n, ... ω, ou 2^ω tels modèles (la conjecture de Vaught est triviale pour une classe pseudo-élémentaire où tous les modèles sont homogènes !) ; ces contre-exemples ont été ensuite modifiés par Daniel Lascar, de manière à être dimensionels ("non multidimensional", comme dirait Shelah), avec, bien sûr, une infinité de dimensions, pour ne pas contrarier Pillay ; ils ont même des rangs U égaux à un : cela indique une théorie des modèles absolument triviale. Pour les premiers de la série, le rang de Morley est 4, mais les types non-récursifs sont de rang de Morley 2, ce qui, dans une théorie récursive, est le minimum possible.

Note technique pour ceux qui savent (et pour donner envie de savoir à ceux qui ne savent point encore) Soit M resplendissant non-dénombrable, dont nous voulons montrer l'homogénéité ; soient A, b, A' dans M, A et A' étant de même type, et de cardinal inférieur à celui de M ; nous voulons b' dans M tel que A^b et A'^b' aient même type ; comme il y a des modèles premiers, nous pouvons supposer que A et A' sont des modèles de T ; en procédant par étapes, on se ramène au cas où tp(b/A) est RK-minimal, et même $f(x,\bar{a})$-régulier, avec en outre tp(b/A) unique extension non-déviante de sa restriction à $\bar{a}$. Soit $\bar{a}'$ le correspondant de $\bar{a}$; par oméga-homogénéité, on peut trouver b' réalisant sur $\bar{a}'$ le type correspondant à $tp(b/\bar{a})$; si on peut en trouver un deuxième, b", b' et b" étant indépendants sur $\bar{a}'$, on a les deux premiers éléments de notre suite de Morley, et, grâce au lemme de Buechler, on force une longue suite indiscernable à être bien la bonne. Sinon, pour tout $\bar{c}$ de A', on peut trouver $b_{\bar{c}}$ avec $tp(b'/\bar{a}') = tp(b_{\bar{c}}/\bar{a}')$, $b_{\bar{c}}$ et $\bar{c}$ étant indépendants au-dessus de $\bar{a}'$ (par oméga-homogénéité) ; comme $b_{\bar{c}}$ et b' sont dépendants au-dessus de $\bar{a}'$, et que $tp(b'/\bar{a}')$ est de poids un, il faut que b' et $\bar{c}$ soient indépendants ; donc A^b et A'^b' ont même type.

5 - LE SOULAGEMENT

Je commence par décrire une innocente petite théorie t_0, où se sent le doigt de Daniel Lascar ; c'est celle de la structure A_0, formée des entiers naturels munis de leur fonction successeur s, certains étant colorés en blanc, d'autres en noir, de la manière suivante :

- si n est pair, tous les x tels que $n^2 \leq x < (n+1)^2$ sont blancs
- si n est impair, tous les x tels que $n^2 \leq x < (n+1)^2$ sont noirs.

Un modèle de t_0 est formé de A_0, plus un certain nombre de copies de la fonction-successeur des entiers relatifs, qui sont soit tous blancs, soit tous noirs, soit blancs puis noirs, soit noirs puis blancs. Voilà une théorie sans mystère : elle est de rang U un, dimensionelle, avec 4 dimensions.

Si r est un ensemble ni fini ni cofini de nombres entiers, on lui associe de même la théorie t_r de la structure A_r définie comme A_0, en remplaçant "n est pair" par "n est dans r", "n est impair" par "n n'est pas dans r" ; c'est une façon très économique de coder un réel dans une structure dont la théorie des modèles est triviale !

Considérons maintenant une théorie T quelconque ; soit T_0 la théorie des structures obtenues en juxtaposant un modèle M de T et un modèle A de t_0 ; il est facile de voir qu'un modèle (M,A) de T_0 est récursivement saturé si et seulement si M et A le sont ; je ne sais si la chose reste vraie en général pour la resplendance, mais c'est vrai en tout cas si T est oméga-stable, à cause de la classification de Buechler.

On fabrique de même la théorie T_r des structures qui sont juxtaposition d'un modèle M de T et d'un modèle B de t_r ; pour la resplendance de (M,B), il faut cette fois la resplendance récursive en r de M.

Il est une façon très simple de passer d'un modèle de T_0 à un modèle de T_r, qui consiste à changer les couleurs des éléments de A_0, et à ne rien toucher d'autre ; on obtient ainsi un foncteur bijectif, qui préserve tout, les cardinalités, les extensions élémentaires, les ensembles de paramètres, les types, la saturation, l'homogénéité, etc... tout sauf la resplendance !

Cet exemple montre clairement que la splendeur ne peut être réduite à des propriétés structurelles, qu'elle restera pour toujours un mélange incongru d'ingrédients disparates : des contraintes structurelles (comme l'homogénéité), et d'autres (comme la saturation récursive) sans valeur modèle-théorique. Ce n'est rien d'autre qu'un gadget, avec lequel les logiciens, les récursivistes, les théoriciens des ensembles, ou les gens comme ça pourront faire joujou, mais qui n'aura jamais de signification pour un mathématicien normal.

REFERENCES

(BUECHLER 1984) Steven Buechler, Expansions of models of omega-stable theories, J. Symb. Logic, 49,

(BUECHLER 198?) Steven Buechler, preprint

(KNIGHT 1982) Julia Knight, Theories whose resplendant models are homogeneous, Israel Journ. Math., 42, 151-161

(KNIGHT 198?) Julia Knight, preprints

(PILLAY 1984) Anand Pillay, Regular types in nonmultidimensional omega-stable theories, J. Symb. Logic, 49, 880-891

(POIZAT 1983) Bruno Poizat, Beaucoup de modèles à peu de frais in Théories Stables III, ed. Poizat, I.H.P., Paris

(POIZAT 1983a) Bruno Poizat, Post-scriptum à "Théories instables" J. Symb. Logic, 48, 239-249

(POIZAT 198?) Bruno Poizat, Cours de Théorie des modèles (600 p.)

LOGIC COLLOQUIUM '84
J.B. Paris, A.J. Wilkie, and G.M. Wilmers (Editors)

On the length of proofs of finitistic consistency statements in first order theories

Pavel Pudlák[†]
Mathematical Institute
Czechoslovak Academy of Sciences
Prague

1. Introduction

By the second incompleteness theorem of Gödel, a sufficiently rich theory cannot prove its own consistency. This leaves open the question, if one can find a feasible proof in T of the statement, say, "there is no proof of falsehood in T whose length is $\leq 10^{10}$". We shall show some bounds to the length of such proofs in some first order theories.

The main results (Theorems 3.1 and 5.5) can be roughly stated as follows: Let $Con_T(x)$ be a reasonable formalization of "there is no proof of contradiction in T whose length is $\leq x$." Then for reasonable T there exist $\varepsilon > 0$ and $k \in \omega$ such that

(1) any proof of $Con_T(\underline{n})$ in T has length $\geq n^\varepsilon$;

(2) there exists a proof of $Con_T(\underline{n})$ in T with length $\leq n^k$.

It had been known that some lower bounds could be derived.[††] In fact we were inspired by a paper of Mycielski [10] and we use an idea of his. The present knowledge of fragments of arithmetic, which is mainly due to Paris and Wilkie, enabled us to reduce the assumptions about T in the lower bound to mere containment of Robinson's arithmetic Q . The upper bound is based on a partial definition of truth. It uses also a technique of writing short formulas, cf. [5], Chapter 7.

[†] This paper was finished while the author was supported by the NSF grant 1-5-34648 at the University of Colorado, Boulder, CO U.S.A.

[††] After the paper had been typed, I learned that H. Friedman had proved a lower bound of the form n^ε , $\varepsilon > 0$.

Our results may be interesting because of the following reasons. (1) The lower and the upper bound are only polynomially distant. (2) Some corollaries of the lower bound, (see Section 4). (3) Relation to some problems in complexity theory (see Theorem 3.2 and Section 6). Perhaps the most interesting application of the lower bound is a more than elementary speed-up for the length of proofs in GB relative to ZF, (Theorem 4.2).

Several important papers that are related to our paper are listed in references. The papers Ehrenfeucht and Mycielski [4], Gandy [6], Gödel [7], Mostowski [9] (last chapter), Mycielski [10], Parikh [11],[12], Statman [16],[17] and Yukami [19] deal with questions about the length of proofs. In Esenin-Volpin [3], Gandy [6], Mycielski [10] and Parikh [11] the reader can find the outlines of some finitistic projects.

In this paper we consider a measure which is different from the measures used in most of the papers mentioned above. Instead of counting just the number of formulas (i.e. proof lines), we include the length of formulas into the complexity. More precisely, we assume that proofs are coded by strings in a finite alphabet and the length of a proof is the length of the correponding string. This is the most realistic measure. We do not know whether a similar lower bound holds also for the number of formulas in the proof. Recently J. Krajíček gave an idea for a lower bound of the number of formulas in the proof of $Con_T(\underline{n})$ in T which is of the form constant$\cdot$log n .

2. Fragments of arithmetic

The weakest fragment of arithmetic that we shall use is Robinson's arithmetic Q . The language of Q consists of $\underline{0}$, S, +, $\cdot$; the axioms are $S(x) = S(y) \rightarrow x = y$; $\underline{0} \neq S(y)$; $x \neq \underline{0} \rightarrow \exists y\, (x = S(y))$; $x + \underline{0} = x$; $x + S(y) = S(x+y)$; $x\cdot\underline{0} = \underline{0}$; $x\cdot S(y) = x\cdot y + x$. $I\Delta_0$ denotes Q plus the scheme of induction for bounded arithmetical formulas, i.e. formulas where all quantifiers are of the form $\exists x \le t$, $\forall x \le t$, t some term in the language of Q (it is sufficient to assume that t is just a variable). $I\Delta_0 + \exp$ is $I\Delta_0$ plus an axiom expressing $\forall xy\, \exists z\, (z = x^y)$. Exponentiation can be introduced naturally without using a function symbol for it (namely, Bennett [1] has shown that exponentiation can be defined by a bounded formula). All the standard theorems of number theory and finite combinatorics are provable in $I\Delta_0 + \exp$, (cf. [2]). Syntax can be arithmetized in a natural way even in some weaker theories, see [13]. The reader can consult these papers for some information. Let us only remark that one can prove the scheme of induction also for exponentially bounded formulas (i.e. formulas with quantifiers of the form $\exists x \le t\ \forall x \le t$, t term in the language of Q plus exponentiation) in $I\Delta_0 + \exp$.

Denote by $\underline{1} = S(\underline{0})$, $\underline{2} = \underline{1} + \underline{1}$. Let $n \ge 1$, $a_i \in \{0,1\}$ and

$$n = \sum_{i=0}^{k} 2^i (a_i + 1) .$$

Then the term

$$(\underline{a}_1 + 1) + \underline{2} \cdot ((\underline{a}_2 + 1) + \underline{2} \cdot (\ldots))$$

will be denoted by $\underline{n}$ and called the nth numeral. The usual definition of the numeral as a term of the form $SS\ldots S(\underline{0})$ is not suitable here, since such a term is too long. $|n|$ denotes the integral part of $\log_2(n+1)$. Hence the length of a numeral $\underline{n}$ is proportional to $|n|$. We shall not introduce new symbols for the formalizations of $+$, $\cdot$, $|\ldots|$, x^y etc. If such a symbol is not in the language of the theory in question, the terms constructed from them should be understood as abbreviations.

When we consider the length of proofs in some theory, it is important to specify the set of axioms. Therefore we shall distinguish two concepts: an <u>axiomatization</u> A is an arbitrary set of sentences, while a <u>theory</u> T is a deductively closed set of sentences. The distinction is more important if T does not have a finite axiomatization, since if T has a finite axiomatization, then the lengths of the shortest proofs in finite axiomatizations differ only by an additive constant (and we usually use only finite axiomatizations). We shall write

$$A \overset{n}{\vdash} \phi$$

to denote that there exists a proof of ϕ in A whose length (including the length of formulas) is $\leq n$.

The aim of this section is to show that, in spite of the fact that Q is much weaker than $I\Delta_0 + \exp$, every numerical instance of a π_1 sentence provable in $I\Delta_0 + \exp$ has a short proof in Q . This is roughly the content of the following lemma.

<u>Lemma 2.1</u>

For every exponentially bounded formula $\phi(x)$ (where x is the only free variable of ϕ), there exists a polynomial p such that if

$$I\Delta_0 + \exp \vdash \forall x\ \phi(x)$$

then, for every $m \in \omega$,

$$Q \overset{p(|m|)}{\vdash} \phi(m) .$$

First we prove another useful lemma. Let $2^x_0, 2^x_1, 2^x_2, \ldots$ denote the $x, 2^x, 2^{2^x}, \ldots$. If I(x) is a formula with the single free variable x , then Cut_I denotes the following sentence

$$I(0) \ \& \ \forall x,y(I(x) \ \& \ y \leq x \rightarrow I(x) \ \& \ I(S(x))) \ .$$

If $A \vdash Cut_I$, then we say that I is a <u>cut</u> in A .

<u>Lemma 2.2</u>

Let I be a cut in A and $Q \subseteq A$. Then there exists a polynomial p such that, for every $k,n \in \omega$

$$A \mathrel{\vdash\!\!\frac{p(|n|,k)}{}} I(2_k^n) \ .$$

Proof:

Given a cut I one can construct another cut I' such that I' is closed under addition and for every x from I' 2^x exits and is in I , cf. [13]. In fact it is possible to find a formula $J_R(x)$ in the language of arithmetic plus a unary predicate R such that

(i) $$Q \vdash Cut_R \rightarrow [Cut_{J_R} \ \& \ \forall x(J_R(x) \rightarrow J_R(2 \cdot x) \ \& \ \exists y(y = 2^x \ \& \ R(y)))] \ .$$

Starting with I instead of R and applying J k-times we get a cut I_k such that

(ii) $$A \vdash I_k(x) \rightarrow \exists y(y = 2_k^x \ \& \ I(y)) \ ,$$

(iii) $$A \vdash I_k(x) \rightarrow I_k(2 \cdot x) \ .$$

Using a technique for writing short formulas which is described in [5], Chapter 7, we can find J_R such that R occurs in it exactly once. Thus the length of I_k will increase only linearly with k . Now it follows from (i) that the lengths of proofs of (ii) and (iii) will be only polynomial in k . Using the fact that I_k is a cut and (iii) we can construct a proof of $I_k(\underline{n})$ in A whose length is polynomial in $|n|$. Combining this proof with the proof of (ii) we obtain the lemma. □

Proof of Lemma 2.1:

If $\phi(x)$ is a bounded formula, then by Corollary 8.8 of [13]

$$I\Delta_0 + \exp \vdash \forall x\, \phi(x)$$

iff for some cut I closed under $+$ and $\cdot$

$$Q \vdash I(x) \rightarrow \phi(x) .$$

By an inessential modification of the proof we get the same theorem also for <u>exponentially</u> bounded formulas. Thus to prove $\phi(\underline{m})$ it is sufficient to prove $I(\underline{m})$. The latter one has a proof with length polynomial in $|m|$ by Lemma 2.2, (where we set $k = 0$) . □

In order to be able to arithmetize syntax in some theory, we have to assume that the theory contains some fragment of arithmetic. Lemma 2.1 enables us to reduce this assumption to Q . This is because (1) the usual syntactical concepts are naturally formalized by exponentially bounded formulas, (2) the basic properties of them are provable in $I\Delta_0 + \exp$, (3) the sentences that we shall consider will be exponentially bounded sentences of the form $\phi(\underline{n})$. Put otherwise, the basic properties of formulas, proofs etc. whose length is assumed to be $\leq n$ have proofs polynomial in $|n|$. The assumption that an axiomatization A contains Q can be weakened by assuming that A only <u>interprets</u> Q , (this is really necessary in case of set theories, e.g. ZF and GB) .

3. The lower bound

In this section we shall prove the lower bound on the length of proofs of finitistic consistency statements. The main theorem will be stated using finitistic counterparts of the well-known derivability conditions for the 2nd Gödel incompleteness theorem. Then we shall argue that they are met by natural arithmetizations. The relation that we shall consider is "y is provable by a proof of length $\leq x$ ". It will be denoted by $P_A(x,y)$, where A is an axiomatization. In this section, however, P_A is not determined by A , it is an arbitrary formula satisfying the derivability conditions. In order to stress this fact, we omit the subscript A in Theorem 3.1. Let $\perp$ denote some standard countradiction, say $\underline{0} = \underline{1}$. Thus $\neg P(\underline{n}, \ulcorner\perp\urcorner)$, which will be denoted by $Con_A(\underline{n})$ later, is a finitistic consistency statement.

It is convenient to assume that formulas and proofs are strings in the two element alphabet $\{0,1\}$. The Gödel numbers of formulas and proofs are the numbers with corresponding diadic expansions. This allows us to use $|\ldots|$ also to denote the length of formulas and proofs. If ϕ is a formula with the Gödel number n , then $\underline{n}$ will be denoted by $\ulcorner\phi\urcorner$. Again the length of ϕ is proportional to $|\phi|$. We shall also use the notation

$$y = \ulcorner\phi(\underset{\sim}{x}_1, \ldots, \underset{\sim}{x}_k)\urcorner$$

for an arithmetization of the function $(n_1,\ldots,n_k) \mapsto$ Gödel number of $\phi(\underline{n}_1, \ldots,\underline{n}_k)$, (thus $x_1,\ldots,x_k$ are free in $\ulcorner\phi(\underset{\sim}{x}_1,\ldots,\underset{\sim}{x}_k)\urcorner$). We shall assume that this arithmetization has the following property: there exists a polynomial p such that

$$Q \overset{p(|n_1|,\ldots,|n_k|)}{\vdash\!\!\!-\!\!\!-\!\!\!-} \ulcorner\phi(\underline{n}_1,\ldots,\underline{n}_k)\urcorner = \phi(\underset{\sim}{n}_1,\ldots,\underset{\sim}{n}_k) \ .$$

Why we can make such an assumption will be explained later.

Theorem 3.1

Let A be a consistent axiomatization, $Q \subseteq A$, let $P(x,y)$ be a formula and let p_1, p_2, p_3, q_1, q_2 be polynomials such that

(0) $$A \vdash x \leq x' \ \& \ P(x',y) \rightarrow P(x,y) \ ;$$

(1) $$A \overset{n}{\vdash} \phi \Rightarrow A \overset{p_1(n)}{\vdash} P(\underline{n}, \ulcorner \phi \urcorner) \ ,$$

(2) $$A \overset{p_2(|n|,|m|)}{\vdash} P(\underline{n},\underline{m}) \rightarrow P(\underline{q_1}(\underline{n}), \ulcorner P(\underline{n},\underline{m}) \urcorner) \ ;$$

(3) $$A \overset{p_3(|n|,|\phi|,|\psi|)}{\vdash} P(\underline{n}, \ulcorner \phi \urcorner) \ \& \ P(\underline{n}, \ulcorner \phi \rightarrow \psi \urcorner) \rightarrow P(q_2(n), \ulcorner \psi \urcorner) \ .$$

Then there exists $\varepsilon > 0$ such that for no $n \in \omega$

$$A \overset{n^{\varepsilon}}{\vdash} \neg P(\underline{n}, \ulcorner \bot \urcorner) \ .$$

Proof:

In order to simplify notation, we shall write

$$\ldots \overset{n}{\underset{*}{\vdash}} \ldots$$

to denote that for some polynomial p

$$\ldots \overset{p(n)}{\vdash} \ldots \ .$$

By Diagonalization Lemma, there exists a formula $D(x)$ such that

$$Q \vdash D(x) \leftrightarrow \neg P(x, \ulcorner D(\dot{x}) \urcorner) \ .$$

Thus

(i) $$Q \overset{|m|}{\underset{*}{\vdash}} D(\underline{m}) \leftrightarrow \neg P(\underline{m}, \ulcorner D(\underline{m}) \urcorner) \ ,$$

hence the same is true also for A. Now, from (1), (i) and the consistency of A we can easily derive for every m

(ii) $$\text{not } A \overset{m}{\vdash} D(\underline{m}) \ .$$

Let $S(m)$ denote $P(m, D(\underline{m}))$. Since

$$S(\underline{m}) \rightarrow (\neg S(\underline{m}) \rightarrow \bot)$$

is a propositional tautology, we get from (0) and (1)

$$A \overset{|m|}{\underset{*}{\vdash}} P(\underline{m}, \ulcorner S(\underline{m}) \rightarrow (\neg S(\underline{m}) \rightarrow \bot)\urcorner) .$$

Now, several applications of (3) and (0) yield

(iii) $$A \overset{|m|}{\vdash} \neg P(\underline{q}_3(\underline{m}), \ulcorner\bot\urcorner) \rightarrow [\neg P(\underline{m}, \ulcorner S(\underline{m})\urcorner) \vee \neg P(\underline{m}, \ulcorner \neg S(\underline{m})\urcorner)] ,$$

for some polynomial q_3, (since $|S(\underline{m})|$ is proportional to $|m|$). By (2) and the definition of $S(\underline{m})$ we have

$$A \overset{|m|}{\underset{*}{\vdash}} S(\underline{m}) \rightarrow P(\underline{q}_1(\underline{m}), \ulcorner S(m)\urcorner) ,$$

which together with (i) implies

$$A \overset{|m|}{\underset{*}{\vdash}} \neg P(\underline{q}_1(\underline{m}), \ulcorner S(\underline{m})\urcorner) \rightarrow D(\underline{m}) .$$

Applying (1) to an implication of (i) we get

$$A \overset{|m|}{\underset{*}{\vdash}} P(q_4(|m|), \ulcorner D(\underline{m}) \rightarrow \neg S(m)\urcorner)$$

for a polynomial q_4 implicitly determined by (i). Thus, if m is sufficiently large, we have by (0)

$$A \overset{|m|}{\underset{*}{\vdash}} P(\underline{m}, \ulcorner D(\underline{m}) \rightarrow \neg S(\underline{m})\urcorner) .$$

By (3) and by the definition of $S(\underline{m})$

$$A \overset{|m|}{\underset{*}{\vdash}} S(\underline{m}) \,\&\, P(\underline{m}, \ulcorner D(\underline{m}) \rightarrow \neg S(\underline{m})\urcorner) \rightarrow P(\underline{q}_2(\underline{m}), \ulcorner \neg S(\underline{m})\urcorner) .$$

Hence, for m sufficiently large,

$$A \overset{|m|}{\underset{*}{\vdash}} S(\underline{m}) \rightarrow P(\underline{q}_2(\underline{m}), \ulcorner \neg S(\underline{m})\urcorner) .$$

By (i),

$$A \overset{|m|}{\vdash} \neg D(\underline{m}) \rightarrow S(\underline{m}) \ .$$

Thus we get, for m sufficiently large,

(v) $$A \overset{|m|}{\vdash} \neg P(\underline{q_2}(\underline{m}), \ulcorner \neg S(\underline{m}) \urcorner) \rightarrow D(\underline{m}) \ .$$

Now (iii), (iv) and (v) implies that for some polynomials p_4 and q_5 and every sufficiently large m

$$A \overset{p_4(|m|)}{\vdash} \neg P(\underline{q_5}(\underline{m}), \ulcorner \bot \urcorner) \rightarrow D(\underline{m}) \ .$$

Thus by (ii)

$$A \overset{m - p_4(|m|) - |D(\underline{m})|}{\vdash} \neg P(\underline{q_5}(\underline{m}), \ulcorner \bot \urcorner)$$

does not hold for any sufficiently large m . The theorem now follows using an easy computation and condition (0) . □

There are several ways in which one can argue that the natural arithmetization meets the conditions (0)-(3). We shall not construct any such particular arithmetization. (For some fragments of arithmetic such an arithmetization is constructed in [13] and can easily be generalized for other axiomatizations). Instead we shall describe some more general properties which look natural and imply the conditions of the Theorem 3.1.

We start by observing that from the finitistic point of view it is too little to know that an axiomatization is recursive. Therefore we shall consider here NP axiomatizations (which means that the set of axioms can be accepted by a nondeterministic polynomial time Turing machine). In particular every finite axiomatization is NP . Now we shall introduce a finitistic counterpart of the concept of numerability.

Definition

Let $\rho(x_1,\dots,x_k)$ be a formula, let A be an axiomatization and let $R \subseteq \omega^k$. We say that ρ polynomially numerates R in A if for some polynomial p and every $n_1,\dots,n_k \in \omega$

$$R(n_1,\dots,n_k) \Leftrightarrow A \overset{p(|n_1|,\dots,|n_k|)}{\vdash\!\!\!\overline{}} \rho(\underline{n}_1,\dots,\underline{n}_k) .$$

Theorem 3.2

Let A be a consistent NP axiomatization such that $Q \subseteq A$ and let $R \subseteq \omega^k$. Then the following are equivalent:

(1) R is NP ;

(2) R is polynomially numerable in Q ;

(3) R is polynomially numerable in A .

Now it is clear that the additional property of the formula $y = \phi(\underset{\sim}{x}_1,\dots,\underset{\sim}{x}_k)$ is just the polynomial numerability. By Theorem 3.2 such a formula exists, since the $k+1$-ary relation

$$m = \text{"the number of } \phi(\underline{n}_1,\dots,\underline{n}_k)\text{"}$$

is NP.

The proofs of (2) => (1) and (3) => (1) are trivial. To prove the converse implications we need first to arithmetize the concept of an NP set. In [13] this was done using so called R_1^+ formulas. Here we briefly sketch another possibility.

<u>Theorem 3.3</u>

There exists an exponentially bounded formula UNP(t,x) such that, for every NP subset R of ω, there exists $k \in \omega$ such that UNP($\underline{k}$,x) polynomially numerates R in Q. (Moreover, there is a fast algorithm to compute k for a given NP Turing machine defining R.)

Proof-sketch:

First consider $I\Delta_0$ + exp instead of Q. In this theory we formalize the computations of a universal nondeterministic Turing machine. Thus UNP(t,x) will mean that the universal nondeterministic Turing machine with the program t accepts the input word x. We also augment the machine with a "clock" so that it runs in time $\leq |x|^t + t$ and still it is universal for NP Turing machines. This enables us to take UNP(t,x) exponentially bounded. The idea is roughly as follows. A word x is accepted with a program t if there exists a matrix M in some finite alphabet such that

(1) the first row consists of t, x and a string of 0's ;

(2) M satisfies finitely many <u>local</u> conditions (which describe relation of m_{ij} to $m_{i-1,j-1}$, $m_{i-1,j}$, $m_{i-1,j+1}$);

(3) the last row codes some accepting configuration (say determined by the occurrence of some particular symbol).

Finally, the matrix is coded by some ℓ adic expansion of a natural number.

Let k be the number which codes an NP Turing machine for R. Then

(i) $I\Delta_o + \exp \vdash UNP(\underline{k},\underline{n}) \Rightarrow R(n)$

since every sentence provable in $I\Delta_0$ + exp is true. To prove, for some polynomial p,

(ii) $R(n) \Rightarrow I\Delta_o + \exp \vdash^{p(|n|)} UNP(\underline{k},\underline{n})$

we have to prove the existence of an accepting computation (the matrix M) via a polynomially long proof. It is enough to take m which codes the accepting computation (the matrix M) and check the conditions (1), (2), (3) for the numeral $\underline{m}$. There are polynomially many in $|m|$ (i.e., also in the length of input) such conditions, hence we are done.

The proof for Q can be obtained by analyzing the above proof and applying Lemma 2.1. We omit the details since the proof for $I\Delta_0$ + exp was only sketched. □

Here we were interested only in the fact that nondeterministic polynomial time corresponds to polynomial length proofs. But it is clear that a more explicit relation between these two measures can be found. One can also bound the length of formulas occurring in proofs using the space bound of the Turing machine.

<u>Proofs of the remaining implications of Theorem 3.2:</u>

(1) => (2) is a direct consequence of Theorem 3.3. To prove (1) =>(3) it is enough to show (i) and (ii) from the proof above for A . (ii) is true, since $Q \subseteq A$ and we have (ii) for Q already. (i) holds since for A consistent, $Q \subseteq A$, every exponentially bounded provable sentence is true. □

<u>Proposition 3.4</u>

Let A be an axiomatization. Suppose $Prf_A(x,y)$ is a polynomial numeration of the relation "z is an A-proof of y" in A , suppose that '$|z| \leq x$' is a polynomial numeration of the relation $|z| \leq x$. Let $P_A(x,y)$ be

$$\exists z('|z| \leq x' \ \& \ Prf_A(z,y)) .$$

Then $P_A(x,y)$ satisfies the conditions (0) and (1) of Theorem 3.1. □

The proof follows immediately from the definition. If A is an NP axiomatization, then the assumption that Prf_A and '$|z| < x$' are polynomial numerations is quite natural, since by Theorem 3.2 there are such formulas.

The second derivability condition is usually proved by formalizing the proof of the first one. This is the case also here. We can use the following theorem, (cf. Theorem 6.4 of [13], where such a theorem is proved for NP formalized by R_1^+ formulas in a weaker theory).

Theorem 3.5

For a suitable polynomial numeration $P_Q(x,y)$ of "there exists a Q-proof of y of length $\leq x$ " and a polynomial q

$$I\Delta_0 + \exp \vdash UNP(t,x) \rightarrow P_Q(q(|x|^t + t), \ulcorner UNP(\underset{\sim}{t},\underset{\sim}{x})\urcorner) . \square$$

This theorem can be proved by formalizing a part of the proof of Theorem 3.3. Using Theorem 3.5 we can prove the derivability condition (2) for P_A if we have the following:

(1) Prf_A and P_A satisfy the assumptions of Proposition 3.4;

(2) A proves that Prf_A is NP; more precisely, for some $k \in \omega$

$$A \vdash Prf_A(z,x) \leftrightarrow UNP(\underline{k},\langle z,x\rangle) ,$$

where $\langle \ldots \rangle$ is the usual pairing function;

(3) A proves that '$|z| \leq x$' is NP (in the same way);

(4) P_A contains P_Q ; more precisely

$$A \mid\!\underline{{}^{p(|n|,|m|)}}\; P_Q(\underline{n},\underline{m}) \rightarrow P_A(\underline{q}(\underline{n}),\underline{m})$$

for some polynomials p,q ;

(5) P_Q satisfies the derivability conditions (0),(1),(3) .

We omit the proofs.

The derivability condition (3) is the simplest one. We can assume, for instance, that if d is a proof of ϕ and e is a proof of $\phi \to \psi$, then the concatenation of d, e, ψ is a proof of ψ. Thus we have in $I\Delta_0 + \exp$

$$P(x,y) \ \& \ P(x,y^{\ulcorner}\to^{\urcorner}z) \to P(3x,z) \ .$$

Hence using Lemma 2.1 and the assumption $Q \subseteq A$ we get (3).

Finally we prove an easy generalization of Theorem 3.1. Let L be some set of closed arithmetical terms. For $t \in L$, let $\underline{t}$ be $\underline{n}$, where n is the value of t in the structure of natural numbers; let $\ell(t)$ denote the length of t, ($|t|$ would be ambiguous).

<u>Theorem 3.6</u>

Let $A \supseteq Q$ be a consistent axiomatization and suppose that there exists a polynomial p such that for every $t \in L$

(i) $A \vdash^{p(\log t,\ell(t))} t = \underline{t}$.

Assuming the derivability conditions of Theorem 3.1 there exists $\delta > 0$ such that for <u>no</u> term $t \in L$

(ii) $A \vdash^{t^{\delta}} Con_A(t)$.

(In (i) we can write the bound also in the form $p'(|t = \underline{t}|)$).

Proof:

Let $\eta > 0$ be so small that

(iii) $$A \vdash^{t^{\eta}} t = \underline{t}$$

and

(iv) $$A \vdash^{t^{\eta}} Con\ (t)$$

would imply

(v) $$A \overset{t^{\varepsilon}}{\vdash} \mathrm{Con}_A(\underline{t})$$

where ε is from Theorem 3.1. Take K so large and δ, $0 < \delta \leq \eta$ so small that

(vi) $$\ell(t) \leq t^{\delta} \ \& \ t > K \Rightarrow p(\log t, \ell(t)) \leq t^{\eta} ,$$

and $K^{\delta} < 1$. Now consider the following three cases.

(a) $\ell(t) > t^{\delta}$. Then the proof of $\mathrm{Con}_A(t)$ must have the length at least

$$|\mathrm{Con}\ (t)| \geq \ell(t) > t^{\delta} .$$

(b) $t \leq K$. Then (ii) is impossible, since the bound is < 1.

(c) $\ell(t) \leq t^{\delta}$ and $t > K$. Then by (i) and (vi) we get (iii). If (ii) were true in this case, then we would get also (iv) and hence (v), which is impossible by Theorem 3.1. □

4. Applications of the lower bound

If A contains a sufficiently strong fragment of arithmetic, then $A + \mathrm{Con}_A$ has a speed-up by an arbitrary recursive function for sentences that are provable in both theories. This theorem goes back to Gödel [7] and Mostowski [9]. Later results of this kind were proved e.g. by Ehrenfencht and Mycielski [4] and Statman [17]. Gandy [6] has shown that if we consider only closed instances of elementary predicates, then the speed-up is still very large. The next corollary shows that such a speed-up is achieved on sentences of the form $\mathrm{Con}_A(\underline{t})$, for some terms t.

We denote $\forall x\ \mathrm{Con}_A(x)$ by Con_A. Recall that $\mathrm{Con}_A(x)$ is $\neg P_A(x, \ulcorner\bot\urcorner)$.

Corollary 4.1

Let $A \subseteq Q$ be a consistent axiomatization. Assume the derivability conditions of Theorem 3.1 for $P_A(x,y)$ and that

$$A \vdash 2^{\underline{0}} = \underline{1} \ \& \ 2^{S(x)} = \underline{2} \cdot 2^x .$$

Then for some constants $\varepsilon > 0$ and c and every $k \in \omega$

(1) $A + Con_A \overset{c\cdot(k+1)}{\vdash} Con_A(2^{\underline{0}}_{\overline{k}})$;

(2) not $A \overset{(2^{0}_{k})^{\varepsilon}}{\vdash} Con_A(2^{\underline{0}}_{\overline{k}})$.

Proof:

The first part is trivial. The second part follows from Theorem 3.6. To this end we should prove condition (i) of Theorem 3.6 for the terms $2^{0}_{\overline{0}}, 2^{0}_{\overline{1}}, \ldots,$ which is an easy exercise. In fact it is not difficult to prove it for any closed term of the alphabet $\{\underline{0}, S, +, \cdot, 2^x\}$. □

Such a speed-up can be achieved also by a conservative extension (cf. Corollary 4.5 of [14]).

Theorem 4.2

There exists $\varepsilon > 0$ and a polynomial p such that for every $k \in \omega$

(1) $GB \overset{p(k)}{\vdash} Con_{ZF}(2^{\underline{0}}_{\overline{k}})$;

(2) not $ZF \overset{(2^{0}_{k})^{\varepsilon}}{\vdash} Con_{ZF}(2^{\underline{0}}_{\overline{k}})$.

Proof:

It is well-known that there is a cut I in GB such that

$$GB \vdash \forall x (I(x) \rightarrow Con_{ZF}(x)) .$$

(This is essentially due to R. Solovay, cf. [14]). Applying Lemma 2.2 we get the first part. The second part is a consequence of the preceding corollary. □

Theorem 4.3

Let $A \subseteq Q$ be a consistent axiomatization and assume the derivability conditions of Theorem 3.1. Then we have:

(1) if I is a cut in A, then

$$A + \exists x(I(x) \ \&\ \neg Con_A(x))$$

is consistent;

(2) if $D(x)$ is Δ_0 (i.e., bounded arithmetical) formula and $D(\underline{0}), D(\underline{1}), \ldots$ are true, then there exists $k \in \omega$ such that

$$A + \exists x(D(x) \ \&\ \neg Con\ (x^k))$$

is consistent.

Proof:

(1) If $A \vdash \forall x(I(x) \rightarrow Con_A(x))$, then in the same way as above we would get

$$A \overset{p(|n|)}{\vdash} Con_A(n)$$

for some polynomial p, which is impossible by Theorem 3.1, since $p(|n|) < n$ for large n.

(2) Let $D(x)$ be a Δ_0 formula and let $D(\underline{0}), D(\underline{1}), \ldots$ be true. Then there is a polynomial p_1 such that for every $n \in \omega$

$$A \overset{p_1(n)}{\vdash} D(\underline{n}) .$$

Hence we can construct a polynomial p_2 with the property that if

(i) $$A \overset{m}{\vdash} \forall x(D(x) \to Con_A(x^k)) ,$$

then

(ii) $$A \overset{p_2(m,n,k)}{\longmapsto} Con_A(\underline{n}) .$$

Take k large so that for ε of Theorem 3.1 we have

(iii) $$p_2(m,n,k)/n^{k\cdot\varepsilon} \to 0 \quad \text{for} \quad n \to \infty .$$

Now suppose that the theory of (2) is inconsistent for this k, i.e. for some m we have (i). Then we get (ii), hence by (iii) we have

$$A \overset{n^{k\cdot\varepsilon}}{\longmapsto} Con\ (n^k) ,$$

for n sufficiently large. But this is prohibited by Theorem 3.1. □

(1) has been proved in [14] (in a different way). It was employed there to show a speed-up by an arbitrary elementary function of the ordinary logic over the logical calculi without cut-rules. (2) is an improvement of a theorem of the same paper.

5. The upper bound

To be able to derive some nontrivial upper bounds to the length of proofs of finitistic consistency statements we have to assume more than we did in section 3. We need that finite pieces of information about the universe are coded in natural numbers. The sequential theories, which we introduced in [15], have this property. The following definition is different from but equivalent to the original one of [15].

Definition

A theory T is called *sequential* if it satisfies the following conditions:

(1) T is a theory with equality,

(2) Q is interpretable in T relativized to N(x), (N(x) is some formula of T),

(3) there exists a formula, which we denote by $x[t] = y$, that defines in T a total function $x[t]$ of two variables x,t such that

$$T \vdash \forall x,y,t\ \exists z(N(t) \to (\forall s < t(z[s] = x[s])\ \&\ z[t] = y)) .$$

Intuitively, (3) means that we have a definition of "y is the t-th element of x" such that for a given t we can always replace the t-th element by an arbitrary one and all the elements which precede it will be preserved.

Examples of sequential theories.

(1) In PA we can take e.g. $x[t] = y_t$ where $x = \prod_{t=1}^{\infty} p_t^{y_t}$, $p_1, p_2, \ldots$ is the series of primes.

(2) In GB we can define

$$X[T] = \emptyset \quad \text{if } T \text{ is a proper class,}$$

$$X[T] = \{w \mid \langle w,T\rangle \in X\} \quad \text{if } T \text{ is a set .}$$

(3) Other examples are $I\Delta_o$, $I\Delta_o$ + exp, ZF, Alternative Set Theory.

It seems to be too difficult even to state the upper bound in such generality as the lower bound. Therefore we shall be more explicit about the logical calculus and its formalization. We shall consider a first order language with finitely many predicate symbols P_d, $d = 1,2,\ldots,e$, one of which is =, with logical symbols $\neg$, $\to$, $\forall$, and with variables $v_o, v_1, \ldots$. (Thus when we speak about theories which contain function symbols, e.g. Q, we assume that the function symbols are treated as relation symbols.) The logical calculus will be the one presented in [8], (it has 5 axiom schemas and the rules of modus ponens and generalization). Formulas and proofs are again strings in $\{0,1\}$ and the strings are arithmetized via diadic expansions. Thus $P_A(x,y)$ (the formalization of "y has an A proof of length $\leq x$") and $Con_A(x)$ is uniquely determined by the numeration of the axiomatization A. We extend the notation $\ulcorner \ldots \urcorner$ to arbitrary strings of symbols. The concatenation will be denoted just by juxtaposition.

Since the complete proofs of the Lemmas which follow would be extremely long and uninteresting, we shall prove only some typical cases, which should demonstrate sufficiently our proof techniques.

In the following three lemmas we assume that A is sequential. In order to simplify our notation let us assume that the language contains just a single, say binary, predicate symbol P. Further, let

$$g =_i f \iff_{df} \forall t \neq i(g[t] = f[t]) \; ;$$

$$Fm_n(x) \iff_{df} \text{"}x \text{ is a formula of length } \leq n\text{"},$$

$n \in \omega$.

Lemma 5.1

There exists a polynomial p such that for every $n \in \omega$ there exists a formula $Sat_n(x,f)$ and there are A-proofs of length $\leq p(n)$ of

(1) $Fm_n(x) \to \{Sat_n(x,f) \leftrightarrow$

$$\leftrightarrow [\exists i,j(x = \ulcorner P(v_i,v_j)\urcorner \;\&\; P(f[i],f[j])) \vee$$

$$\vee\; \exists y,z(x = y\ulcorner\to\urcorner z \;\&\; (Sat_n(y,f) \to Sat_n(z,f))) \vee$$

$$\vee\; \exists y(x = \ulcorner\neg\urcorner y \;\&\; \neg Sat_n(y,f)) \vee$$

$$\vee\; \exists i(x = \ulcorner\forall v_i\urcorner y \;\&\; \forall g(g =_i f \to Sat_n(y,g)))]\} \; ;$$

(2) $Fm_n(x) \to (Sat_n(x,f) \leftrightarrow Sat_{n+1}(x,f))$.

Proof:

Sat_0 can be an arbitrary formula, since there are no formulas of length ≤ 0. Denote by $\Sigma(Sat_n)$ the right hand side of the equivalence in (1). In order to avoid exponential growth of the length of Sat_n, we replace Σ by Σ', using a technique of [5], Chapter 7, so that Sat_n occurs in $\Sigma'(Sat_n)$ only once and the equivalence

(i) $$\Sigma(R) \leftrightarrow \Sigma'(R) ,$$

where R is a new predicate, is provable in the predicate calculus. Now we can define by induction for $n > 1$

$$Sat_n(x,f) \leftrightarrow_{df} \Sigma'(Sat_{n-1}) .$$

Then the length of such formulas is linear in n, hence also

(ii) $$Sat_n(x,f) \leftrightarrow \Sigma(Sat_{n-1})$$

has a proof of length linear in n by (i).

Let Φ_n denote the universal closure of (2). We shall describe a proof of $\Phi_n \to \Phi_{n+1}$ whose <u>shape</u> does not depend on n (i.e. these proofs will be instances of a proof schema). Hence the lengths of these proofs will increase also only linearly. Arguing in A, assume that Φ_n is true and x is a formula of length $\leq n+1$. We have to show that

(iii) $$Sat_{n+1}(x,f) \leftrightarrow Sat_{n+2}(x,f) .$$

We can distinguish the cases: x is atomic, x is an implication etc. E.g. let $x = \ulcorner\neg\urcorner y$. Then $|y| \leq n$, thus by Φ_n

$$Sat_n(y,f) \leftrightarrow Sat_{n+1}(y,f) .$$

Applying (ii) to n+1 and n+2 we get

$$\mathrm{Sat}_{n+1}(x,f) \leftrightarrow \neg \mathrm{Sat}_n(y,f) \ ,$$

$$\mathrm{Sat}_{n+2}(x,f) \leftrightarrow \neg \mathrm{Sat}_{n+1}(y,f) \ .$$

The last three equivalences yield (iii). Since Φ_0 is trivial and we have the proofs of $\Phi_n \rightarrow \Phi_{n+1}$ of linear length we get a proof of Φ_n, i.e. of (2), of polynomial length.

Now we can construct a polynomial proof of (1), i.e. of

$$\mathrm{Fm}_n(x) \rightarrow (\mathrm{Sat}_n(x,f) \leftrightarrow \Sigma(\mathrm{Sat}_n)) \ .$$

This follows easily from (ii) and (2), since all the formulas to which Sat_n is applied in $\Sigma(\mathrm{Sat}_n)$ are of length $< n$. □

<u>Lemma 5.2</u>

Sat_n preserves the logical axioms, i.e. there are A-proofs of lengths polynomial in n of

(1) $\mathrm{Fm}_n(x)$ & "x is a logical axiom" $\rightarrow \mathrm{Sat}_n(x,f)$;

Sat_n preserves the logical rules, i.e. there are proofs of lengths polynomial in n of

(2) $\mathrm{Fm}_n(x \ulcorner\rightarrow\urcorner y)$ & $\mathrm{Sat}_n(x,f)$ & $\mathrm{Sat}_n(x\ulcorner\rightarrow\urcorner y,f) \rightarrow \mathrm{Sat}_n(y,f)$;

(3) $\mathrm{Fm}_n(\ulcorner\forall v_i\urcorner x)$ & $\forall f\ \mathrm{Sat}_n(x,f) \rightarrow \forall f\ \mathrm{Sat}_n(\ulcorner\forall v_i\urcorner x, f)$. □

Proofs of (1) for propositional axioms and the proofs of (2) and (3) follow directly from Lemma 5.1 (1). The proof of (1) for quantifier axioms requires an additional lemma, therefore is omitted.

Lemma 5.3

There exists a polynomial p such that for every $n \in \omega$ and every formula $\phi(v_{i_1},\ldots,v_{i_m})$ of length $\leq n$ (where all free variables of ϕ are displayed) there exists an A-proof of length $\leq p(n)$ of

$$\mathrm{Sat}_n(\ulcorner\phi\urcorner, f) \leftrightarrow \phi(f[i_1],\ldots,f[i_m]) .$$

Proof:

Let $\Psi(\phi)$ be the formula above. For ϕ atomic we have such a proof of $\Psi(\phi)$ from Lemma 5.1 (1). Now it is sufficient to show that there exists a polynomial p such that all the following implications have proofs of lengths $\leq p(n)$:

$$\Psi(\phi)\ \&\ \Psi(\psi) \rightarrow \Psi(\phi \rightarrow \psi) , \quad \text{for } |\phi \rightarrow \psi| \leq n ;$$

$$\Psi(\phi) \rightarrow \Psi(\neg\phi) , \quad \text{for } |\neg\phi| \leq n ;$$

$$\Psi(\phi) \rightarrow \Psi(\forall v_i \phi) , \quad \text{for } |\forall v_i \phi| \leq n .$$

This can be easily derived from Lemma 5.1 (1). E.g. consider the second implication, then we have, by Lemma 5.1 (1), a polynomial proof of

$$\mathrm{Sat}_n(\ulcorner\neg\phi\urcorner, f) \leftrightarrow \neg\mathrm{Sat}_n(\ulcorner\phi\urcorner, f) ,$$

which, together with $\Psi(\phi)$, yields $\Psi(\neg\phi)$. □

Let α be a sentence. Then $P_{\{\alpha\}}(x,y)$ and $\mathrm{Con}_{\{\alpha\}}(y)$ will denote the arithmetizations of provability and consistency where the axiomatization $\{\alpha\}$ is numerated by the formula $x = \ulcorner\alpha\urcorner$.

Theorem 5.4

Let A be sequential. Then there exists a polynomial p such that for every $n \in \omega$ and every sentence α , $|\alpha| \leq n$.

$$A \vdash^{p(n)} \alpha \rightarrow \mathrm{Con}_{\{\alpha\}}(\underline{n}) .$$

Proof:

By the preceding lemma we have $\neg \mathrm{Sat}_n(\ulcorner \bot \urcorner, f)$. Hence it is sufficient to show that

$$\alpha \mathbin{\&} P_{\{\alpha\}}(\underline{n}, x) \rightarrow \forall f \; \mathrm{Sat}_n(x,f)$$

has an A-proof of polynomial length. Denote this formula by θ_n . θ_0 is trivial, therefore we need only polynomial proofs of $\theta_n \rightarrow \theta_{n+1}$. So assume that θ_n and α hold true and w is a proof of x , $|w| \leq n+1$. We have to prove

(i) $$\forall f \; \mathrm{Sat}_{n+1}(x,f) \; .$$

Now w is a sequence of formulas where the last one is x . For every formula of this sequence, except of x , we have

$$\forall f \; \mathrm{Sat}_n(y,f)$$

by θ_n . Using (2) of Lemma 5.1 we get the same for Sat_{n+1} . Now we consider the following two cases:

(a) x is a logical axiom or follows from the preceding formulas of w by some logical rule. Then (i) has a polynomial proof by Lemma 5.2;

(b) $x = \ulcorner \alpha \urcorner$. Then we get a polynomial proof of (i) using the assumption α and Lemma 5.3. □

By Theorem 5.4, if A is a finite axiomatization of a sequential theory, then in A we have proofs of length polynomial in n of $\mathrm{Con}_A(\underline{n})$, (assuming that the numeration of A is reasonable). This theorem could be easily generalized to infinite axiomatizations which are sparse, i.e. for every n there are only polynomially many axioms of length $\leq n$. However this would not include the theories that we are interested in (PA,ZF) , since they are not sparse. Therefore we shall prove a different theorem. The proof of this theorem is based on the fact that the axiomatizations in question can be replaced by sparse ones.

Theorem 5.5

Let $A = \{\forall y\ \Phi(\phi(y,z)) \mid \phi(y,z)$ formula with two free variables $y,z\}$ be an axiomatization of a sequential theory. Suppose that the variable y is not bounded in the schema Φ. Suppose that a numeration of A in A is chosen so that it is provable in A that "α is an axiom iff α is of the form $\forall y\ \Phi(\phi(y,z))$". Then for some polynomial p and every $n \in \omega$.

$$A \overset{p(n)}{\vdash} Con_A(\underline{n}) \ .$$

Proof:

Define

$$Tr_n(x,y,z) \leftrightarrow_{df} \forall f\ (f[\underline{0}] = y \ \&\ f[\underline{1}] = z \rightarrow Sat_n(x,f)) \ .$$

Let α_n denote

$$\forall y\ \Phi(Tr_n(y[\underline{0}],y[\underline{1}],z)) \ .$$

Then α_n is an instance of the schema, hence A proves that α_n is an axiom of A. In fact this proof has length polynomial in the length of Tr_n, thus also polynomial in n (we know that Sat_n has polynomial length). Let β be the conjunction of finitely many sentences provable in A which ensure the sequentiality of A. Let β_n be $\alpha_n \ \&\ \beta$. Since α_n is an axiom of A, β_n has an A-proof of polynomial length. Using Theorem 5.4 we can construct for every polynomial q another polynomial p such that for every $n \in \omega$

$$A \overset{p(n)}{\vdash} Con_{\{\beta_n\}}(\underline{q(n)}) \ .$$

It suffices to prove now that for a suitable polynomial q we have a polynomial proof of

(i) $P_A(\underline{n},x) \to P_{\{\beta_n\}}(\underline{q(n)},x)$

in A. First we shall argue in the metatheory. Let ϕ be a formula $|\phi| \leq n$, let v_0, v_1 be the free variables of ϕ. By Lemma 5.3 we can derive from β

$$Sat_n(\ulcorner\phi\urcorner, f) \leftrightarrow \phi(f[\underline{0}], f[\underline{1}]) ,$$

using a proof of polynomial length, (since the theory axiomatized by β is sequential). Further, β also implies

$$\forall y,z\ \exists f\ (f[\underline{0}] = y\ \&\ f[\underline{1}] = z) .$$

Thus we get a polynomial proof from β of

$$Tr_n(\ulcorner\phi\urcorner, y, z) \leftrightarrow \phi(y,z) .$$

Again using the sequentiality of β we find an x such that $x[\underline{0}] = \ulcorner\phi\urcorner$ and $x[\underline{1}] = y$. For this x we have then

$$\Phi(Tr_n(x[\underline{0}], x[\underline{1}], z)) \to \Phi(\phi(y,z)) .$$

If we assume moreover α_n then we can derive $\forall y\ \Phi(\phi(y,z))$. Thus we have shown that the instance of the schema for arbitrary ϕ, $|\phi| < n$ is derivable from β_n (which is $\alpha_n\ \&\ \beta$) via a proof of polynomial length.

Now let a proof w in A be given and let $|w| < n$. We can transform this proof into a proof w' from a single axiom β_n in such a way that we replace every axiom of A by the proof of this axiom from β_n. Thus we have $|w'| < q(|w|)$ for some polynomial q.

In order to get (i) we have to formalize the above argument in A . It is clear that this argument can be formalized in $I\Delta_o$ + exp , A contains Q , hence we can apply Lemma 2.1. □

The usual axiomatizations of PA and ZF are not exactly of this form, since instead of a single parameter (which is y in $\forall y\ \Phi(\phi(y,z))$) they allow arbitrarily many parameters. Since all sequential theories have a pairing function, this is an inessential difference. The fact that PA and ZF are axiomatized by such schemas is provable (for reasonable numerations) already in $I\Delta_o$ + exp . Thus the polynomial upper bounds are true also for PA and ZF .

The theorem of Vaught [18] implies that every recursively axiomatizable sequential theory is axiomatizable by a schema. We would like to know if it can be axiomatized by a schema of the form described in Theorem 5.5, (i.e., with y free in Φ).

6. Some problems related to NP = coNP?

So far we have studied only the question of the size of the shortest proof of $Con_A(\underline{n})$ in A . But what about the proofs of $Con_A(\underline{n})$ in weaker theories? The best that we can say is the following informal proposition.

Proposition 6.1

Let A be a consistent co-NP axiomatization, let $Q \subseteq B$. Then for a reasonable numeration of A in B there exists a polynomial p such that for every $n \in \omega$

$$B \overset{2^{p(n)}}{\vdash} Con_A(\underline{n}) \ .$$

Proof-sketch:

Working in B enumerate all sequences of length $\leq n$ and check that none of them is a proof of contradiction in A . □

Problem 1

Is there a finite consistent axiomatization A , $Q \subseteq A$ and a polynomial p such that

(1) $A \overset{p(m)}{\vdash} Con_{A+Con_A}(\underline{n})$?

(2) or $A \overset{p(m)}{\vdash} Con_{A+Con_A(2^n)}(\underline{n})$? (Mycielski)

We conjecture that the answer is no. We have added the finiteness assumption in order to avoid possible pathological examples, but we do not know any such example.

The quantifier complexity of the formulas occurring in the proofs of $Con_A(\underline{n})$ that we have constructed in the preceding section increased with n . A truly finitistic proof should have limited quantifier complexity. Such proofs were used in the proof-sketch of Proposition 6.1, but they were exponentially long.

Problem 2

Is there a consistent finite axiomatization A , $Q \subseteq A$, a number k and a polynomial p such that for every $n \in \omega$ there exists a proof of $Con_A(\underline{n})$ in A of length $\leq p(n)$ which uses only formulas of complexity Σ_k ?

Again we conjecture that the answer is no. But we have the following proposition.

Proposition 6.2

A negative answer to any of the two problems above would imply $NP \neq coNP$, (hence also $P \neq NP$) .

Proof:

We shall show that such an answer would imply the stronger inequality NEXP ≠ coNEXP . Sets of numbers which belong to NEXP are exactly those sets X for which there exists an NP algorithm which accepts 2^n iff $n \in X$. If A is finite, then $Con_A(\underline{n})$, as a predicate on ω , is in coNEXP .

Proving the counterpositive implications assume that NEXP = coNEXP. Then for sufficiently large finite part A of the true arithmetical sentences we have

$$A \vdash Con_A(x) \leftrightarrow UNP(\underline{k}, 2^x)$$

for some $k \in \omega$; (see the definition of UNP in Section 3). Thus to prove $Con_A(\underline{n})$ in A it is sufficient to find a computation of the Turing machine with the number k on the input 2^n and check in A that it is such a computation (for the corresponding numeral). The length of this computation is polynomial in the length of the input, which is $|2^n| = n$. Thus the proof of $Con_A(x)$ in A has polynomial length and bounded quantifier complexity.

Acknowledgement. I would like to thank Jeff Paris, Jan Krajíček, Jan Mycielski and W.N. Reinhardt for their comments about this paper.

References

[1] J.H Bennett, On Spectra, P .D. dissertation, Princeton University, 1962.

[2] C. Dimitracopoulos, Matijasevič's Theorem and Fragments of Arithmetic, Ph.D. thesis, University of Manchester, (1980).

[3] A.S. Esenin-Volpin, The ultraintuitionistic criticism and the anti-traditional programme for foundations of mathematics, in Intuitionism and Proof Theory, Ed. A. Kino, J. Myhill & R.E. Vesley, NHPC (1970), pp. 3-45.

[4] A. Ehrenfencht and J. Mycielski, Abbreviating proofs by adding new axioms, Bulletin of the A.M.S. 77 (1971), pp. 366-67.

[5] J. Ferrante, Ch. W. Rackoff, The Computational Complexity of Logical Theories, Springer-Verlag LNM 718, (1979).

[6] R.O. Gandy, Limitations to mathematical knowledge, in Logic Colloquium '80, Ed. D. Van Dalen, D. Lascar, T. J. Smiley, NHPC (1982), pp. 129-146.

[7] K. Gödel, Uber die Länge der Beweise, Ergebuisse eines mathematischen Kolloquiums, 7 (1936), pp. 23-24, (English translation in The Undecidable, Ed. M. Davis, Raven Press, (1965), pp. 82-83).

[8] E. Mendelson, Introduction to Mathematical Logic, D. Van Nostrand Co., 1964.

[9] A. Mostowski, Sentences Undecidable in Formalized Arithmetic, NHPC, (1952).

[10] J. Mycielski, Finitistic intuitions supporting the consistency of ZF and ZF + AD. (manuscript).

[11] R. Parikh, Existence and feasibility in arithmetic, J.S.L. 36 (1971), pp. 494-508.

[12] R. Parikh, Some results on the lengths of proofs, T.A.M.S. 177 (1973), pp. 29-36.

[13] J. Paris and A. Wilkie, On the scheme of induction for bounded formulae, manuscript.

[14] P. Pudlák, Cuts, consistency statements and interpretations, to appear in J.S.L.

[15] P. Pudlák, Some prime elements in the lattice of interpretability types, T.A.M.S. 280 (1983), pp. 255-275.

[16] R. Statman, Bounds for proof-search and speed-up in the predicate calculus, Annals of Math. Logic 15 (1978), pp. 225-287.

[17] R. Statman, Speed-up by theories with infinite models, Proceedings of the A.M.S. 81 (1981), pp. 465-469.

[18] R.L. Vaught, Axiomatizability by a schema, J.S.L. 32 (1967), pp. 473-479.

[19] J. Yukami, Some results on speed-up, The Annals of the Japan Association for Philosophy of Science 6 (1984), pp. 195-205.

LOGIC COLLOQUIUM '84
J.B. Paris, A.J. Wilkie, and G.M. Wilmers (Editors)

ON CATEGORICAL THEORIES

JÜRGEN SAFFE

Ringseisstr. 12, 8000 München 2
Germany

Abstract

In this talk we collect some old and new results on categorical theories as well as some (famous) open problems.

0. INTRODUCTION

In this talk we are mainly concerned with some applications (or implications) of the (so far known) structure theory of ω-stable ω-categorical theories. This structure theory is the cornerstone for the solutions of two old conjectures:

1. a totally categorical theory is not finitely axiomatizable;
2. an ω-categorical ω-stable theory has finite Morley-rank.

The solutions of these can be found in [CHL] - in fact they proved more, namely that in 1. the assumption "ω-categorical ω-stable" is sufficient as well.

Unfortunately, I am not sure that they really got the "natural" proofs of their results - the reason probably beeing that the structure theory still has "missing links". Here I shall try to provide you with one of these missing links which I suppose to be a step towards "natural proofs".

But even there are "unnatural conjectures" in this field. For example, 1. above is not the statement we should try to prove (although it is proved). As Cherlin has pointed out, 1. should read like its opposite, i. e.

3. if T is totally categorical, then there exists a sentence φ of T such that the models of T are exactly the infinite models of φ.

Clearly, 3. implies 1. - but it contains much more information (unfortunately, it is still unsolved).

We need to fix some assumptions. Throughout T denotes a (fixed) ω-categorical ω-stable theory, acl(A) denotes the algebraic closure of the set A, and R(-) denotes Morley-rank.

Although the result(s) will be understandable with very rudimentary knowledge of model theory, for checking details you need to know the [CHL]-paper and (not too few) stability theory.

1. L'HYPOTHESE DE MAZOYER

I won't fall into French the next moment - but I have to recall some ideas from [Po]. L'hypothèse de Mazoyer is the following

1.1 Conjecture.

Let p be a strongly minimal type. Then there exists a natural number n such that the following holds:

if c realizes p, $c \in acl(A \cup b)$ and A is algebraically closed, then there exists a subset A' of A of cardinality at most n such that $c \in acl(A' \cup b)$.

If you are not convinced that this is an interesting problem, I have to sketch a proof of

1.2 Theorem.

If T satisfies conjecture 1.1 for p, then T is not finitely axiomatizable.

Proof sketch (Lascar): Let C^p collect all realizations of p. For a set A call E(A) a p-envelop of A if E(A) is a maximal superset of A with the following property:

if $b \in acl(E(A)) \cap C^p$, then $b \in acl(A) \cap C^p$.

The following are easy to see:

1. a p-envelop $E(A)$ is a model of T iff $E(A) \cap C^p$ is infinite;
2. for all $n < \omega$ there exists (a finite set) A such that $n < |E(A) \cap C^p| < \omega$.

(Carefully reading [CHL], you will detect that this kind of p-envelop "essentially" coincides with their kind.)

If we now adjoin a new predicate Y to our language, by our assumption we can express that Y is a p-envelop. From 1. and 2. we then see that T can't be finitely axiomatizable.

I am now going to show that conjecture 1.1 is true if A is "sufficiently large" - that means A has to be m-saturated (i. e. every 1-type of T over a set of cardinality less than m is realized in A) for an m depending on T (not on A).

The proof has a good and a bad feature: the existence of m-saturated, algebraically closed sets which are not models is not easy to establish - that's bad; but the proof uses a good amount of stability theory what for my taste is clearly good.

For the beginning, I have to quote some results from [CHL].

2. RESULTS FROM [CHL]

That theorem of the paper that will be most usefull here is

2.1 Theorem (T^{eq}).

For any type $q \in S(\emptyset)$ of finite rank there exists a type $p \in S(\emptyset)$ of rank one such that for any $\bar{a}$ realizing q $acl(\bar{a}) \cap C^p \neq \emptyset$.

The proof of this theorem very strongly relies on the main structure theorem for strongly minimal types in ω-categorical theories, i. e.

2.2 Theorem.

Let p be a strongly minimal type such that for all a realizing p $acl(a) \cap C^p = a$. Then the dependence relation on C^p is either

1. trivial, i. e. $a \in acl(a_0, \ldots, a_{n-1})$ iff $a = a_i$ for some $i < n$;
2. a projective space over a finite field; or
3. an affine space over a finite field

(in 2. and 3. we interprete $a \in C^p$ as points an the sets acl(ab) for $a \neq b$ as lines).

In cases 1. and 2. the dependence relation is "modular", i. e. we have dim(A B) + dim(A B) = dim(A) + dim(B), where "dim" is the cardinality of a maximal independent set.

In the affine case we can build the so-called "associated" modular type in T^{eq} by factoring the lines by the equivalence relation "parallelism".

We need two further implications of this important theorem.

2.3 Theorem

$R(\bar{x} = \bar{x})$ is finite. So theorem 2.1 is applicable to all types.

The last result needed from [CHL] will be stated in a slightly different way to come around into stability theory.

2.4 Theorem

Suppose that p and q are strongly minimal modular types over A which are not orthogonal. Then p and q are not weakly orthogonal.

This last theorem is a particular instance of a much more general

2.5 Conjecture (for any stable T).

Let p and q be types over A such that p and q are not orthogonal. Let $(B_i)_{i<\omega}$ and $(C_i)_{i<\omega}$ be the Morley-sequences of p and q respectively. Then $t(B_1B_2,A)$ and $t(C_1C_2,A)$ are not weakly orthogonal.

It is not too difficult to see that theorems 2.1 to 2.4 imply the conjecture 2.5 beeing true for ω-categorical ω-stable theories, and it is well known that his conjecture is true replacing the sequence of length two by the sequence of length ω.

Finally we have to remark that theorems 2.2 to 2.4 are "essentially" true not only for strongly minimal types but for all complete types of rank one - we shall use this without explicit mention.

3. THE THEOREM AND ITS PROOF

Here I shall really prove conjecture 1.1 for m-saturated A for sufficiently large m. To make the proof (and its notation) easier understandable, I shall not prove it as it stands - the plan of the proof is as follows:

we construct a finite extension T^e of T, where T T^e T^{eq}, and prove a slightly different theorem for T^e. It should be obvious from the construction of T^e and the proof of that theorem that this really implies our statement.

The construction of T^e is as follows: we iterate theorem 2.1 and get a finite extension of T such that for all a there exist a sequence $a_0,\ldots,a_{n-1}$ satisfying $R(a_i,(a_j)_{j<i}) = 1$ and $acl(a) = acl((a_j)_{j<n})$. This is already almost T^e, but we need to take care of the possibility that the type $t(a_i,(a_j)_{j<i})$ is "affine" - in this case we shall throw in the associated modular type as well. So we find a finite extension T_n^e of T with the following property: if $t(a,\bar{b})$ is an affine type of rank one and $|\bar{b}| < n$, then the associated modular type is in T_n^e. We finally put $T^e = T^e_{R(x=x)}$.

Our theorem reads as follows:

<u>3.1 Theorem</u> (T_n^e).

For all $i < n$ there exists $m_i < \omega$ such that for all m_i-saturated and algebraically closed A and for all sequences $\bar{a} = a_0 \ldots a_i$ satisfying $j \leq i \rightarrow R(a_j,(a_k)_{k<j}) \leq 1$ there exists $B \subset A$ of cardinality at most m_i such that $t(\bar{a},A)$ does not fork over B.

<u>Proof</u>: This will be proved by induction on i. For $i = 0$, we are concerned with a single element a_0 such that $R(a_0,\emptyset) \leq 1$. If $t(a_0,A)$ does not fork over $\emptyset$, we can take $B = \emptyset$, otherwise $a_0 \in A$ because A is algebraically closed and we can take $B = a_0$. So $m_0 = 1$ does the job.

Now we assume that the theorem is proved for i, i. e. we have found m_i, and we shall construct m_{i+1}.

Using ω-categoricity, we find s with the following property: If $\bar{b}$ is a sequence of length m_i, $a_0 \ldots a_i$ a sequence such that $R(a_j,\bar{b} \cup (a_k)_{k<j}) = 1$ for $j \leq i+1$ and $t(a_{i+1},\bar{b} \cup (a_k)_{k\leq i})$ is not almost orthogonal to $\bar{b}$, then there exists an indiscernible sequence $(\bar{a}^j)_{j\leq s}$ such that $\bar{a}^0 = a_0 \ldots a_{i+1}$; $(\bar{a}^j)_{j<s}$ is independent; $t(a_0^s \ldots a_i^s,\bar{b} \cup (\bar{a}^j)_{j<s})$ does not fork over $\bar{b}$, but $a_{i+1}^s \in acl(\bar{b} \cup (\bar{a}\ )_{j<s}\ (a_j^s)_{j\leq i})$.

We shall prove that we can take $m_{i+1} = 2 \cdot m_i + s \cdot (i+2) + 1$.

So let $a_0 \ldots a_{i+1}$ be a sequence such that $j \leq i+1$ implies $R(a_j,(a_k)_{k<j}) \leq 1$ and A be an algebraically closed, m_{i+1}-saturated set.

We shall construct $\bar{a}, \bar{b}, \bar{c} \subset A$ such that $|\bar{a}| \leq s \cdot (i+2)$, $|\bar{b}| \leq m_i$, $|\bar{c}| \leq m_i$ and $t(a_0 \ldots a_{i+1},A)$ does not fork over $\bar{a}\bar{b}\bar{c}$.

By induction hypothesis we find $\bar{b} \subset A$ such that $|\bar{b}| \leq m_i$ and $t(a_0 \ldots a_i,A)$ does not fork over $\bar{b}$. If $R(a_{i+1},A \cup a_0 \ldots a_i) = R(a_{i+1},\bar{b} \cup a_0 \ldots a_i)$, we are done for $\bar{a}\bar{c} = \emptyset$.

Otherwise $a_{i+1} \in acl(A \cup a_0 \ldots a_i)$ but $a_{i+1} \notin acl(\bar{b} \cup a_0 \ldots a_i)$. This implies that $t(a_{i+1},\bar{b} \cup a_0 \ldots a_i)$ is not almost orthogonal to $\bar{b}$. so by the definition of s and A's saturativity we find $\bar{a} = (\bar{a}^j)_{1 \le j \le s} \subset A$ such that for $\bar{a}^0 = a_0 \ldots a_{i+1}$, $(\bar{a}^j)_{j \le s}$ is an independent indiscernible sequence, and we find b^s_{i+1} such that $t(b^s_{i+1}, \bar{b}a^s_o \ldots a^s_i) = t(a^s_{i+1}, \bar{b}a^s_o \ldots a^s_i)$;

$b^s_{i+1} \in acl(\bar{b}\bar{a}a_0 \ldots a_{i+1})$ and $a_{i+1} \in acl(\bar{b}\bar{a}a_0 \ldots a_i b^s_{i+1})$ (1).

Hence $b^s_{i+1} \in acl(A \cup a_0 \ldots a_i)$, and $t(b^s_{i+1}, \bar{b}\bar{a})$ is a modular type. If now $b^s_{i+1} \in acl(A \cup a_0 \ldots a_{i-1})$ we can use the induction hypothesis to find $\bar{c} \subset A$ such that $t(a_0 \ldots a_{i-1} b^s_{i+1}, A)$ does not fork over $\bar{c}$ and $|\bar{c}| \le m_i$, whence $b^s_{i+1} \in acl(\bar{c}a_0 \ldots a_{i-1})$. It follows using (1) that $a_{i+1} \in acl(\bar{a}\bar{b}\bar{c}a_0 \ldots a_i)$ whence $t(a_0 \ldots a_{i+1}, A)$ does not fork over $\bar{a}\bar{b}\bar{c}$.

So suppose $b^s_{i+1} \notin acl(A \quad a_0 \ldots a_{i-1})$. We can apply theorem 2.4 to get some b_i such that $st(b_i, \bar{a}\bar{b}a_0 \ldots a_{i-1}) = st(a_i, \bar{a}\bar{b}a_0 \ldots a_{i-1})$;

$b_i \in acl(\bar{a}\bar{b}a_0 \ldots a_i b^s_{i+1})$ and $b^s_{i+1} \in acl(\bar{a}\bar{b}a_0 \ldots a_i b_i)$ (2).

So $b_i \in acl(A \quad a_0 \ldots a_i)$. If now $b_i \in acl(A \cup a_0 \ldots a_{i-1})$, we can use the induction hypothesis as above to find $\bar{c} \subset A$ such that $t(a_0 \ldots a_{i-1} b_i, A)$ does not fork over $\bar{c}$ and $|\bar{c}| \le m_i$. Using (1) and (2), we are done as in the case "$b^s_{i+1} \in acl(A \cup a_0 \ldots a_{i-1})$".

So suppose $b_i \notin acl(A \cup a_0 \ldots a_{i-1})$. Looking at the Morley-sequence of $t(a_i b_i, A \cup a_0 \ldots a_{i-1})$ we find $c \in acl(A \cup a_0 \ldots a_{i-1})$ such that $b_i \in acl(\bar{a}\bar{b}a_0 \ldots a_{i-1} c)$ (3)

and $t(c, \bar{b}a_0 \ldots a_{i-1})$ is either $t(a_i, \bar{b}a_0 \ldots a_{i-1})$ or the associated modular type.

We then find $\bar{c} \quad A$ such that $t(a_0 \ldots a_{i-1}, A)$ does not fork over $\bar{c}$ and $|\bar{c}| \le m_i$, what using (1) - (3) implies that $t(a_0 \ldots a_{i+1}, A)$ does not fork over $\bar{a}\bar{b}\bar{c}$ as in the previous cases.

Glueing this together with the proof of 1.2, we see that we have found a "natural proof" of the conjecture 3. from the introduction - if we could prove the following

3.2 Conjecture.

There exists m_0 such that all m_0-saturated A are algebraically closed.

This would remove the hypothesis "A algebraically closed" from theorem 3.1 and provide the "natural" induction beginning.

So I returned to my belief in "natural proofs" and make a new section of it.

4. ON NATURAL PROOFS

The most important theorem quoted here is clearly theorem 2.2 determining (essentially) all possible strongly minimal - categorical theories. Unfortunately, both known proofs (see [CHL] and [Zi3]) rely heavily on finite combinatorics - so there is no hope generalizing these proofs. But we want to generalize the theorem - for example dropping the ω-categoricity (and even more). My favourite version would read as follows:

4.1 Conjecture.

Suppose that p is a stationary regular type in a stable theory T. Then either

1. p is modular or
2. p is affine or
3. we can define an infinite field in C^p.

This conjecture would clearly imply (much more than) theorems 2.2 and 2.3, and provide a "counting-free" proof of theorem 2.2. This is somehow related to another conjecture that is a "better" version of conjecture 2.5 and that I call the

4.2 "Four-Parameter-Conjecture".

Every structure possessing a real structure theory is already determined by its four-place relations.

This last conjecture is rather vague and general and I would be interested in any counterexample you can give me (please together with the structure theory ...); the important part of the conjecture clearly is the existence of an uniform bound not depending on the theory.
I can provide you with a more concrete version of the conjecture good in this context:

4.3 Conjecture.
Suppose that T is a first-order (super)stable theory in the language L. Then there is a language L' consisting of predicates having at most four places and a theory T' in L' such that T and T' are equivalent in the following sense: every predicate (or function) of T is Ø-definable in T' and vice versa.
(The last phrase is formally not correct - but hopefully understandable.)

Until now, this written exposition has followed the original talk as closely as it reasonably could. Unfortunately, I ended the talk with a statement that was so misstated (to be polite to myself) that I decided to leave it out in a "final" version. Instead, I have good reasons to supply

5. ADDED TO THE TALK.
After this talk, I have learned a lot during discussions with Ahlbrandt, Cherlin, Lachlan, Poizat, Ziegler ... (?).
These things are collected in the follwing.

5.1 - Conjecture 4.3 is wrong as it stands - for a counter-example see [Ah]. But I am still convinced that conjecture 4.2 can be given a precise and correct meaning.

5.2 - The figure 2 appearing in the statement of conjecture 2.5 should be replaced by a number ≥ 4 - so as for 4.2 concentrate on the important part: prove the existence of an uniform finite bound.

5.3 - My opinion that the proof of 1.2 together with 3.1 and 3.2 would give a "natural" (short) proof of the third conjecture of the introduction is wrong - to see the insuffiency consult [AZ].

References:

[Ah] G.Ahlbrandt: Totally Categorical Theories of Modular Type, Ph.D.-Thesis, Chicago 1984

[AZ] G.Ahlbrandt/M.Ziegler: Quasi finitely axiomatizable totally categorical theories, preprint 1984

[CHL] G.Cherlin/L.Harrington/A.H.Lachlan: ω-categorical ω-stable structures, preprint 1981

[Zi1] B.I.Zil'ber (Б.U.Зильбер): Totally categorical theories: Structural properties and the non-finite axiomatizability, Proceedings Karpacz 1979, Springer Lecture Notes 834, p. 381 - 410

[Po] B.Poizat: Review of [Zi1], MR 82m:03045

[Zi2] Б.U.Зильбер: Сильно минимальные счетно категоричные структуры, Сибирский математический журналь, 21(1980), 98 - 112

[Zi3] Б.U.Зильбер: Сильно минимальные счетно категоричные стуктуры II, III

LOGIC COLLOQUIUM '84
J.B. Paris, A.J. Wilkie, and G.M. Wilmers (Editors)

FINITE HOMOGENEOUS RINGS OF ODD CHARACTERISTIC

Dan Saracino and Carol Wood

Colgate University, Hamilton, NY 13346
Wesleyan University, Middletown, CT 06457

0. Introduction. We complete here the classification of finite homogeneous rings of odd characteristic begun in [5] and based on work of Berline and Cherlin in [1]. A countable ring R is homogeneous if every isomorphism between finitely generated subrings of R extends to an automorphism of R. Although we draw heavily on results in [1] and [5] for our present analysis, we give (without proof) enough background results in Section 1 to allow the present work to be read independently.

In [1] the classification of all quantifier-eliminable (QE) rings is reduced to the classification of the Jacobson radicals of such rings. Assuming finiteness (hence QE is equivalent to homogeneous), we listed in [5] all possible radicals for the characteristic p^2 case, p an odd prime. Previous work [2] provided a list for characteristic p, p any prime. We now extend this to finite homogeneous rings of characteristic p^n, $n > 2$ and p odd. Since any QE ring of finite characteristic is a product of QE rings of prime power characteristic, this gives a list of all finite QE rings of odd characteristic. (Actually, using [2] we can get characteristic $2m$, m odd.) As usual, the prime 2 behaves differently, and we do not include it here. Much of the analysis in [5] has however been done for $p = 2$, and we expect to complete the picture for finite homogeneous rings in a subsequent note.

In light of fascinating developments involving homogeneous structures (e.g., [3], [4], and unpublished work of Cherlin on homogeneous digraphs) the present enterprise might become part of a special case analysis of some quite general phenomenon involving homogeneous structures. It is our hope that at least we may have found some clues to understanding either QE or homogeneous rings.

In Section 1 we state our main theorem and background results. The analysis splits roughly into two cases (roughly, large and small rings, but with a special role in the latter for $p = 3$). In Sections 2 and 3 we obtain basic information about the two cases, respectively, and in the final two sections we obtain the classification.

We thank Greg Cherlin for giving our manuscript a careful reading and offering valuable comments.

1. Background and main theorem.

Throughout this paper, R will denote a QE ring (with 1) of characteristic p^n where p is an odd prime and $n \geq 1$, and J will denote its Jacobson radical. We let R_1 be the annihilator of p^{n-1} in R, and $\mathbf{F}_q$ the field with q elements, q a power of p.

We begin by listing facts from [1], [2], [5] which will be used to reduce the problem at hand.

Fact 1 (odd p case of [2]). If R is a finite QE ring of characteristic p, p odd, then $J^3 = 0$ and either J is trivial (i.e., $\mathbf{Z}_p \oplus \cdots \oplus \mathbf{Z}_p$, with all products 0) or J is C_p or $D_p(t)$, where

$C_p = \langle x,y : px = py = x^3 = y^3 = xy = yx = y^2 + tx^2 = 0\rangle$, t a nonsquare in $\mathbf{F}_p$, and $D_p(t) = \langle x,y : px = py = x^3 = y^3 = xy = 0,\ yx = x^2,\ y^2 = tx^2\rangle$, $1 - 4t$ a non-square.

Henceforth we assume $n > 1$.

Fact 2 (Theorems I, II, [1]). (p odd, $n > 1$)

(i) J is nilpotent of order at most $2n + 1$ and J has QE(p), i.e., QE for the language based on $(0, p, +, \cdot)$.

(ii) $R_1 = J$ or $J \oplus \mathbf{F}_q$ or $J \oplus C(X, x; \mathbf{F}_q, 0)$ where q is a power of p, and $C(X,x; \mathbf{F}_q,0)$ is the subring of $C(X; \mathbf{F}_q)$ consisting of functions vanishing at $x \in X$, X a Boolean space without isolated points.

(iii) $R/R_1 \approx \mathbf{F}_p$.

Thus Fact 2 reduces the classification of QE rings of characteristic p^n to that of their radicals (and similar results hold for $p = 2$). For R not semisimple, one finds many such J (cf. [2], [6]), hence many R, and the structure of such rings is at present unclear. Some information is however available from the analysis in [1], where J is studied "in layers".

Notation ([1]). (i) For $1 \leq i \leq n-1$, let J_i be the annihilator of p^i in J. (Notice J_i is QE (p^{n-i}) of characteristic p^i.)

(ii) Let k be the least integer such that

$$J = \langle p\rangle + J_k.$$

(iii) Let $V = \{x \in J_1 : x^2 = 0\}$, $V' = V - \langle p^{n-1}\rangle$, $J_k' = J_k - \langle J_{k-1}, p^{n-k}\rangle$.

Intuitively, the kth layer of J is the last place where anything noteworthy occurs. Our analysis succeeds by showing that J_1 and J_k are the only relevant layers for determining J. The set V' appears again and again, since for homogeneous R there is only one type realized in V'.

We focus first on the bottom layer J_1:

Fact 3 (Main Theorem, [5]). Let J_1 be a finite $QE(p^{n-1})$ ring of characteristic p, p odd. Then J_1 is isomorphic to one of the following: (Here we write J_1 as a vector space over $\mathbf{F}_p$ and determine the multiplication; t is a fixed nonsquare in $\mathbf{F}_p$.)

I. $J_1 = V = \langle p^{n-1}\rangle \oplus \langle a_1\rangle \oplus \cdots \oplus \langle a_r\rangle$ with trivial multiplication.

II. $J_1 = V \oplus \langle a\rangle$, V of type I, $V\cdot a = a\cdot V = 0$, $a^2 = p^{n-1}$ or tp^{n-1}.

III. $J_1 = V \oplus \langle a\rangle \oplus \langle b\rangle$, V of type I, $V\cdot a = V\cdot b = a\cdot V = b\cdot V = 0$,

$$ab = \ell p^{n-1} = -ba,\ \ell \in \mathbf{F}_p,\ a^2 = p^{n-1},\ \text{and } b^2 = \begin{cases} tp^{n-1}, & p \equiv 1 \bmod 4 \\ p^{n-1}, & p \equiv 3 \bmod 4. \end{cases}$$

IV. $J_1 = \langle p^{n-1}\rangle \oplus \langle a_1\rangle \oplus \cdots \oplus \langle a_r\rangle$, $a_ia_j = a_ja_i = 0$, $i \neq j$,

$$a_1^2 = \ldots = a_{r-1}^2 = p^{n-1},\ a_r^2 = p^{n-1} \text{ or } tp^{n-1}.$$

V. $J_1 = V = \langle p^{n-1}\rangle \oplus \langle a_1\rangle \oplus \langle b_1\rangle \oplus \cdots \oplus \langle a_r\rangle \oplus \langle b_r\rangle$ where

$a_ib_i = p^{n-1} = -b_ia_i$, $1 \leqq i \leqq n$, and $a_ib_j = a_i = b_i = b_ja_i = 0$, $i \neq j$.

VI. ($p = 3$ only) $J_1 = \langle 3^{n-1}\rangle \oplus \langle a\rangle \oplus \langle b\rangle$ where $a^2 = 0$, $ab = 3 = -ba$, $b^2 = \pm 3$. Moveover, any J_1 as in I - VI is QE.

Thus if $k = 1$ we are done: $J = J_1 + \langle p\rangle$, and J_1 is known by Fact 3. If $k > 1$ we can get additional information from [1]:

Fact 4 ([1] or trivial). If $k > 1$ then $p^{k-1} J_k = V$ and so $VJ_1 = J_1V = 0$. Moreover $V \supsetneqq \langle p^{n-1}\rangle$.

In the list in Fact 3, only I, II, and III (and cases of IV already included in II and III) satisfy $VJ_1 = J_1V = 0$, so we have

Fact 5. If $k > 1$, then J_1 is of type I, II, or III.

(We shall see that all three possibilities can occur.)

Additional technical information about multiplication on J_k is summarized in:

Fact 6. ([1] or trivial). If $k > 1$ then $J_k^2 \subseteq \langle J_{k-1}, p^{n-k}\rangle$. If in addition the dimension of V over $\mathbf{F}_p$ is greater than 2 or $p > 3$, then $J_k^2 \subseteq J_{k-1}$ and $J_iJ_{i-1}, J_{i-1}J_i \subseteq J_{i-2}$ for $3 \leqq i \leqq k$, $J_1J_k, J_kJ_1 \subseteq V$, and $VJ_k = J_kV = 0$.

We remark that more is claimed in [1] than what we state above, namely that $J_k^2 \subseteq J_{k-1}$ whenever $V \neq \langle p^{n-1}\rangle$. However, in the proof of Lemma 19, p. 150 of [1], the additional hypothesis that $p > 3$ is required when V is two- dimensional over $\mathbf{F}_p$.

As we see in the theorem below, it can in fact happen for $p = 3$ that $J_k^2 \nsubseteq J_{k-1}$.

The goal of this paper, then, is the following:

Classification Theorem. Let J be the Jacobson radical of a finite QE ring of characteristic p^n, $n > 1$, p odd. Suppose $k > 1$ (i.e., $J \neq J_1 + \langle p\rangle$). Then J is isomorphic to one of the following (all of which are QE):

A. ($\mathbf{F}_p$-dim $V > 2$ or $p > 3$) $J = \langle J_1, x_1, \ldots, x_s, p\rangle$, where

$x_i J_k = J_k x_i = 0$, J_1 of type I, II, or III, $k \leq n/2$,

$V = \langle p^{n-1}\rangle \oplus \langle v_1\rangle \oplus \cdots \oplus \langle v_s\rangle$, $v_i = p^{k-1} x_i$, $i = 1, \ldots, s$.

or

B. ($\mathbf{F}_p$-dim $V = 2$, $p = 3$)

(i) ($J_k^2 \subseteq \langle 3^{n-1}\rangle$) $J = \langle J_1, x, 3\rangle$, $x^2 = j3^{n-1}$, $j = 0, 1$, or -1,

$xJ_1 = J_1 x = 0$,

J_1 type I, II, or III, $k \leq n/2$, $V = \langle 3^{k-1}x\rangle \oplus \langle 3^{n-1}\rangle$.

or

(ii) ($J_k^2 \nsubseteq \langle 3^{n-1}\rangle$) $J = \langle J_1, x, 3\rangle$ where $x^2 \in \langle 3^{n-2}\rangle - \langle 3^{n-1}\rangle$,

$xJ_1 = J_1 x = 0$, $k = 2$, J_1 of type I, II, or of type III with the additional restriction that $ab = ba = 0$, $V = \langle 3x\rangle \oplus \langle 3^{n-1}\rangle$.

This, then, together with Facts 1, 2, 3 provides the classification of finite QE rings of characteristic p^n, hence of all odd characteristics.

The proof breaks into cases in two different ways: according to the size of V and according to whether or not $p = 3$. There is considerable overlap, but at little or no cost, in some of the cases--in particular, for small V and $p > 3$. We choose to include arguments in some generality, with an eye to possible future analysis.

2. Case A ($\mathbf{F}_p$-dimension $V > 2$ or $p > 3$).

Let ν be the dimension of V over $\mathbf{F}_p$; by Fact 4, $\nu \geq 2$. We assume $\nu > 2$ or $p > 3$, and so by Fact 4, ν is also the dimension of J_k/J_{k-1} over $\mathbf{F}_p$.

We assume $k > 1$, J finite as before. Our goal in this section is to show that in Case A, multiplication on J_k goes into $\langle p^{n-1}\rangle$, hence is quite trivial. We work our way through the layers of J via a series of lemmas.

Definitions. (i) Let $x, y \in J_k'$ ("real J_k-elements"). We say x and y are independent if $p^{k-1}x$, $p^{k-1}y$, and p^{n-1} are linearly independent over $\mathbf{F}_p$.

(ii) Given $x, y \in J$, $x * y = xy - yx$.

(iii) Given $x_1,\dots,x_r, y_1,\dots,y_r \in J$, we say $(x_1,\dots,x_r) \sim (y_1,\dots,y_r)$ (read "has the same type as") provided the map $x_i \to y_i$, $i = 1,\dots,r$, $p \to p$ determines an isomorphism of the corresponding subrings of J (generated by $\{p, x_1, \dots, x_r\}$, $\{p, y, \dots, y_r\}$, resp.).

Lemma 1. ($\nu > 2$ or $p > 3$) If $k \geq 3$ and $x \in J_k$, then $x^2 \in J_{k-2} + \langle p^{n-(k-1)}\rangle$.

Proof: By Fact 6, the squaring map ϕ from J_k/J_{k-1} to J_{k-1}/J_{k-2} is well-defined. If there is $x \in J_k'$ with $x^2 \in J_{k-2} + \langle p^{n-(k-1)}\rangle$ then $p^{k-2}x^2 \in V'$ and by homogeneity of V' we get that every coset of J_{k-1}/J_{k-2} except possibly ones of $\langle p^{n-(k-1)}\rangle$ is in the image of ϕ. Thus $p^\nu - p \leq p^\nu/2$, $p^\nu \leq 2p$, a contradiction. □

Lemma 2. ($\nu > 2$ or $p > 3$). Suppose $k \geq 3$. If $x \in J_k'$ then $x^2 \in J_{k-2}$.

Proof: Suppose not. Let $p^{k-1}x \in V'$, $p^{k-2}x^2 = \ell p^{n-1}$, $\ell \not\equiv 0 \bmod p$. Then by homogeneity, for every $v \in V'$ there is $y \in J'$, $p^{k-1}y = v$, $p^{k-2}y^2 = \ell p^{n-1}$. But now for any $z \in J_k'$ we have $y \in J_k'$ with $p^{k-1}(y-z) = 0$, $p^{k-2}y^2 = \ell p^{n-1}$. Thus $z = y + c$, $c \in J_{k-1}$, and so

$$p^{k-2}z^2 = p^{k-2}(y+c)^2 = p^{k-2}y^2 + p^{k-2}(yc + cy + c^2)$$
$$= p^{k-2}y^2 \quad \text{by Fact 6.}$$

This gives $p^{k-2}z^2 = \ell p^{n-1}$ for all $z \in J_k'$. If $p > 3$ we get an immediate contradiction by comparing z and 2z : $p^{k-2}(2z)^2 = 4p^{k-2}z^2 = 4\ell p^{n-1} \neq \ell p^{n-1}$. If $p = 3$ we use $\nu > 2$ to choose x and y independent. Now $p^{k-2}(x+y)^2 = p^{k-2}(x-y)^2 = \ell p^{n-1} = p^{k-2}x^2 = p^{k-2}y^2$. Thus $p^{k-2}(x * y) = -\ell p^{n-1}$ and $p^{k-2}(-x * y) = -\ell p^{n-1}$, contradicting $\ell \not\equiv 0 \bmod p$. Thus $x^2 \in J_{k-2}$ in both cases. □

Lemma 3. $(\nu > 2$ or $p > 3)$. If $k \geq 3$ then $J_k^2 \subseteq J_{k-2}$.

Proof: If $\nu = 2$ and $p > 3$, then Lemma 2 implies $J_k^2 \subseteq J_{k-2}$ easily. Suppose now $\nu > 2$, and choose x and y independent. If $p^{k-2}xy \in \langle p^{n-1}\rangle$ then $\langle p^{n-1}, p^{k-1}x\rangle \cap \langle p^{n-1}, p^{k-1}y\rangle = \langle p^{n-1}\rangle$ implies that $p^{k-2}xy \notin \langle p^{n-1}, p^{k-1}x\rangle \cap \langle p^{n-1}, p^{k-1}y\rangle$, say $p^{k-2}xy \notin \langle p^{n-1}, p^{k-1}x\rangle$. By homogeneity, then, for every $v \in V - \langle p^{n-1}, p^{k-1}x\rangle$ there exist x' and y' so that $p^{k-1}x' = p^{k-1}x$, $p^{k-2}x'y' = v$. But now $x \equiv x' \bmod J_{k-1}$, so $p^{k-2}xy' = p^{k-2}x'y' = v$. We fix x and count y''s: there are at most p^ν elements $p^{k-2}xy'$, since for $c \in J_{k-1}$, $p^{k-2}xc = 0$. These elements must cover the $p^\nu - p^2$ elements of $V - \langle p^{n-1}, p^{k-1}x\rangle$. Since $2(p^\nu - p) > p^\nu$, it must be that for a given v and x, exactly one y' exists mod J_{k-1} with $p^{k-2}xy' = v$. But $p^{k-2}x(x + y') = p^{k-2}xy'$ while $x + y' \not\equiv y' \bmod J_{k-1}$, a contradiction. Thus $p^{k-2}xy \in \langle p^{n-1}\rangle$ for independent x and y.

We now show $xy \in J_{k-2}$. Let $p^{k-2}xy = \ell p^{n-1}$. Since x and y are independent, $(p^{k-1}x, p^{k-1}y) \sim (p^{k-1}x, 2p^{k-1}y)$. Thus there is y' with $(p^{k-1}x, p^{k-1}y, y) \sim (p^{k-1}x, 2p^{k-1}y, y')$; in particular $p^{k-1}y' = 2p^{k-1}y$ and $p^{k-2}xy' = \ell p^{n-1}$. The first equation implies $y' \equiv 2y \bmod J_{k-1}$, so $\ell p^{n-1} = p^{k-2}xy' = p^{k-2}x \cdot 2y = 2\ell p^{n-1}$. Thus $\ell \equiv 0 \bmod p$, and so $xy \in J_{k-2}$ for independent x and y. Now fix any $x \in J_k'$. If x, y are independent so are $x, y + p^{n-k}$, and we get that $xy \in J_{k-2}$, $x(y + p^{n-k}) \in J_{k-2}$ implies $x \cdot p^{n-k} \in J_{k-2}$. By Lemma 2, $x^2 \in J_{k-2}$. Thus $x \cdot J_k \subseteq J_{k-2}$. Also $x + p^{n-k}$ and $y + p^{n-k}$ are independent, so

$$(x + p^{n-k})(y + p^{n-k}) \in J_{k-2}, \text{ giving } (p^{n-k})^2 \in J_{k-2}.$$

This proves that all products from J_k are in J_{k-2}, as desired. □

Lemma 4. $J_{k-1} = J_1 + pJ_k$.

Proof: We proceed inductively to show that $J_{i-1} = J_1 + pJ_i$, $2 \leq i \leq k$. For $i = 2$ this is trivial: $J_1 = J_1 + pJ_2$. Suppose $J_{i-1} = J_1 + pJ_i$. Multiplication by p maps J_{i+1}/J_i onto J_i/J_{i-1} and so $pJ_{i+1} + J_{i-1} = J_i$, hence $J_i = pJ_{i+1} + J_1 + pJ_i = J_1 + pJ_{i+1}$. When $i = k$ we are finished. □

Lemma 5. ($\nu > 2$ or $p > 3$). $J_k^2 \subseteq J_1$.

Proof: For $k = 2$ this is contained in Fact 6. For $k = 3$, it is Lemma 3. For $k > 3$, we repeat the proofs of Lemmas 1-3, using Lemma 4 for the induction, to go from $J_k^2 \subseteq J_{k-2}$ to squares in $\langle J_{k-3}, p^{n-(k-2)}\rangle$, etc., until we get to $J_k^2 \subseteq J^{k-3}$ and then repeat down to J_1. We omit the gory details. □

Lemma 6. ($\nu > 2$ or $p > 3$). $J_k^2 \subseteq V$.

Proof: By Fact 6, $J_1J_k \subseteq V$, so $J_k^3 \subseteq J_1J_k \subseteq V$ by Lemma 5. Thus $xyx \in V$ for all $x, y \in J_k$. Since $VJ_k = 0$ by Fact 6, $xyxy = 0$. Since $xy \in J_1$ by Lemma 5, this shows $xy \in V$. □

Lemma 7. ($\nu > 2$ or $p > 3$). $(J_{k-1})^2 \subseteq \langle p^{n-1}\rangle$.

Proof: $J_{k-1} = J_1 + pJ_k$ by Lemma 4, hence $(J_{k-1})^2 \subseteq J_1^2 \subseteq \langle p^{n-1}\rangle$. □

Lemma 8. ($\nu > 2$ or $p > 3$). $J_{k-1}\cdot J_k \subseteq \langle p^{n-1}\rangle$ and $J_k\cdot J_{k-1} \subseteq \langle p^{n-1}\rangle$.

Proof: Since $J_{k-1} = J_1 + pJ_k$ and $J_k^2 \subseteq V$, it suffices for $J_{k-1}J_k \subseteq \langle p^{n-1}\rangle$ to know that $J_1 J_k \subseteq \langle p^{n-1}\rangle$.

Suppose there is $u \in J_1 - V$, $x \in J_k'$ with $ux \notin \langle p^{n-1}\rangle$.

Case 1: $ux \notin \langle p^{n-1}, p^{k-1}x\rangle$. In this case for all $v \in V - \langle p^{n-1}, p^{k-1}x\rangle$ there is y with $uy = v$, $p^{k-1}x = p^{k-1}y$. But then $x \equiv y \bmod J_{k-1}$ and from $J_{k-1} = J_1 + pJ_k$ this implies $ux \equiv uy \bmod \langle p^{n-1}\rangle$. Choosing $v = -ux$ we get $v = -v$, a contradiction. Thus Case 1 cannot occur.

Case 2: For all u and x, $ux \in \langle p^{n-1}, p^{k-1}x\rangle$. Fix u. Let $ux = \ell p^{k-1}x$ mod $\langle p^{n-1}\rangle$. Then for all $v \in V'$ there is y with $p^{k-1}y = v$, $uy \equiv \ell p^{k-1}y$ mod $\langle p^{n-1}\rangle$. Since $uJ_{k-1} \subseteq \langle p^{n-1}\rangle$, this says that whenever $uz \notin \langle p^{n-1}\rangle$, $uz \equiv \ell p^{k-1}z$ mod $\langle p^{n-1}\rangle$. But now replace u by $-u$ (since $u \sim -u$) and get again that $(-u)z \notin \langle p^{n-1}\rangle$ implies $(-u)z \equiv \ell p^{k-1}z$ mod$\langle p^{n-1}\rangle$. This says $-\ell p^{k-1}x \equiv (-u)x \equiv \ell p^{k-1}x$. Thus $\ell \equiv 0$ mod p, and so $ux \in \langle p^{n-1}\rangle$, showing $J_1J_k \subseteq \langle p^{n-1}\rangle$.

Similarly, $J_kJ_1 \subseteq \langle p^{n-1}\rangle$, hence $J_kJ_{k-1} \subseteq \langle p^{n-1}\rangle$. □

Now we are ready to prove

Proposition A. ($\nu > 2$ or $p > 3$). $J_k^2 \subseteq \langle p^{n-1}\rangle$ and $k \leq n/2$.

Proof: Case 1: $\nu > 2$. Take x and y independent. If $v = xy \notin \langle p^{n-1}, p^{k-1}x, p^{k-1}y\rangle$ then for all $v' \in V - \langle p^{n-1}, p^{k-1}x, p^{k-1}y\rangle$ there exist x', y' with $(p^{k-1}x, p^{k-1}y, v, x, y) \sim (p^{k-1}x, p^{k-1}y, v', x', y')$, hence $x'y' = v'$, $x' \equiv x$ mod J_{k-1}, $y' \equiv y$ mod J_{k-1}. Now by Lemma 8 we have $x'y' \equiv xy$ mod $\langle p^{n-1}\rangle$. But for $v' = v + p^{k-1}x$ we see this is impossible: $v \not\equiv v + p^{k-1}x$ mod $\langle p^{n-1}\rangle$. Thus $xy \in \langle p^{n-1}, p^{k-1}x, p^{k-1}y\rangle$, say $xy = \alpha p^{k-1}x + \beta p^{k-1}y$ mod $\langle p^{n-1}\rangle$. Since $(x,y) \sim (x,-y)$, this implies $x(-y) = \alpha p^{k-1}x + \beta p^{k-1}(-y)$. But $x(-y) = -xy = -\alpha p^{k-1}x - \beta p^{k-1}y$. This implies $\alpha \equiv 0$ mod p. Similarly $\beta \equiv 0$ mod p, and so $xy \in \langle p^{n-1}\rangle$ for all independent x and y. Now $x(x + y) \in \langle p^{n-1}\rangle$, since x and $x + y$ are also independent, and this implies $x^2 \in \langle p^{n-1}\rangle$. Likewise $x(p^{n-k} + y) \in \langle p^{n-1}\rangle$ implies $xp^{n-k} \in \langle p^{n-1}\rangle$. This gives $x\cdot J_k \subseteq \langle p^{n-1}\rangle$; similarly $J_k^2\cdot x \subseteq \langle p^{n-1}\rangle$. Thus $p^{n-k}(p^{n-k} + x) \in \langle p^{n-1}\rangle$ and so $(p^{n-k})^2 \in \langle p^{n-1}\rangle$. This shows $J_k^2 \subseteq \langle p^{n-1}\rangle$. From $p^{n-k}x \in \langle p^{n-1}\rangle$ we deduce that $p^{n-k}x = 0$, since all multiples of x lie outside $\langle p^{n-1}\rangle$ except 0. Thus $n - k \geq k$, and so $k \leq n/2$, finishing Case 1.

Case 2: $p > 3$, $\nu = 2$. If $x^2 \equiv \alpha p^{k-1}x \bmod \langle p^{n-1}\rangle$, $\alpha \in \mathbf{F}_p^*$, then for all $v \in V'$ there is y with $p^{k-1}y = v$, $y^2 \equiv \alpha p^{k-1}y \bmod \langle p^{n-1}\rangle$. By Lemma 8 we conclude that for all $z \in J_k'$, $z^2 = \alpha p^{k-1}z \bmod \langle p^{n-1}\rangle$. But for $z = \gamma x$, $\gamma \in \mathbf{F}_p - \{0,1\}$ we get a contradiction:

$$z^2 = \gamma^2\alpha p^{k-1}x = \gamma(\alpha p^{k-1}z) \not\equiv \alpha p^{k-1}z \bmod \langle p^{n-1}\rangle.$$

Since $\nu = 2$, $J_k = \langle x, p^{n-k}, J_{k-1}\rangle$ for any $x \in J_k'$ and using Lemma 8 it follows readily that $J_k^2 \subseteq \langle p^{n-1}\rangle$. The argument that $k \leq n/2$ is as in Case 1. □

3. Small V.

In what follows we consider J_k such that $\nu = 2$, ν the dimension of V over $\mathbf{F}_p$, and show that all products from J_k lie in the prime subring, hence in $\langle p^{n-k}\rangle$. In particular, we prove this in case $\nu = 2$, $p = 3$, which is all we really add, in light of Proposition A. The proof for arbitrary p is apparently no harder than for $p = 3$, and does give an alternate route to showing that for $p > 3$ and $\nu = 2$, $J_k^2 \subseteq \langle p^{n-1}\rangle$, as an easy corollary to $J_k^2 \subseteq \langle p^{n-k}\rangle$.

Notation. We fix some notation for this section. Since $\nu = 2$, we write $V = \langle p^{n-1}\rangle \oplus \langle v\rangle$, and fix $x \in J_k'$ with $p^{k-1}x = v$. Using Fact 5 we have J_1 is of type I, II, or III, and we write $J_1 = V$ or $\langle V,a\rangle$ or $\langle V,a,b\rangle$ as in Fact 3, hence $J_k = \langle p^{n-k},x\rangle$ or $\langle p^{n-k},x,a\rangle$ or $\langle p^{n-k},x,a,b\rangle$. Throughout the proofs in this section we argue for the type III case only, and note that the obvious modifications (where b or both a and b are absent) give the proof for the other two types. We set $a^2 = p^{n-1}$, $ab = -ba = \ell p^{n-1}$, where $b^2 = tp^{n-1}$ or p^{n-1} according to $p \equiv 1$ or $3 \bmod 4$.

Lemma 1 ($\nu = 2$). $xa, ax, xb, bx \in V$.

Proof: Since $pa = 0$ we know $pxa = 0$, hence $xa \in J_1$ and so $axa \in \langle p^{n-1}\rangle$. Thus $xaxa = 0$, giving that $xa \in V$. Similarly for ax, etc. □

Lemma 2 ($\nu = 2$). $xa, ax, xb, bx \in \langle p^{n-1}\rangle$.

Proof: Suppose $xa \in V'$. Since $(a,v) \sim (-a,v)$, we conclude that there exists $y \in J_k$ with $(a,v,x) \sim (-a,v,y)$. In particular, $y(-a) = xa$ and $p^{k-1}y = v$; from the latter we conclude that $y \equiv x \bmod J_{k-1}$. By Lemma 2.4, $J_{k-1} = J_1 + pJ_k$, and so $xa = y(-a) \equiv x(-a) \bmod \langle p^{n-1}\rangle$, contradicting $xa \in V'$. Thus $xa \in \langle p^{n-1}\rangle$; we argue similarly for $ax, xb, bx \in \langle p^{n-1}\rangle$. □

Lemma 3 ($\nu = 2$). $x^2 \in \langle px, p^{n-k}\rangle$.

Proof: Write $x^2 = \alpha p^r x + \beta p^{n-k} + \gamma a + \delta b$ where $\alpha, \beta, \gamma, \delta \in \mathbf{Z}$, $\alpha \not\equiv 0 \bmod p$. Here $r \geq 1$, since by Fact 6 we know that $J_k^2 \subseteq \langle J_{k-1}, p^{n-k}\rangle$. Now by Lemma 2, $xa \in \langle p^{n-1}\rangle$, and this implies that $x^2a = 0$, so $0 = \gamma a^2 + \delta ba = (\gamma - \delta \ell)p^{n-1}$. Also $ax^2 = 0$ by Lemma 2, and so $(\gamma + \delta \ell)p^{n-1} = 0$. This gives p divides 2γ, hence $\gamma \equiv 0 \bmod p$. By a similar argument with $x^2b = bx^2 = 0$ we have $\delta \equiv 0$, and so $x^2 \in \langle px, p^{n-k}\rangle$. □

Proposition B. ($\nu = 2$). $J_k^2 \subseteq \langle p^{n-k}\rangle$ and $k \leq n/2$.

Proof: First we show that $x^2 \in \langle p^{n-k}\rangle$. Suppose $x^2 = \alpha p^r x + \beta p^{n-k}$, where $\alpha, \beta \in \mathbf{Z}$, $\alpha \not\equiv 0 \bmod p$, and $1 \leq r \leq k-1$. Then $p^{k-r-1}x^2 = \alpha v + \beta p^{n-r-1}$. Now $p^{k-r-1}J_{k-1}J_k = p^{k-r-1}(J_1 + pJ_k)J_k = p^{k-r-1}J_1J_k + p^{k-r} J_k^2 \subseteq \langle p^{n-k}\rangle$. We use $v \sim -v$ to produce y with $(v,x) \sim (-v,y)$, getting $y^2 = \alpha p^r y + \beta p^{n-k}$, $p^{k-1}y = -v$, and $p^{k-r-1}y^2 = \alpha(-v) + \beta p^{n-r-1}$. From $y \equiv -x \bmod J_{k-1}$ we have $p^{k-r-1}y^2 = p^{k-r-1}x^2 \bmod \langle p^{n-k}\rangle$. This implies $\alpha(-v) \equiv \alpha v \bmod \langle p^{n-k}\rangle$, contradicting $\alpha \not\equiv 0 \bmod p$. Thus $x^2 \in \langle p^{n-k}\rangle$.

To see $k \leq n/2$ we look at $(x + p^{n-k})^2 \in \langle p^{n-k}\rangle$ and conclude $2p^{n-k}x \in \langle p\rangle$. But this can only happen if $n - k \geq k$, so $k \leq n/2$.

By Lemma 2, then, it follows from $x^2 \in \langle p^{n-k}\rangle$ and $k \leq n/2$ that $J_k^2 \subseteq \langle p^{n-k}\rangle$. □

4. Large V Classified.

We assume in this section that $\nu > 2$, and we show the Classification Theorem's claims for this case. Again we use the notation of Fact 3 for possible J_1's.

Lemma 1 ($\nu > 2$). If J_1 is of type I, then $J_k^2 = 0$.

Proof: If $J_1 = V$, then $J_{k-1} = J_1 + pJ_k = V + pJ_k$. Moreover, $VJ_k = J_kV = 0$ by Fact 6. Therefore $J_{k-1}J_k = J_kJ_{k-1} = pJ_k^2 = 0$, by Proposition A. Choose x and y independent, and let $xy = sp^{n-1}$. Since $(p^{k-1}x, p^{k-1}y, sp^{n-1}) \sim (p^{k-1}x, -p^{k-1}y, sp^{n-1})$, there exist x', y' with $p^{k-1}x' = p^{k-1}x$, $p^{k-1}y' = -p^{k-1}y$, $xy = x'y'$. Now $y' + y$ and $x' - x \in J_{k-1}$, hence $x(y' + y) = (x' - x)y = 0$, giving $x'y' = xy' = x(-y)$. But $x'y' = xy$, so $xy = 0$. Now x and $x + y$ are also independent, so $x(x + y) = 0$, giving $x^2 = 0$. Thus $J_k^2 = 0$. □

This settles the picture for J_1 of type I completely. We turn to types II and III:

Lemma 2. ($\nu > 2$). If J_1 is of type II, then there is $x \in J_k'$ with $x * J_1 = 0$ and $x^2 = 0$.

Proof: Since $\nu > 2$ we can find y, z independent so that $y * J_1 = = z * J_1 = 0$. By adding a suitable multiple of y to z we can assume that $y * z = 0$ also. If either y^2 or z^2 is 0, fine. Otherwise we can represent any sp^{n-1} for $s \not\equiv 0 \bmod p$ as w^2 where $w = \alpha y + \beta z \in J_k'$ for some α, β. By definability of J_1, this says that all elements w of J_k' with nonzero square satisfy $w * J_1 = 0$. Taking w with $w^2 = a^2$ we get $(w + a)^2 = 2a^2 \neq 0$ and so $(w + a) * a = 0$. But $(w + a) * a = a * a = 2a^2 \neq 0$, a contradiction. □

Lemma 3. ($\nu > 2$). If J_1 is type II or III and $x \in J_k'$ then $x * J_1 = 0$ if and only if $x^2 = 0$.

Proof: Case 1: J_1 of type II. Here we know $x^2 = 0$ and $x \in J_k'$ implies $x * J_1 = 0$ by Lemma 2 and homogeneity. To prove the other direction, suppose $y * J_1 = 0$ with $y^2 \neq 0$. Since $(x + a)^2 = a^2$ and $(x + a) * J_1 \neq 0$ we know (again by homogeneity) that $y^2 = ta^2$, t a non-square mod p.

If $p > 3$ we can choose t so that both t and $t + 1$ are non-squares. Then $(y + a)^2 = (t + 1)a^2$ says that $2a^2 = (y + a) * a = 0$, a contradiction.

If $p = 3$ then $y^2 = -a^2$, so $(y + a)^2 = 0$, giving $2a^2 = a * (y + a) = 0$ by homogeneity, again a contradiction. Thus $y * J_1 = 0$ implies $y^2 = 0$.

Case 2: J_1 of type III. Suppose there is $x \in J_k'$, $x * a = x * b = 0$, $x^2 = sp^{n-1} \neq 0$. Since $J_kV = VJ_k = 0$ we know for any $x' = \delta x + \alpha a + \beta b$ that $(x')^2 = \delta^2 sp^{n-1} + \alpha^2 a^2 + \beta^2 b^2$. If $p > 3$ we can find δ with $\delta^2 \not\equiv 0, 1 \bmod p$ and solve for α, β so that $(x')^2 = sp^{n-1}$. Necessarily α and β are not both zero, since $\delta^2 \neq 0, 1$, and so $x' * a$ and $x' * b$ are not both 0. But $(x')^2 = x^2$ implies $x' \sim x$, so $x' * a = x' * b = 0$, a contradiction. Thus $x^2 = 0$ if $p > 3$.

If $p = 3$ we use $\nu > 2$ to produce x, y independent so that $x * J_1 = y * J_1 = 0$. Recall that here $a^2 = b^2 = 3^{n-1}$. If $x^2 = -3^{n-1}$ we get $(x + a)^2 = 0$, $(x + a + b)^2 = 3^{n-1}$, so the only elements of J_k' which satisfy $z * J_1 = 0$ have $z^2 = -3^{n-1}$. Similarly, if $x^2 = 3^{n-1}$ then $z \in J_k'$ and $z * J_1 = 0$ imply $z^2 = 3^{n-1}$. Notice that $x \pm y \in J_k'$, $(x \pm y) * J_1 = 0$. But $(x + y)^2 = x^2$ says $x * y = -x^2$ and $(x - y)^2 = x^2$ says $x * y = x^2$. This contradicts $x^2 \neq 0$, and so $x^2 = 0$ is the only possibility when $x * J_1 = 0$.

This will complete our argument for Case 2 provided we can find some $x \in J_k'$ with $x * J_1 = 0$. To do this we intersect the $*$-annihilators of a and b in J_k/pJ_k: each is a $(\nu + 1)$-dimensional subspace, and so the $*$-annihilator of J_1 in J_k/pJ_k has dimension at least ν. Since $\nu > 2$, this subspace gives us suitable $x \in J_k'$. □

Lemma 4. $(\nu > 2)$. If J_1 is of type II or III and $x \in J_k'$ such that $x^2 = 0$, then $xJ_1 = J_1x = 0$.

Proof: Take x as in the hypothesis, and let $v = p^{k-1}x$. Notice $x * J_1 = 0$ by Lemma 3. Then for any $v' \in V'$ we have $(v, a, b, p^{k-1}) \sim (v', a, b, p^{n-1})$ (or $(v, a, p^{n-1}) \sim (v, a, p^{n-1})$). Thus there is x' so that $p^{k-1}x' = v'$, $x'a = xa$, $(x'b = xb)$, $(x')^2 = 0$. By choosing $v' \notin \langle v, p^{n-1} \rangle$ we can be sure that x and x' are independent. Thus $x - x' \in J_k'$. But $(x - x') * J_1 = 0$, and so $(x - x')^2 = 0$ by Lemma 3. Moreover $(x - x')J_1 = 0$. But now $x - x' \sim x$ gives $xJ_1 = 0$. Similarly $J_1x = 0$. □

Lemma 5. $(\nu > 2)$. If J_1 is of type II or III, and x and y are independent with $x^2 = y^2 = 0$, then $xy = yx = 0$.

Proof: Take x and y as above and notice that $(x + y) * J_1 = 0$, hence $(x + y)^2 = 0 = x * y$. Let $p^{k-1}x = v$, $p^{k-1}y = v'$. Then $(v,v') \sim (v',v)$, and so there are x', y' with $(v,v',x,y) \sim (v',v,x',y')$. In particular, $p^{k-1}x' = v'$, $p^{k-1}y' = v$, $x'y' = xy$, $(x')^2 = (y')^2 = 0$. But then $x' \equiv y \bmod J_{k-1}$, $y' \equiv x \bmod J_{k-1}$, and so by Lemma 4 and Lemma 2.4, $xy = x'y' = yx = -xy$. Thus $xy = 0$, and from $x * y = 0$ it follows that $yx = 0$ also. □

We now prove the Classification Theorem, Part A, for the Case $\nu = \mathbf{F}_p\text{-dim } V > 2$. For $J_1 = V$, Lemma 1 is all we need. For J_1 of type II or III, we see that J_k is generated by J_1, p^{n-k}, and elements with square 0, by modifying any element y of J_k' by a suitable multiple of a (or of a and b) to get $(y + \alpha a + \beta b) * J_1 = 0$, then applying the previous lemmas.

The verification that the resulting J's are homogeneous is routine.

5. Small V Classified.

In this final section we consider V of dimension 2 over $\mathbf{F}_p$, using results in Sections 2 and 3 to obtain the rest of our Classification Theorem. Recall that from Proposition B we know that $J_k^2 \subseteq \langle p^{n-k}\rangle$, $k \leq n/2$, hence $p^{n-k}J_k = 0$. If $p > 3$ we known more by Proposition A, namely $J_k^2 \subseteq \langle p^{n-1}\rangle$, $VJ_k = J_kV = 0$.

Lemma 1 ($\nu = 2$, $p > 3$). If $J_1 = V$ (type I) then $J_k^2 = 0$.

Proof. Notice that $J_{k-1} = V + pJ_k$ annihilates J_k. Let $x \in J_k'$. It suffices to show $x^2 = 0$. From $(x^2, p^{k-1}x) \sim (x^2, 2p^{k-1}x)$ we know there is $y \in J_k'$ such that $(p^{k-1}x, x^2, x) \sim (2p^{k-1}x, x^2, y)$. But then $p^{k-1}y = 2p^{k-1}x$ implies $y \equiv 2x \mod J_{k-1}$ and so $x^2 = y^2 = 4x^2$. For $p > 3$ this implies $x^2 = 0$. □

Thus we have again that for type I J_1's, multiplication is trivial, and so we turn to types II and III.

Lemma 2 ($\nu = 2$, $p > 3$). Suppose J_1 is type II or III, and suppose $x \in J_k'$ such that $x * J_1 = 0$. Then $J_1x = xJ_1 = 0$.

Proof. (i) J_1 of type II: Let $J_1 = \langle V, a\rangle$, $a * x = 0$. Since $x + a \sim x - a$ and $\{a, -a\}$ is definable modulo V, we get $a(x + a) = \pm a(x - a)$. If $a(x + a) = a(x - a)$, then $a^2 = -a^2$, which is impossible. From $a(x + a) = -a(x - a)$ we conclude that $ax = -ax$, hence $ax = 0$. Now $a * x = 0$ gives $xa = 0$ as well.

(ii) J_1 of type III: Suppose $J_1 * x = 0$. Since $J_1x \subseteq \langle p^{n-1}\rangle$ there exists $c \in J_1 - V$ such that $cx = 0$. Choose $d \in J_1$ with $c * d = 0$, $J_1 = \langle V, c, d\rangle$, $d^2 = c^2$ or tc^2. Since $\nu = 2$, $\pm x$ is definable mod $\langle V, pJ_k, p^{n-k}\rangle$, and so if $d^2 = c^2$ then $dx = 0$ also, giving $J_1x = 0$. Suppose then $d^2 = tc^2$ (and so $p \equiv 1 \mod 4$). Now every element of $\mathbf{F}_p^*$

can be written as $\mu^2 + \nu^2 t$ with $\mu\nu \neq 0$, and in particular there exist $\mu, \nu \neq 0$ with $\mu c + \nu d \sim c$. But then $(\mu c + \nu d)x = 0$, and so $dx = 0$, giving $J_1 x = 0$. It follows from $J_1 * x = 0$ that $xJ_1 = 0$ also. □

We are now ready to prove the small V part of Case A of our Theorem ($\nu = 2$, $p > 3$):

If J_1 is of type II, then choose $x \in J_k'$ so that $a * x = 0$, $J_1 = \langle V,a\rangle$. From Lemma 2 we know $ax = xa = 0$. Let $x^2 = \alpha p^{n-1}$, $\alpha \in \mathbf{F}_p$, and choose $\beta \neq 0, 1, -1$, $\beta \in \mathbf{F}_p$. Since $p^{k-1}x \sim \beta p^{k-1}x$, there is $y \in J_k'$, $(p^{k-1}x, x^2, x) \sim (\beta p^{k-1}x, x^2, y)$. This gives $p^{k-1}y = \beta p^{k-1}x$, $y^2 = x^2$, $y * a = 0$. Write $y = \gamma a + \delta x \bmod \langle p^{n-k}, pJ_k\rangle$ for $\gamma, \delta \in \mathbf{F}_p$. Then $\delta = \beta$ and from $y * a = 0$ we get $\gamma = 0$, so $y = \beta x$, $x^2 = y^2 = \beta^2 x^2$. Since $\beta^2 \neq 1$, we conclude that $x^2 = 0$. Thus $J = \langle a, x, p\rangle$ as in the theorem.

If J_1 is of type III, then choose $x \in J_k'$, $x * J_1 = 0$ and argue as above, again using Lemma 2, to get that

$$J = \langle a, b, x, p\rangle,\ xJ_k = J_k x = 0,\ k \leqq n/2.$$

Again, the checks that these are homogeneous are routine and we omit them.

We now turn to Part B of our theorem. Notice that a complication arises here since we do not know that $VJ_k = 0$.

Lemma 3. ($v = 2$, $p = 3$). Let $x \in J_k'$.

(i) If $J_1 = \langle V,a\rangle$ is of type II, and $x * a = 0$ then $ax = xa = 0$.

(ii) If $J_1 = \langle V,a,b\rangle$ is of type III, and $x * a = x * b = 0$, then $ax = xa = bx = xb = 0$. Moreover, if $Vx \neq 0$, then $ab = ba = 0$.

Proof: (i) If $Vx = 0$ we argue exactly as in Lemma 2. If not, then $V = \langle 3^{n-1}, 3^{k-1}x\rangle$, so $vx = xv$ for all $v \in V$. Choose $v \in V'$ with $vx = xv = a^2$. Since $(v,a) \sim (v,v-a)$, there exists $y \in J_k$ with $(v, a, x) \sim (v, v - a, y)$, hence

$$vy = vx,\ (v - a)y = ax,\ (v - a) * y = 0.$$

Now $y \equiv kv + \ell a + mx \bmod \langle 3^{n-k}\rangle$ for some $k, \ell, m \in \mathbf{Z}$. Since $vy = vx$ we have $m \equiv 1 \bmod 3$. Thus from $(v - a)y = ax$ it follows that

$vx - \ell a^2 - ax = ax$, hence $a^2 - \ell a^2 = -ax$. Finally, from $(v - a) * y = 0$ we get $v * x - \ell a * a = 0$, $a^2 - \ell a^2 = 0$. Thus $ax = 0$, and so $xa = 0$.

(ii) As in (i) we need only consider when $Vx \neq 0$. Choose $v \in V'$ with $vx = xv = a^2$. Then $(v, v - a, -b) \sim (v, a, b)$, so there is y with $(v, v - a, -b, y) \sim (v, a, b, x)$. In particular, $vy = vx$, $(v - a)y = ax$, $(v - a) * y = 0$, $-b * y = 0$, $-by = bx$. Writing $y = kv + ra + sb + mx$ mod $\langle 3^{n-k} \rangle$ with $m = 1$ mod 3 we get $(v - a)y = ax$ and so $a^2 - ra^2 - sab = -ax$, while $(v - a) * y = 0$ yields $a^2 - ra^2 = 0$. Now $b * y = 0$ implies $sb * b = 0$, so $s \equiv 0$ mod 3. Combining all this gives $ax = 0$, hence $xa = 0$. Similarly $bx = xb = 0$.

Now from $-by = bx$ we get

$$-bra - sb^2 - bx = bx, \quad -bra - sb^2 = -bx.$$

Since $s = 0$, $r = 1$ this becomes $-ba = -bx$, hence $ba = 0$. From $a * b = 0$ it follows that $ab = 0$ also, when $Vx \neq 0$.

We now prove Case B of our theorem. If $J_k^2 \subseteq \langle 3^{n-1} \rangle$ then the description as in Case B (i) is exactly as for Case A, except that now $x^2 = \pm 3^{n-1}$ is also possible.

If $J_k^2 \subseteq \langle 3^{n-1} \rangle$ then we choose $x \in J_k'$ with $x * a = x * b = 0$, hence $xa = xb = ax = bx = 0$ by Lemma 3. Let $x^2 = \alpha 3^{n-r}$, $1 < r \leq k$, $\alpha \not\equiv 0$ mod 3. If $r > 2$ consider $m = 3^{r-2} - 1$: $m^2 \equiv 1$ mod 3^{r-2} but $m^2 \not\equiv 1$ mod 3^{r-1}. Observe that $(3x, a, b) \sim (3mx, a, b)$ and so there is y with $(3x, a, b, x) \sim (3mx, a, b, y)$, hence

$$y^2 = x^2, \; 3y = 3mx, \; ya = yb = ay = by = 0.$$

From $3y = 3mx$ we get $y - mx \in J_1$, and from $ya = yb = 0$, etc. we get $y - mx \in V$, so $y = mx + v$. Thus $x^2 = y^2 = m^2x^2 + 2mxv$. But now $(1 - m^2)x^2 = 2mxv \in \langle 3^{n-1} \rangle$ and so $(1 - m^2) = 0$ mod $\langle 3^{n-1-(n-r)} \rangle$, contradicting $m^2 \not\equiv 1$ mod 3^{r-1}. Thus $r = 2$, $x^2 = \alpha 3^{n-2}$, $\alpha \equiv 0$ mod 3.

Now apply the above argument with $m = 2$ to get $y = 2x + v$ such that $x^2 = y^2 = 4x^2 + 4xv$, $-3x^2 = 4xv$. If $4xv = 0$ then $-3 \cdot 3^{n-2} = 0$, contradicting characteristic 3^n. If $4xv \neq 0$ then $x \cdot 3^{k-1}x = \pm 3^{n-1}$, and so $3^{k-1} \cdot \alpha 3^{n-2} = \pm 3^{n-1}$, hence $n - 2 + k - 1 = n - 1$, so $k = 2$. Thus we have $k = 2$, $x^2 \in \langle 3^{n-2} \rangle - \langle 3^{n-1} \rangle$ The rest of Case B (ii) follows from Lemma 3.

Again we omit verification of homogeneity, as tedious and routine.

This proves the Classification Theorem, and thus completes our description of finite homogeneous rings of odd characteristic.

References:

[1] C. Berline and G. Cherlin, QE rings in characteristic p^n, J. Symbolic Logic 48 (1983), 140-162.

[2] C. Berline and G. Cherlin, QE rings in characteristic p, in Logic Year 1979-80 (Storrs), Lecture Notes in Math. 859 (Springer, Berlin, 1981).

[3] G. Cherlin and A. Lachlan, Stable finite homogeneous structures, preprint.

[4] A. Lachlan, On countable stable structures which are homogeneous for a finite relational language, preprint.

[5] D. Saracino and C. Wood, Finite QE rings in characteristic p^2, to appear in Annals of Pure and Applied Logic.

[6] D. Saracino and C. Wood, QE commutative nilrings, J. Symbolic Logic 49 (1984), 644-651.

LOGIC COLLOQUIUM '84
J.B. Paris, A.J. Wilkie, and G.M. Wilmers (Editors)
© Elsevier Science Publishers B.V. (North-Holland), 1986

SUBSTRUCTURE LATTICES OF MODELS OF PEANO ARITHMETIC

James H. Schmerl*

Department of Mathematics
University of Connecticut
Storrs, Connecticut 06268
U.S.A.

1. INTRODUCTION

The intersection of an arbitrary collection of elementary substructures of a model $\mathcal{N}$ of Peano arithmetic (PA) is again an elementary substructure of $\mathcal{N}$. Hence, the collection of elementary substructures of $\mathcal{N}$ forms a lattice, denoted by $\mathrm{Lt}(\mathcal{N})$ and referred to as the <u>substructure lattice</u> of $\mathcal{N}$. The basic question concerning substructure lattices is: Which lattices can be realized as substructure lattices? That is, for which lattices L is there a model $\mathcal{N}$ of PA such that $L \simeq \mathrm{Lt}(\mathcal{N})$? The question can also be asked for models of specific completions T of PA: Which lattices can be realized as substructure lattices of models of T? At present, there is no known way that this question distinguishes between different completions T of PA, provided $T \neq TA$ (where TA is true arithmetic). This peculiarity of TA is due to the fact that each model of TA is an end extension of the minimal model of TA, whereas the minimal models of all other completions of PA have a rich collection of cofinal extensions. Furthermore, we know of certain finite lattices which cannot occur as $\mathrm{Lt}(\mathcal{N})$ when $\mathcal{N}$ is an end extension of its minimal elementary substructure, but no such restriction is known for cofinal extensions. A specific example is the 1-3-1 lattice, also known as M_5: if $\mathcal{N} \models TA$, then $\mathrm{Lt}(\mathcal{N}) \not\simeq M_5$; however (as will be proved here) if $T \neq TA$, then there is some $\mathcal{N} \models T$ such that $\mathrm{Lt}(\mathcal{N}) \simeq M_5$.

Therefore, it seems appropriate to study the more general <u>intermediate structure lattice</u> $\mathrm{Lt}(\mathcal{N}/\mathcal{M})$, where $\mathcal{M} \prec \mathcal{N}$, which is the principal filter of $\mathrm{Lt}(\mathcal{N})$ generated by $\mathcal{M}$. That is, $\mathrm{Lt}(\mathcal{N}/\mathcal{M}) = \{\mathcal{M}_0 \in \mathrm{Lt}(\mathcal{N}) : \mathcal{M} \prec \mathcal{M}_0 \prec \mathcal{N}\}$. In this context the basic question has the following reformulation: For each model $\mathcal{M}$ of PA, which lattices can be realized as intermediate structure lattices $\mathrm{Lt}(\mathcal{N}/\mathcal{M})$? Refinements to this question can now be made by restricting $\mathcal{N}$ to be an end extension, a cofinal extension, or some other type of extension of $\mathcal{M}$.

The main question to be considered in this paper is: For a given countable,

*Research was supported in part by NSF Grant No. 8301603.

nonstandard model M of PA , which finite lattices can be realized as $Lt(N/M)$ where N is a cofinal extension of M ? It is still not known that for any such M there are any finite lattices which cannot be realized in this manner.

It will be proved, however, that for any countable, nonstandard model M of PA , those finite lattices which can be realized as $Lt(N/M)$, where N is a cofinal extension of M , depend only on $Th(M)$. Many finite lattices will be shown to be realized in this way. For example, every countable, nonstandard M has a cofinal extension N such that $Lt(N/M)$ is the 1-3-1 lattice. Moreover, for any lattice L and any nonstandard model M , there is a cofinal extension N of M such that $Lt(N/M)$ embeds L .

From now on, all structures are models of PA . Throughout this paper we will be working with a formalization of PA that includes terms for all definable functions. Thus, all extensions and substructures of models will be assumed to be elementary. We adhere to the convention that $M \prec M$. If $M \prec N$ and $a \in N$, then $M(a)$ is the substructure of N generated by $M \cup \{a\}$.

In this paragraph the previous developments on the problem of characterizing the possible substructure lattices will be discussed. If $M \prec N$, then there is a unique $M_0 \in Lt(N/M)$ such that M_0 is a cofinal extension of M and N is an end extension of M_0 . Therefore, it follows that every nonstandard M has a proper cofinal extension. The fundamental theorem of MacDowell and Specker [9] is that every M has a proper end extension. Gaifman [5] extended the MacDowell-Specker theorem by showing that every M has a minimal end extension N (so that $Lt(N/M)$ is the 2-element lattice), and Blass [1] showed that every countable nonstandard model has a minimal cofinal extension. Later in [7] Gaifman showed that other distributive lattices, such as all finite Boolean lattices, some infinite Boolean lattices, and some chains, could also be realized. Paris [11] characterized those countable distributive lattices which can occur as substructure lattices as those which are complete and compactly generated. Schmerl [15] proved that for any finite distributive lattice L every model M has an end extension N such that $Lt(N/M) \cong L$. Then Mills [10] generalized this, showing that for any complete ω_1-like-compactly generated distributive lattice L and any M there is $N \succ M$ such that $Lt(N/M) \cong L$. Paris [12] showed that $Lt(N)$ need not be distributive by proving that every model M has an extension N such that $Lt(N/M)$ embeds the 1-3-1 lattice (or even the 1-ω-1 lattice). This contrasts with an earlier observation of Gaifman [7] and Paris [11] that if N is an end extension of M , then $Lt(N/M)$ is not the 1-n-1 lattice whenever $3 \leq n \leq \omega$. Wilkie [17] showed that the lattice $Lt(N)$ need not even be modular; specifically, he proved that every model M has an end extension N such that $Lt(N/M)$ is the pentagon lattice. He also showed in [17] that the hexagon lattice (and other related lattices) cannot occur as $Lt(N/M)$ when N is an end extension of M.

In outline the contents of this paper are as follows. In §2 we discuss a certain type of lattice representation. This section is lattice-theoretic and contains the crucial definitions which will be used later. Some examples of these types of representations of lattices are presented in §3. All of these examples come from canonical partition theorems. In §4 we prove the main result which relates substructure lattices and the lattice representations of §2. Then applying the examples from §3 we obtain in §5 all of our results about specific lattices which can be realized as substructure lattices. We conclude with §6 with some conjectures.

2. REPRESENTATIONS OF LATTICES

Given a set A let $\Pi(A)$ be the set of all partitions of A. If $a,b \in A$ and $\pi \in \Pi$, then we write

$$a \approx b \pmod{\pi}$$

if $\{a, b\} \subseteq C$ for some $C \in \pi$. There are two extreme partitions $\mathbf{0}_A$ and $\mathbf{1}_A$ in $\Pi(A)$, where

$$a \approx b \pmod{\mathbf{1}_A} \quad \text{iff} \quad a = b ,$$

and

$$a \approx b \pmod{\mathbf{0}_A} \quad \text{for every} \quad a,b \in A .$$

We partially order $\Pi(A)$ so that if $\pi_1,\pi_2 \in \Pi$, then $\pi_1 \le \pi_2$ iff π_2 is a refinement of π_1. Then we have for each $\pi \in \Pi(A)$ that

$$\mathbf{0}_A \le \pi \le \mathbf{1}_A .$$

With this partial order, $\Pi(A)$ becomes a complete lattice (which is the dual of the lattice one usually considers on $\Pi(A)$). In particular, if $\pi_1,\pi_2 \in \Pi(A)$, then $\pi_1 \vee \pi_2 = \{A_0 \subseteq A : A_0 = A_1 \cap A_2 \neq \emptyset$ for some $A_1 \in \pi_1$ and $A_2 \in \pi_2\}$.

Let $\Theta : A \to B$ be an injection. Then there is an induced function $\Theta^* : \Pi(B) \to \Pi(A)$ such that for any $a,b \in A$ and $\pi \in \Pi(B)$, $a \approx b \pmod{\Theta^*(\pi)}$

iff $\theta(a) \approx \theta(b) \pmod{\pi}$. One easily checks that $\theta^*(\mathbb{0}_B) = \mathbb{0}_A$, $\theta^*(\mathbf{1}_B) = \mathbf{1}_A$, and $\theta^*(\pi_1) \vee \theta^*(\pi_2) = \theta^*(\pi_1 \vee \pi_2)$ whenever $\pi_1, \pi_2 \in \Pi(B)$. In particular, if $A \subseteq B$ and $\theta_{AB} : A \to B$ is the inclusion function, then for each $\pi \in \Pi(B)$, we let $\pi | A = \theta^*_{AB}(\pi)$.

Let $L = (L, \wedge, \vee)$ be any finite lattice. We denote the minimum and maximum elements of L by 0_L and 1_L respectively (or simply by 0 and 1) . A <u>representation</u> α <u>of</u> L <u>in</u> $\Pi(A)$ is an injection $\alpha : L \to \Pi(A)$, such that

$$\alpha(0) = \mathbb{0}_A ,$$

$$\alpha(1) = \mathbf{1}_A ,$$

and

$$\alpha(x \vee y) = \alpha(x) \vee \alpha(y) \quad \text{for any} \quad x, y \in L .$$

Notice that we do not require that a representation satisfy $\alpha(x \wedge y) = \alpha(x) \wedge \alpha(y)$. Thus, if we consider both L and $\Pi(A)$ as bounded upper semilattices, then a representation is merely an embedding.

Let $\alpha : L \to \Pi(A)$ be a representation. Then α is <u>nontrivial</u> if $|\alpha(x)| \neq 2$ whenever $x \in L$. If $B \subseteq A$, then we let $\alpha | B : L \to \Pi(B)$ where $(\alpha | B)(x) = \alpha(x) | B$. Notice that $\alpha | B$ may fail to be a representation since it may not be an injection; however, if it is an injection then it is a representation.

We now come to the crucial new definition.

<u>Definition 2.1</u>. Let $\alpha : L \to \Pi(A)$ be a representation of the finite lattice L . We say that α has the 0-<u>canonical partition property</u> if whenever $A = X \cup Y$, then $\alpha | X$ or $\alpha | Y$ is a representation of L . Proceeding recursively, we say that α has the (n+1)-<u>canonical partition property</u> if whenever $\pi \in \Pi(A)$, then there is some $X \subseteq A$ such that $\alpha | X$ is a representation of L having the n-canonical partition property and $\alpha(x) | X = (\alpha | X)(x) = \pi | X$ for some $x \in L$. (For brevity, we say that α is an n-CPP representation whenever it is a representation with the n-canonical partition property.)

<u>Lemma 2.2</u>. If $\alpha : L \to \Pi(A)$ is an n-CPP representation for some n , then α is nontrivial.

Proof: For a contradiction suppose $\alpha(x) = \{X,Y\}$, where $X \neq A \neq Y$, so that $x \neq 0$. We proceed by induction on n. Assume $n = 0$. By the definition of 0-CPP, either $\alpha|X$ or $\alpha|Y$ is a representation, yet $(\alpha|X)(0) = \alpha(0)|X = \alpha(x)|X = (\alpha|X)(x)$ and $(\alpha|Y)(0) = \alpha(0)|Y = \alpha(x)|Y = (\alpha|Y)(x)$. Now suppose $n > 0$, and consider the partition $\pi = \{X,Y\}$. Then there is $B \subseteq A$ such that $\pi|B$ is an $(n-1)$-CPP representation of L and $\pi|B = \alpha(y)|B$ for some $y \in L$. By the inductive hypothesis $|\alpha(y)|B| \neq 2$, so either $B \subseteq X$ or $B \subseteq Y$. Hence $(\alpha|B)(x) = (\alpha|B)(0)$, and this contradicts $\alpha|B$ being a representation of L. □

Lemma 2.3. If $\alpha : L \to \Pi(A)$ is an n-CPP representation and $m \leq n$, then α is also an m-CPP representation.

Proof: Clearly it suffices to consider just $m = 0$ and $n > 0$. Suppose $A = X \cup Y$, and without loss of generality, let $\pi = \{X,Y\}$ be a partition of A. Let $B \subseteq A$ be such that $\pi|B$ is an $(n-1)$-CPP representation of L and $\pi|B = (\alpha|B)(x)$ for some $x \in L$. By Lemma 2.2, either $B \subseteq X$ or $B \subseteq Y$, so either $\pi|X$ or $\pi|Y$, respectively, is a representation of L. Thus α is a 0-CPP representation. □

Lemma 2.4. If $\alpha : L \to \Pi(A)$ is an n-CPP representation, then for some finite $B \subseteq A$, $\alpha|B$ is an n-CPP representation.

Proof: Use the Compactness Theorem and induction on n. □

We will be interested in those finite lattices which, for each $n < \omega$, have n-CPP representations. In practice, such representations exhibit more uniformity than is required, suggesting the following definition.

Definition 2.5. Suppose $\alpha_i : L_i \to \Pi(A_i)$ are representations for $i = 1,2$. Then α_1 arrows α_2 (in symbols: $\alpha_1 \to \alpha_2$) iff whenever $\pi \in \Pi(A_1)$, then there is an injection $\theta : A_2 \to A_1$ such that

$$\alpha_2 = \theta^*\alpha_1$$

and

$$\theta^*(\pi) = \alpha_2(x) \quad \text{for some } x \in L_2 .$$

Notice that if $\alpha \to \beta$ and β is nontrivial, then α is a 0-CPP representation.

The procedure for showing that a finite lattice has, for each $n < \omega$, an n-CPP representation will typically be as follows. Suppose that α_n is a representation of L for each $n < \omega$ and that

$$\cdots \to \alpha_{n+1} \to \alpha_n \to \cdots \to \alpha_1 \to \alpha_0 ,$$

where α_0 is nontrivial. Then, by induction each α_{n+1} is an n-CPP representation. In particular, if $\alpha \to \alpha$, where α is a nontrivial representation of L, then α is an n-CPP representation of L for each $n < \omega$.

3. SOME EXAMPLES

In this section we will give some examples of n-CPP representations of lattices, and also of representations which arrow other ones.

Example 3.1. For $1 \leq j < \omega$, let B_j be the set of all subsets of $j = \{0, 1, \ldots, j-1\}$, so that B_j is the Boolean lattice with 2^j elements. For $X \subseteq \omega$ we let $[X]^j$ denote the set of increasing j-tuples from X. Now let $j < n \leq \omega$, and define

$$\alpha_n : B_j \to \Pi([n]^j)$$

so that if $x \in B_j$ and $a,b \in [n]^j$, then

$$a \approx b \ (\text{mod } \alpha_n(x)) \ \text{ iff } \ a_i = b_i \ \text{ for each } i \in x .$$

It is easily checked that α_n is a representation of B_j, and that it is nontrivial provided $n \geq 3$. The theorem of Erdös and Rado [4], which is the canonical version of Ramsey's Theorem, implies that $\alpha_\omega \to \alpha_\omega$. In particular, the Erdös-Rado theorem asserts that if $\pi \in \Pi([\omega]^j)$, then there is an infinite $X \subseteq \omega$ and $x \in B_j$ such that $\pi | [X]^j = \alpha_\omega(x) \mid [X]^j$. Thus B_j has a k-CPP representation for each $k < \omega$. Indeed, for each $n < \omega$ there is $m < \omega$ such that $\alpha_m \to \alpha_n$. Thus, if $k < \omega$ then α_n is a k-CPP representation of B_j for all sufficiently large finite n.

Example 3.2. Let D be a finite distributive lattice with $|D| \geq 2$, and let $J \subseteq D$ be the set of join-irreducible elements of D. For $2 \leq n \leq \omega$, define

$$\alpha_n : D \to \Pi(n^J)$$

so that if $x \in D$ and $a,b \in n^J$, then

$$a \approx b \pmod{\alpha_n(x)} \text{ iff } a_j = b_j \text{ whenever } j \in J \text{ and } j \leq x .$$

It is easily checked that α_n is a representation of D, and that it is nontrivial provided $n \geq 3$. There is a rather straightforward proof, by induction on $|J|$, that $\alpha_\omega \to \alpha_\omega$. Thus, D has a k-CPP representation for each $k < \omega$. Indeed, whenever $2 \leq n < \omega$ there is $m < \omega$ such that $\alpha_m \to \alpha_n$. Thus, if $k < \omega$ then α_n is a k-CPP representation of D for all sufficiently large $n < \omega$.

Example 3.3. We consider representations of $\Pi(k)$, where $2 \leq k < \omega$. For $1 \leq n < \omega$, define

$$\alpha_n : \Pi(k) \to \Pi(k^n)$$

so that if $\pi \in \Pi(k)$ and $a,b \in k^n$, then

$$a \approx b \pmod{\alpha_n(\pi)} \text{ iff } a_j \approx b_j \pmod{\pi} \text{ for each } j < n .$$

(Here, k^n denotes the set of functions from n to k.) It is easily checked that α_n is a representation of $\Pi(k)$, and that it is nontrivial provided $n \geq 2$. A special case of the theorem of Prömel and Voigt [13] which is the canonical version of the Hales-Jewett Theorem (see [8]) implies that for each $n < \omega$ there is $m < \omega$ such that $\alpha_m \to \alpha_n$. Thus, if $m < \omega$, then α_n is an m-CPP representation of $\Pi(k)$ for all sufficiently large $n < \omega$.

Example 3.4. By using the full theorem of Prömel and Voigt referred to in Example 3.3, we can find more examples of finite lattices which have m-CPP representations for all $m < \omega$. These lattices tend to get rather complicated. The simplest one, other than the partition lattices $\Pi(k)$, is the 10-element lattice described at the end of [13] and pictured at right. Notice that the pentagon lattice P is an ideal of this lattice, so it also has an m-CPP representation for each $m < \omega$.

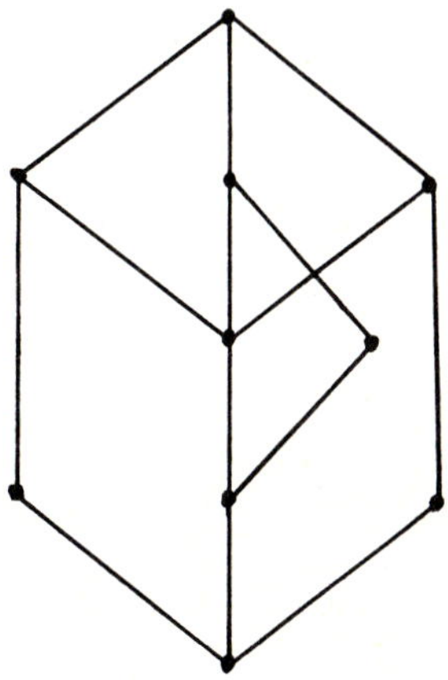

Example 3.5. Wilkie [17] shows, essentially, that there is a nontrivial representation $\alpha : P \to \Pi(\omega)$ of the pentagon lattice P such that $\alpha \to \alpha$. Therefore, P has an m-CPP representation for each $m < \omega$.

Example 3.6. For each n, $3 \leq n < \omega$, let M_{n+2} be the 1-n-1 lattice. For each $k < \omega$ we obtain a certain sublattice of $\Pi(k\times k)$ which besides $\mathbf{0}_{k\times k}$ and $\mathbf{1}_{k\times k}$, consists of all partitions of $k\times k$ of the following sort. Think of $k\times k$ as a subset of the plane R^2, and for any line ℓ in R^2 such that $|\ell \cap k\times k| \geq 2$, define the partition associated with ℓ to be such that if $(i_1,j_1),(i_2,j_2) \in k\times k$, then (i_1,j_1) and (i_2,j_2) are equivalent iff there is a line through (i_1,j_1) and (i_2,j_2) which is parallel (or equal) to ℓ. Then some M_{n_k} is isomorphic to this sublattice. Let $\alpha_k : M_{n_k} \to \Pi(k\times k)$ be an isomorphism. Clearly, α_k is a nontrivial representation of M_{n_k} provided $k \geq 3$. The theorem of Deuber, Graham, Prömel and Voigt [2], which is the canonical version of the 2-dimensional version of van der Waerden's theorem, implies that for each $k < \omega$ there exists $r < \omega$ such that $\alpha_r \to \alpha_k$. Thus, if $k < \omega$ then α_r is a k-CPP representation of M_{n_r} for all sufficiently large $r < \omega$. The theorem of [2] has higher-dimensional versions, which yield additional examples.

Example 3.7. Finally we mention that the theorem of Deuber, Prömel and Voigt [3] yields still more examples. From this theorem there is an increasing sequence $L_0 \subset L_1 \subset L_2 \subset \ldots$ of finite lattices such that each L_k has an m-CPP representation for each $m < \omega$. It is noteworthy that every finite lattice is embeddable in some L_k.

4. THE MAIN THEOREM

The principal result of this section is Theorem 4.1 which gives a first-order characterization of those countable nonstandard models M that have cofinal extensions N such that $Lt(N/M)$ is some given finite lattice.

Theorem 4.1. Let $M \models PA$ be countable and nonstandard, and let L be a finite lattice. Then the following are equivalent.

(1) M has a cofinal extension N such that $Lt(N/M) \cong L$.

(2) For each $n < \omega$, $M \models$ "L has an n-CPP representation".

Proof: Throughout this proof any term which we refer to is permitted to have parameters from M.

(1) $\Longrightarrow$ (2) . (The proof of this direction does not require that M be countable.) Let $h : L \to Lt(N/M)$ be an isomorphism. For each $i \in L$ let $b_i \in M$ be such that $h(i) = M(b_i)$, and then let $b = b_1$. For each $i \in L$ there is a term $t_i(x)$ such that $N \models t_i(b) = b_i$; and for each $i,j \in L$ there is a term $t_{ij}(x,y)$ such that $N \models t_{ij}(b_i,b_j) = b_{i\vee j}$. There is a bounded definable $X \subseteq M$ such that $b \in X^N$ (where by X^N is meant that unique subset of N definable in N by the same formula defining X) and in M each of the following sentences holds:

$$\forall x \in X[t_{ij}(t_i(x),t_j(x)) = t_{i\vee j}(x)] ,$$

$$\forall x, y \in X[t_0(x) = t_0(y)] ,$$

$$\forall x \in X[t_1(x) = x] .$$

For each $i \in L$, let π_i be the partition of X induced by t_i ; that is, $x \approx y \pmod{\pi_i}$ iff $M \models t_i(x) = t_i(y)$. Let $\alpha : L \to \Pi(X)$ be defined by $\alpha(i) = \pi_i$. This definition of α can be made inside M . We now want to show, working inside M , that α is an n-CPP representation for each $n < \omega$.

First we show that α is a representation. Clearly, $\alpha(0) = \pi_0 = \mathbf{0}_X$ and $\alpha(1) = \pi_1 = \mathbf{1}_X$. So we need only show that α is an injection and that $\alpha(i\vee j) = \alpha(i)\vee\alpha(j)$, or, equivalently, that if $x \approx y \pmod{\pi_i}$ and $x \approx y \pmod{\pi_j}$, then $x \approx y \pmod{\pi_{i\vee j}}$. To prove the latter statement, suppose $x \approx y \pmod{\pi_i}$ and $x \approx y \pmod{\pi_j}$. Then $t_i(x) = t_i(y)$. But $t_{i\vee j}(x)$ $= t_{ij}(t_i(x),t_j(x)) = t_{ij}(t_i(y),t_j(y)) = t_{i\vee j}(y)$, so $x \approx y \pmod{\pi_{i\vee j}}$.

To prove α is an injection, suppose $i,j \in L$ and $\pi_i = \pi_j$. Thus, for $x,y \in X$, $t_i(x) = t_i(y)$ iff $t_j(x) = t_j(y)$. Define a term $t(w)$ as follows: $t(w) = z$ iff there is $x \in X$ such that $t_i(x) = w$ and $t_j(x) = z$. Clearly, $N \models t(b_i) = b_j$, so $b_j \in M(b_i)$. Similarly, $b_i \in M(b_j)$, so that $M(b_i) = M(b_j)$, implying $i = j$. This proves that α is a representation of L .

This representation is nontrivial. For, if $|\pi_i| = 2$, then clearly $i \neq 0$ yet $b_i \in M$.

To show that α is an n-CPP representation for each standard n , it suffices to show that for each definable partition π of X there is a definable $Y \subseteq X$ such that $b \in Y^N$ and $\alpha(i)|Y = \pi|Y$ for some $i \in L$. Let $t(x)$ be a term which induces π . There is some $i \in L$ such that $M(b_i) = M(t(b))$. Thus, there are terms $t'(z)$ and $t''(z)$ such that $t'(t(b)) = b_i$ and $t''(b_i) = t(b)$.

So there is a definable $Y \subseteq X$ such that $b \in Y^N$ and

$$\mathcal{M} \models \forall x \in Y[t'(t(x)) = t_i(x)]$$

and

$$\mathcal{M} \models \forall x \in Y[t''(t_i(x)) = t(x)] \quad .$$

Therefore,

$$\mathcal{M} \models \forall x,y \in Y[t(x) = t(y) \leftrightarrow t_i(x) = t_i(y)]$$

so that $\alpha(i)|Y = \pi|Y$.

(2) $\Longrightarrow$ (1) . Our object is to construct a type over M such that if $\mathcal{N} = \mathcal{M}(b)$ and b realizes this type, then $\mathcal{N}$ is a cofinal extension of $\mathcal{M}$ and $\mathrm{Lt}(\mathcal{N}/\mathcal{M}) \simeq L$.

Let $\langle \pi_k : k < \omega \rangle$ be an enumeration of all the definable partitions of M . Such an enumeration exists because $\mathcal{M}$ is countable. (This is the only place in the proof where the countability of $\mathcal{M}$ is used.)

Suppose $\mathcal{M} \models$ "L has an n-CPP representation" for each standard n . Since $\mathcal{M}$ is nonstandard, by overspill there is a nonstandard $r \in M$ such that $\mathcal{M} \models$ "L has an r-CPP representation". Let $X_0 \subseteq M$ be a bounded definable subset and let $\alpha \in M$ be such that $\mathcal{M} \models$ "$\alpha : L \to \Pi(X_0)$ is an r-CPP representation". We will define a decreasing sequence $X_0 \supseteq X_1 \supseteq X_2 \supseteq \ldots$ of subsets of M such that in $\mathcal{M}$, $\alpha_k : L \to \Pi(X_k)$ is an (r-k)-CPP representation of L , where $\alpha_k = \alpha|X_k$. Suppose we already have such an X_k . Since α_k is an (r-k)-CPP representation, there is a definable $X_{k+1} \subseteq X_k$ such that $\alpha_k|X_{k+1}$ is an (r-k-1)-CPP representation and $\alpha_k(i) = \pi_k|X_{k+1}$ for some $i \in L$.

Let $\Sigma(x)$ be the set of formulas $\phi(x)$ (allowing parameters from M) such that for some $k < \omega$, $\mathcal{M} \models \forall x \in X_k\ \phi(x)$. Certainly, $\Sigma(x)$ is a type since each $X_k \neq \emptyset$. Furthermore, $\Sigma(x)$ is complete. For, consider any formula $\phi(x)$ and let π_k be the partition of M where $x \approx y \pmod{\pi_k}$ iff $\mathcal{M} \models \phi(x) \leftrightarrow \phi(y)$. Since α_k is, in particular, a 0-CPP representation, it follows that $\pi_k|X_{k+1} = \mathbf{0}_{X_{k+1}}$. Thus, either $\phi(x)$ or $\neg\phi(x)$ is in $\Sigma(x)$. Notice that if $\phi(x)$ is the formula $x \neq a$, then since $|X_{k+1}| > 1$, it must be that $\phi(x) \in \Sigma(x)$. Therefore, $\Sigma(x)$ is nonprincipal.

Let b realize $\Sigma(x)$ in an extension of M, and let $N = M(b)$. Since X_0 is bounded, N is a cofinal extension of M. We will now show that $Lt(N/M) \simeq L$. For each $i \in L$ let $t_i(x)$ be a term such that

$$M \models \forall x \in X_0[x \in t_i(x) \in \alpha(i)] .$$

Let $b_i = t_i(b)$. The following two things need to be shown:

(a) If $M \prec M_0 \prec N$, then $M_0 = M(b_i)$ for some $i \in L$.

(b) If $i,j \in L$, then $M(b_i) \prec M(b_j)$ iff $i \leq j$.

To prove (a), let us suppose that there is some $M_0 \in Lt(N/M)$ such that $M_0 \neq M(b_i)$ for each $i \in L$. Since L is finite, we can find M_0 of the form $M(c)$. Let $t(x)$ be a term such that $N \models c = t(b)$. There is a partition π_k which is induced by $t(x)$. But then for some $i \in L$, $\pi_k | X_{k+1} = \alpha_{k+1}(i)$. Clearly, $M \models \forall x,y \in X_{k+1}[t(x) = t(y) \leftrightarrow t_i(x) = t_i(y)]$. Therefore, there are terms $t'(z)$ and $t''(z)$ such that

$$M \models \forall x \in X_{k+1}[t'(t(x)) = t_i(x) \wedge t''(t_i(x)) = t(x)] ,$$

so that $N \models t'(c) = b_i \wedge t''(b_i) = c$. Hence, $M(b_i) = M(c)$, proving (a).

For (b), suppose $i \leq j$. Let $t(x)$ be a term such that for any $z \in \alpha(j)$, $t(z)$ is the equivalence class of $\alpha(i)$ which contains z. Then,

$$M \models \forall x \in X_0[t(t_j(x)) = t_i(x)] ,$$

so $N \models t(b_j) = b_i$. Therefore, $M(b_i) \prec M(b_j)$.

Conversely, suppose $M(b_i) \prec M(b_j)$ and let $N \models t(b_j) = b_i$. Then for some $k \in \omega$, $M \models \forall x \in X_k[t(t_j(x)) = t_i(x)]$. Clearly, then, $\alpha(j)|X_k$ is a refinement of $\alpha(i)|X_k$, so that $i \leq j$. This proves (b) and completes the proof of the theorem. □

The following variant of Theorem 3.1 will be useful in the proof of Corollary 5.8.

<u>Theorem 4.2.</u> Let $M \models PA$ be countable and nonstandard, and let L be a finite lattice. Then the following are equivalent.

(1) For each nonstandard $a \in M$, $\mathcal{M}$ has a cofinal extension $\mathcal{N} = \mathcal{M}(b)$ such that $\mathcal{N} \models b < a$ and $\mathrm{Lt}(\mathcal{N}/\mathcal{M}) \simeq L$.

(2) For each nonstandard $a \in M$ there is a cofinal extension $\mathcal{N}$ of $\mathcal{M}$ and an isomorphism $h : L \to \mathrm{Lt}(\mathcal{N}/\mathcal{M})$ such that for each $i \in L$ there is $b_i \in N$, where $\mathcal{N} \models b_i < a$ and $h(i) = \mathcal{M}(b_i)$.

(3) For each $n < \omega$, L has an n-CPP representation.

<u>Proof</u>: (1) $\Longrightarrow$ (3). Fix some $n < \omega$. We know from Theorem 4.1 that $\mathcal{M} \models$ "L has an n-CPP representation". In fact, from condition (1) it follows that in $\mathcal{M}$ there are arbitrarily small nonstandard n-CPP representations of L. By overspill, there is some standard one, and this one must be an n-CPP representation of L in the real world.

(3) $\Longrightarrow$ (2). Just mimic the proof of (2) $\Longrightarrow$ (1) in Theorem 4.1, starting out with a sufficiently small X_0. □

There is an analogue of half of Theorem 4.1 for end extensions. Before stating this analogue we will need some definitions.

For a lattice L let $\mathbf{1} \oplus L$ be the lattice obtained from L by adjoining to L a new least element. If $\mathcal{M} \prec \mathcal{N}$, then (following [16]) $\mathcal{N}$ is an <u>almost-minimal</u> extension of $\mathcal{M}$ if $\mathcal{N}$ is an end extension of $\mathcal{M}$ and for any $a \in N \setminus M$, $\mathcal{M}(a)$ is cofinal in $\mathcal{N}$. Gaifman [6] observed (see (6.1.1) of Lemma 6.1) that if $\mathcal{N}$ is an almost-minimal extension of $\mathcal{M}$ and $\mathrm{Lt}(\mathcal{N}/\mathcal{M})$ is finite, then $\mathrm{Lt}(\mathcal{N}/\mathcal{M}) \simeq \mathbf{1} \oplus L$ for some lattice L.

The following definitions are due to Gaifman [6], and are made with respect to some fixed completion T of Peano arithmetic. A type (by which we mean a complete 1-type) $\Sigma(x)$ is <u>definable</u> if for every formula $\phi(u,x)$ there is a formula $\sigma_\phi(u)$ such that whenever t is a constant term,

$$\phi(t,x) \in \Sigma(x) \quad \text{iff} \quad T \vdash \sigma_\phi(t) .$$

For any model $\mathcal{M} \models T$ each definable type $\Sigma(x)$ has a canonical extension to a type $\Sigma_{\mathcal{M}}(x)$ over $\mathcal{M}$, where $\phi(a,x) \in \Sigma_{\mathcal{M}}(x)$ iff $\mathcal{M} \models \sigma_\phi(a)$. Then $\mathcal{N}$ is a $\Sigma(x)$-<u>extension</u> of $\mathcal{M}$ if $\mathcal{N} = \mathcal{M}(c)$, where c is an element realizing $\Sigma_{\mathcal{M}}(x)$. A definable type $\Sigma(x)$ is said to be <u>end-extensional</u> if every $\Sigma(x)$-extension of each model of T is almost-minimal.

The following theorem is proved in much the same manner as was (2) $\Longrightarrow$ (1) in Theorem 4.1.

Theorem 4.3. Suppose $T \supseteq PA$ is a complete theory and L a finite lattice such that $T \models \forall x$("L has an x-CPP representation"). Then there is an end-extensional type $\Sigma(x)$ such that whenever $M \models T$ and N is a $\Sigma(x)$-extension of M, then $Lt(N/M) \simeq \mathbf{1} \oplus L$.

Proof: We will only construct the definable type $\Sigma(x)$, leaving to the reader the verification that it does what it is supposed to do.

We work inside the minimal model M of T. Since $M \models \forall x$("L has an x-CPP representation"), we are able to find a formula $\phi(u,x)$ and a term $t(u,v)$ which have the following properties. For each u, let $X_u = \{a \in M : M \models \phi(u,a)\}$. Then $u < w \wedge a \in X_u \wedge b \in X_w$ implies $a < b$. For each u and each $i \in L$, $t(u,i)$ is a partition of X_u. Let $\alpha_u : L \to \Pi(X_u)$, where $\alpha_u(i) = t(u,i)$. Then α_u is a u^2-CPP representation of L.

Our object is to define a sequence $\langle \phi_n(x) : n < \omega \rangle$ of formulas, and then let $\Sigma(x)$ be the type these formulas generate.

Let $\langle \psi_n(x,y,z) : n < \omega \rangle$ be a sequence of all formulas whose free variables are among x, y, z, such that for each z the formula $\psi_n(\cdot\,,\cdot\,,z)$ defines an equivalence relation on M.

Along with the formulas $\phi_n(x)$, we are also going to define formulas $\theta_n(u)$. (The sequence $\langle \theta_n(u) : n < \omega \rangle$ will generate a minimal type, and the construction of this sequence will be exactly the same as the construction of a minimal type.) Let $\theta_0(u)$ be $u = u$. Each of the following sentences should be in T:

$$\forall u(\theta_{n+1}(u) \to u > n) ;$$

$$\forall w \exists u > w\ \theta_{n+1}(u) ;$$

$$\forall u(\theta_{n+1}(u) \to \theta_n(u)) ;$$

$$\forall z \exists w[(\forall x > w \forall y > w(\theta_{n+1}(x) \wedge \theta_{n+1}(y) \to \psi_n(x,y,z)))$$
$$(\forall x > w \forall y > w(\theta_{n+1}(x) \wedge \theta_{n+1}(y) \to \neg\psi_n(x,y,z)))] .$$

Let $\phi_0(x)$ be the formula $\exists u\, \phi(u,x)$. Suppose we already have $\phi_n(x)$ and $\theta_n(u)$, and that the following sentences are in T :

$$\theta_n(u) \leftrightarrow \exists x\ \phi_n(u,x) \ ;$$

$$\forall u[\theta_n(u) \rightarrow \text{"}\alpha_u | \{x : \phi_n(u,x)\} \text{ is a } (u^2\text{-}nu)\text{-CPP representation of } L\text{"}] \ .$$

In order to get $\phi_{n+1}(x)$, first obtain $\theta_{n+1}(u)$. Let $\pi_{n,z}$ be the partition of M defined by $\psi_n(\cdot, \cdot, z)$. For each u such that $\theta_{n+1}(u)$, there is $Y_u \subseteq \{x : \phi_n(u,x)\}$ such that for each $z < u$ there is $i \in L$ such that $\pi_{n,z}|Y_u = \alpha_u(i)|Y_u$ and $\alpha_u|Y_u$ is a $(u^2-(n+1)u)$-CPP representation of L . Then $\phi_{n+1}(x)$ iff $x \in Y_u$ for some u . □

5. CONSEQUENCES

Using the theorems in §4 and the examples in §3 we can obtain interesting results about intermediate structure lattices. Before stating these we make two remarks about the examples from §3, the first being relevant when applying Theorem 4.1 and the second when applying Theorem 4.3.

Remark 5.1. Suppose $n < \omega$ and L is a finite lattice. The statement "L has an n-CPP representation" can be formalized in the language of PA as a Σ_1 sentence. Thus, if in fact L has an n-CPP representation, then PA $\vdash$ "L has an n-CPP representation". For example, we get from Example 3.3 that for each $n,k < \omega$, PA $\vdash$ "$\Pi(k)$ has an n-CPP representation". It would be quite remarkable if "L has an n-CPP representation" were consistent with PA without actually being true.

Actually, more than what is indicated in Remark 5.1 is true, and this is the point of the second remark.

Remark 5.2. An inspection of the proof of the theorem of Prömel and Voigt [13] which was used in Example 3.3 reveals that it can be carried out on the basis of PA . Thus, whenever $2 \leq k < \omega$, then PA $\vdash \forall x$("$\Pi(k)$ has an x-CPP representation") . Similarly, the proof of the theorem involved in Example 3.2 can be carried out on the basis of PA . Thus, we get that for any finite distributive lattice D , PA $\vdash \forall x$("D has an x-CPP representation") .

As consequences of Example 3.2 and Theorems 4.1 and 4.3, we obtain the following two results. The second one is explicitly stated in [15]; the first is not but could also be derived by the method of [15].

Corollary 5.3. Let M be countable and nonstandard, and let L be a finite distributive lattice. Then M has a cofinal extension N such that $\mathrm{Lt}(N/M) \simeq L$.

Corollary 5.4. Let M be any model and L a finite distributive lattice having a unique atom. Then M has an almost-minimal end extension N such that $\mathrm{Lt}(N/M) \simeq L$.

From Example 3.3 and Theorem 4.1 we can derive the next corollary.

Corollary 5.5. Let M be countable and nonstandard, and let $2 \leq k < \omega$. Then M has a cofinal extension N such that $\mathrm{Lt}(N/M) \simeq \Pi(k)$.

In particular, we can improve upon Paris [12] by taking $k = 3$ to realize the 1-3-1 lattice.

Using Theorem 4.3 with Example 3.3 yields the following.

Corollary 5.6. Let M be any model and $2 \leq k < \omega$. Then M has an almost-minimal end extension N such that $\mathrm{Lt}(N/M) \simeq \mathbf{1} \oplus \Pi(k)$.

The celebrated theorem of Pudlák and Tůma [14] says that every finite lattice is embeddable in some $\Pi(k)$. From this theorem and Corollary 5.6 we get the following corollary. (Example 3.7 could have been used instead of Example 3.3, but the Pudlák-Tůma theorem would still be necessary.)

Corollary 5.7. Let M be any model and L any finite lattice. Then M has an almost-minimal end extension N such that $\mathrm{Lt}(N/M)$ is finite and embeds L .

Using Theorem 4.2 and the Compactness Theorem, we can obtain the following corollary.

Corollary 5.8. Let M be nonstandard and L be any lattice. Then M has a cofinal extension N such that $Lt(N/M)$ embeds L .

Let L be a lattice. A sublattice L_0 of L is convex if whenever $a,b \in L_0$ and $x \in L$ are such that $a < x < b$, then $x \in L_0$.

Corollary 5.9. Let M be a countable structure, L a finite lattice, and L_0 a convex sublattice of L . If M has a cofinal extension N such that $Lt(N/M) \simeq L$, then M has a cofinal extension N_0 such that $Lt(N_0/M) \simeq L_0$.

Proof: Let $L_0 = \{x \in L : a \le x \le b\}$ and let $h : L \to Lt(N/M)$ be an isomorphism. Then $Lt(h(b)/h(a)) \simeq L_0$. But since $M \equiv h(a)$, we get from Theorem 4.1 that M also has a cofinal extension N_0 such that $Lt(N_0/M) \simeq L_0$. □

The lattice pictured at right is a convex sublattice of the lattice pictured in Example 3.4. Therefore, every countable nonstandard M has a cofinal extension N such that $Lt(N/M)$ is isomorphic to this lattice.

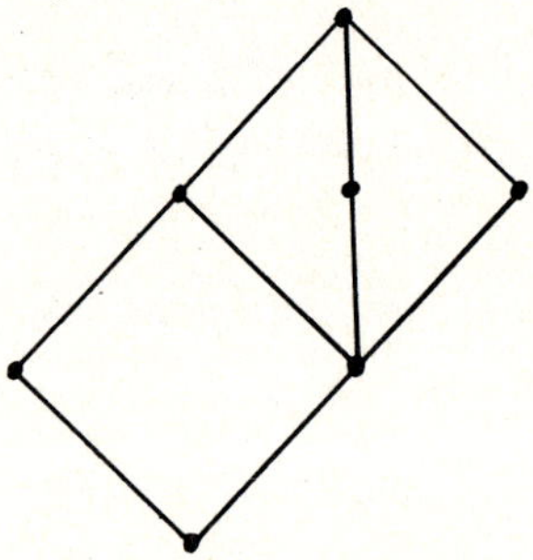

We finally mention that Example 3.6 together with the technique of the proof of Theorem 4.1 and with an additional overspill argument yields the following.

Corollary 5.10. Let M be countable and nonstandard. Then M has a cofinal extension N such that $Lt(N/M)$ is the 1-ω-1 lattice.

6. CONJECTURES

The problem of which lattices, or even which finite lattices, can occur as $Lt(N)$ remains unsettled. As an example, we do not know if the 1-n-1 lattices, where $4 \le n < \omega$, can be realized as substructure lattices, nor do we know about the hexagon lattice. Yet, there is no finite lattice which is known to be impossible to realize. There are finite lattices that cannot be realized as $Lt(N)$, when $N \models TA$. In regard to the question of which finite lattices can occur as $Lt(N)$, where $N \models TA$, Wilkie [17] says "... there is not even an obvious conjecture." We propose here to go out on a limb and make such a conjecture.

Let L be a finite lattice and let $C \subseteq L$ be a linearly ordered subset such that $1 \varepsilon C$. We say that (L,C) is a <u>ranked lattice</u>. Ranked lattices arise in the following manner. Let $M \prec N$, where $Lt(N/M)$ is finite. Let

$$D = \{M_0 \varepsilon Lt(N/M) : N \text{ is an end extension of } M_0\}$$

(so, in particular, $N \varepsilon D$) . Then let $Lt^*(N/M) = (Lt(N/M),D)$, which is a ranked lattice.

Suppose $h : (L,C) \simeq Lt^*(N/M)$ and $x < y \varepsilon L$. Whether or not h(y) is a cofinal extension of h(x) , and whether or not h(y) is an end extension of h(x) , is a property of (L,C) . To see this make the following definition: for each $z \varepsilon L$, set $\bar{z} = \min(\{c \varepsilon C : z \leq c\})$. Then h(y) is a cofinal extension of h(x) iff $\bar{x} = \bar{y}$; and h(y) is an end extension of h(x) iff $x = y \wedge \bar{x}$.

<u>Lemma 6.1</u>. Suppose (L,C) is a finite ranked lattice and $(L,C) \simeq Lt^*(N/M)$. Then the following two properties hold:

(6.1.1) If $x,y \varepsilon L$ and $\bar{x} = \bar{y}$, then $\overline{x \wedge y} = \bar{x}$.

(6.1.2) If $w,x,y,z \varepsilon L$ are such that $x \vee w = y \vee w = z$, $x \wedge w = y \wedge w$ and $w = \bar{w} \wedge z$, then $x = y$.

<u>Proof</u>: The proofs are easy and can be found in [15]. For (6.1.1) see Proposition 1.5 of [15], and for (6.1.2) see Lemma 2.6 of [15]. □

As an example of how Lemma 6.1 is implemented, we will show that the diagrammed 6-element lattice L cannot be realized as $Lt(N/M)$, where N is an end extension of M . For, suppose $(L,C) \simeq Lt^*(N/M)$, where $0,1 \varepsilon C$. Either $x \varepsilon C$ or $w \varepsilon C$ since otherwise (6.1.1) is contradicted. Similarly, $y \varepsilon C$ or $w \varepsilon C$. But C is linearly ordered, so either $x \notin C$ or $y \notin C$. Therefore $w \varepsilon C$, and so $C = \{0,w,1\}$. But then (6.1.2) is contradicted.

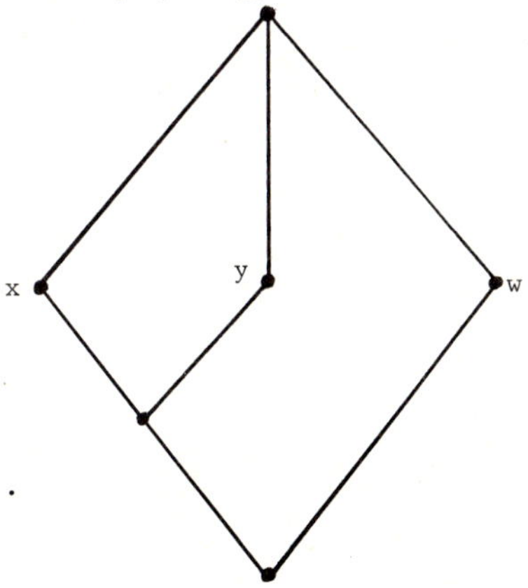

The main result of [15] is that if (L,C) is a finite ranked distributive lattice satisfying (6.1.1) such that $0 \in C$, then each model M has an elementary extension N such that $Lt^*(N/M) \simeq (L,C)$. Only minor changes in that proof are needed to prove the following theorem.

Theorem 6.2. Suppose that (L,C) is a finite ranked distributive lattice satisfying (6.1.1), and suppose that M is a countable nonstandard model. Then M has an elementary extension N such that $Lt^*(N/M) \simeq (L,C)$.

Consequently, for ranked distributive lattices (6.1.2) implies (6.1.1). It is quite easy to give a direct proof of this.

All known examples of finite lattices that cannot be realized by $Lt(N)$, where $N \models TA$, fail because of Lemma 6.1. Encouraged by this and by Theorem 6.2, we make the following bold conjectures.

Conjecture 6.3. Suppose that (L,C) is a finite ranked lattice satisfying (6.1.1) and (6.1.2), and suppose that M is a countable nonstandard model. Then M has an elementary extension N such that $Lt^*(N/M) \simeq (L,C)$.

For any finite lattice L, $(L,\{1\})$ is a ranked lattice satisfying (6.1.1) and (6.1.2). So, in particular, Conjecture 6.3 would imply that for any finite lattice L, every nonstandard countable M has a cofinal extension N such that $Lt(N/M) \simeq L$.

Conjecture 6.4. Suppose that (L,C) is a finite ranked lattice satisfying (6.1.1) and (6.1.2) where $0 \in C$, and suppose that M is any model of PA. Then M has an elementary end extension N such that $Lt^*(N/M) \simeq (L,C)$.

REFERENCES

[1] Blass, A., On certain types and models for arithmetic, J. Symb. Logic 39 (1974), 151-162.

[2] Deuber, W., Graham, R.L., Prömel, H.J. and Voigt, B., A canonical partition theorem for equivalence relations on Z^t, J. Comb. Th. (A) 34 (1983), 331-339.

[3] Deuber, W., Prömel, H.J. and Voigt, B., A canonical partition theorem for chains in regular trees, in: Combinatorial Theory, Lecture Notes in Mathematics 969, Springer-Verlag, 1983, pp. 115-132.

[4] Erdös, P. and Rado, R., A combinatorial theorem, J. London Math. Soc, 25 (1950), 249-255.

[5] Gaifman, H., On local arithmetic functions and their applications for constructing types of Peano's arithmetic, in: Mathematical Logic and Foundations of Set Theory (North-Holland, Amsterdam, 1970), 105-121.

[6] Gaifman, H., A note on models and submodels of arithmetic, in: Conference in Mathematical Logic, London '70, Lecture Notes in Mathematics 255, Springer-Verlag, 1972, pp. 128-144.

[7] Gaifman, H., Models and types of Peano's arithmetic, Annals Math. Logic 9 (1976), 223-306.

[8] Graham, R.L., Rothschild, B.L., and Spencer, J.H., Ramsey Theory, Wiley, New York, 1980.

[9] MacDowell, R., and Specker, R., Modelle der Arithmetic, in: Infinitistic Methods (Pergamon Press and PWN, Warsaw, 1961), 257-263.

[10] Mills, G., Substructure lattices of models of arithmetic, Annals Math. Logic 16 (1979), 145-180.

[11] Paris, J., On models of arithmetic, in: Conference in Mathematical Logic, London '70, Lecture Notes 255, Springer-Verlag, 1972, pp. 252-280.

[12] Paris, J., Models of arithmetic and the 1-3-1 lattice, Fund. Math. 95 (1977), 195-199.

[13] Prömel, H.J., and Voigt, B., Canonical partition theorems for parameter sets, J. Comb. Th.(A) 35 (1983), 309-327.

[14] Pudlák, P., and Tůma, J., Every finite lattice can be embedded in the lattice of all equivalences over a finite set, Algebra Universalis 10 (1980), 74-95.

[15] Schmerl, J.H., Extending models of arithmetic, Annals Math. Logic 14 (1978), 89-109.

[16] Shelah, S., End extensions and number of countable models, J. Symb. Logic 43 (1978), 550-562.

[17] Wilkie, A., On models of arithmetic having non-modular substructure lattices, Fund. Math. 95 (1977), 223-237.

LOGIC COLLOQUIUM '84
J.B. Paris, A.J. Wilkie, and G.M. Wilmers (Editors)
Elsevier Science Publishers B.V. (North-Holland), 1986

DECIDABLE THEORIES OF VALUATED ABELIAN GROUPS

P. H. Schmitt

Universität Heidelberg
Im Neuenheimer Feld 294
6900 Heidelberg
W. Germany

We introduce the class of tamely p-valuated abelian groups and prove its decidability by a relative quantifier elimination procedure.

SECTION 0: INTRODUCTION

Somewhere in the seventies the concept of a valuated Abelian group emerged and quickly established itself as a ubiquiteous and promising research topic in Abelian group theory. For a prime p a p-valuation v is a function from an Abelian group G into the ordinals plus the symbol ∞ (this situation will be refined a little in Section 1 below) satisfying the axioms:

(1) $v(pg) > v(g)$ if $v(g) < \infty$

(2) $v(g-h) \geqq \min\{v(g), v(h)\}$.

These are generalizations of the axioms characterizing the p-height function. A good idea of this concept is conveyed by the theorem proved in [6] that for every p-valuation v on G there is a supergroup H of G such that v coincides on G with the p-height function on H. For more information see [3], [5], [6].

We want to start in this paper the model theoretic investigations of p-valuated Abelian groups, which we view as two-sorted structures $(G, \alpha \cup \{\infty\}, v)$ with α an ordinal and v a function from G into $\alpha \cup \{\infty\}$. In [7] we already obtained some undecidability results. Here we introduce the notion of a <u>tamely p-valuated</u> Abelian group (G,α,v) by the requirements:

(1) G is torsion free

(2) $v(pg) = v(g) + 1$ for $v(g) < \infty$

(3) for all primes q and all $s \geqq 1$ q^sG is a nice subgroup of G i.e. for every $g \in G$ there is some h such that $v(g+q^sh)$ is maximal among $v(g+q^sh')$.

These axioms were chosen in a minimal way to prevent an undecidability proof by the methods of [7] and we obtain indeed as Theorem 5.3 below:

The elementary theory of tamely p-valuated Abelian groups is decidable.

By adding countably many unary predicates to the value sort we can eliminate group quantifiers in favor of value quantifiers (relative quantifier elimination Theorem 3.1). Decidability is obtained from this by observing the decidability of the theory of well-orderings with countably many unary predicates (Theorem 5.2) and by determining which of these structures arise from tamely p-valuated Abelian groups (Lemma 5.1). The relative quantifier elimination result also leads to an axiomatization of the class of direct sums of p-valuated rank one groups satisfying $v(pg) = v(g) + 1$ for $v(g) < \infty$ and of the class of free valuated Z_p-modules (introduced in [6]).

We also consider the class of valuated p^s-groups which are defined as direct sums of copies of cyclic groups of order p^s with a p-valuation w satisfying:

If for $0 < m < s$ and $w(p^m g) = \gamma$
then there is some h satisfying $p^m h = p^m g$ and $\gamma = w(h) + m$.

Valuated p^s-groups occur basically as quotients G/p^sG of tamely p-valuated Abelian groups. We obtain in the course of the proof of Theorem 3.1 also a relative quantifier elimination Theorem for this class and subsequently decidability of its theory.

SECTION 1: PRELIMINARIES

We use standard notation concerning group theory as established e.g. in L. Fuchs' monograph on infinite Abelian groups.

$|X|$ denotes the cardinality of the set X. □ signals the end of a proof.

Let p be a prime.

Definition: A p-valuated Abelian group is a two-sorted structure $(G, \alpha \cup \{c_0, c_1\}, v)$ where

G is an Abelian group

$\alpha \cup \{c_0, c_1\}$ is a linearly ordered set with α a well-ordering and for all $\gamma \in \alpha : \gamma < c_0 < c_1$

v is a mapping from G into $\alpha \cup \{c_0, c_1\}$ satisfying the axioms:

(V1) $v(pg) > v(g)$ for all g such that $v(g) < c_0$

(V2) $v(g-h) \geqq \min\{v(g), v(h)\}$

(V3) $v(g) = c_1$ iff $g = 0$.

In the following "group" will always mean Abelian group. We will write (G,α,v) for p-valuated groups the presence of the additional elements c_0, c_1 being implicitely understood; or even (G,v) when α is clear from the context.

The two-sorted language L_V which we will be using contains

(I) group variable $x,y,z,\ldots$
group function symbols and constant $+,-,0$

(II) value variables ξ, υ, η
binary relation symbol $<$, constant symbols c_0, c_1

(III) the function symbol v.

Terms built up from group function and group constant symbols are called group terms. The constants c_0, c_1 and strings $v(t)$ for any group term t are called value terms. We denote by $L_<$ the language of linear orders containing the symbols listed under (III).

As an immediate consequence of the axioms we obtain:

Lemma 1.1.

(a) if $v(g) < v(h)$ then $v(g+h) = v(g)$

(b) $v(n\cdot g) = v(g)$ for $(n,p) = 1$. □

The most important notion in analysing the elementary properties of p-valuated groups G is the m-part v_m of v for every natural number $m > 1$.

Definition: $v_m(g) = \min\{\gamma :$ there is no $h \in G$ such that $v(g+mh) \geqq \gamma\}$.

In order that $v_m(g)$ be defined for all elements g we add a new value denoted by c_2 such that $c_1 < c_2$.

Lemma 1.2.

(a) $v_m(g) > v(g)$

(b) $v_m(g) = c_2$ iff g is divisible by m

(c) $v_m(g+mh) = v_m(g)$

(d) $v_m(g-h) \geqq \min v_m(g), v_m(h)$

(e) If $v_m(g) < v_m(h)$ then $v_m(g+h) = v_m(g)$

(f) If $(n,p) = 1$ then $v_{m\cdot n}(n\cdot g) = v_m(g)$

(g) If $(m,n) = 1$ then $v_m(ng) = v_m(g)$

(h) If $(m,n) = 1$ then $v_{n\cdot m}(g) = \min\{v_n(g), v_m(g)\}$

(i) $v_m(g)$ is a limit number or one of the elements c_0, c_1, c_2 if $(m,p) = 1$.

Proofs:

(a) - (e) are immediate from the definition.

(f) By Lemma 1.1 (b) we have $v_m(g) \geqq v_{m\cdot n}(n\cdot g)$ while the reverse inequality is obvious.

(g) We have in any case $v_m(n\cdot g) \geqq v_m(g)$. If now $v(ng + mh) \geqq \gamma$ then choose k,r such that $1 = kn + rm$ and we obtain from $v(nk\cdot g + mk\cdot h) \geqq \gamma$: $v(g + m(k\cdot h - r\cdot g)) \geqq \gamma$.

(h) Of course $v_{n\cdot m}(g) \leqq v_n(g), v_m(g)$. If on the other hand there are elements h_1, h_2 such that $v(g + nh_1) \geqq \gamma$ and $v(g + mh_2) \geqq \gamma$ then we get for $kn + rm = 1$ $\;v(g + nm(rh_1 + kh_2)) \geqq \gamma$.

(i) It suffices to show for each $\beta < \alpha$ that $v(g + mh) \geqq \beta$ implies the existence of some h' such that $v(g + mh') \geqq \beta+1$. Let $1 = kp + rm$ then $v(pkg + mpkh) > \beta$ and we may take $h' = pkh - rg$.

<u>Remark</u>: The m-part v_m of a p-valuation need not be again a p-valuation, axiom (V1) may be violated, even for $m = p$.

<u>Notation</u>: We write $v_{p,s}$ instead of v_m for $m = p^s$.

<u>Definition</u>: A subgroup H of a p-valuated group (G,v) is <u>nice</u> if for every $g \in G$ there is some $h \in H$ such that for all $h' \in H$ $v(g+h) \geqq v(g+h')$.

<u>Remark</u>: Nice subgroups are exactly the kernels of homomorphisms in the category of p-valuated groups.

<u>Definition</u>: For a p-valuated group (G,v), $m > 1$ and γ a value, we set:

$$G_v(\gamma) = \{g \in G : v(g) \geqq \gamma\}$$

$$G_v(m,\gamma) = \{g \in G : v_m(g) \geqq \gamma\} .$$

When there is no danger of confusion we will simply write $G(\gamma)$ and $G(m,\gamma)$.

Lemma 1.3.

(i) p^sG is a nice subgroup of (G,α,v)
iff
$v_{p,s}(g)$ is never a limit value.

(ii) For m with $(m,p) = 1$ mG is a nice subgroup of (G,α,v)
iff
$v_m(g)$ is never a limit of limits.

Proofs. obvious. □

Remark: The element c_0 is considered a limit value, c_1, c_2 are not. c_0 is considered a limit of limits if there is no greatest limit number in α.

Definition: A p-valuated group (G,α,v) is called tamely p-valuated if

(T1) G is torsionfree

(T2) $v(pg) = v(g) + 1$ for $v(g) < c_0$

(T3) q^sG is a nice subgroup of (G,α,v) for all primes p and $s \geqq 1$.

Here $\gamma+1$ denotes the successor of the ordinal γ, with the conventions $c_0+1 = c_1$ and $c_1+1 = c_2$ and $\gamma+1 < c_0$ for every $\gamma < c_0$.

Lemma 1.4. Let G be a tamely p-valuated group.

(i) $v_{p\cdot m}(pg) = v_m(g) + 1$ if $v_{p\cdot m}(pg) < c_0$
$v_{p\cdot m}(pg) = v_m(g)$ if $v_{p\cdot m}(pg) \geqq c_0$.

(ii) $v_{p,s}(pg) > v_{p,s}(g)$ if $v_{p,s}(g) < c_0$.

(iii) For $g \in \cap\{p^nG : n \geqq 1\}$ we have $v(g) \geqq c_0$.

(iv) For all primes q different from p
if $v_{q,s}(q^kg) = \gamma$ and $k < s$
then there exists some h satisfying $v_{q,s}(h) = \gamma$ and
$q^k(g-h) \in q^sG$.

(v) If $v_{p,s}(p^kg) = \gamma$ and $k < s$
then there exists some h satisfying $v_{p,s}(h) + k = \gamma$ and
$p^k(g-h) \in p^sG$.

Proofs:

(i) $v_{p\cdot m}(pg) \geqq v_m(g) + 1$ is obvious. Now let $v_{p\cdot m}(pg) = \gamma + 1$ which is by (T3) and Lemma 1.3 (i) always the case for $v_{p\cdot m}(pg) < c_0$. Thus we have for some h: $v(pg + phm) = \gamma$ and therefore by (T2)

$v(g+mh) + 1 = \gamma$. Thus $v_m(g) \geqq \gamma$. If $v_{p\cdot m}(pg) \geqq c_0$ then we must already have $v_{p\cdot m}(pg) = c_2$ by (T3) and (T1).

(ii) immediate consequence of (T3), Lemma 1.3 (ii) and (V1).

(iii) Assume $v(g) < c_0$, say $v(g) = \lambda+k$ for λ a limit number and $0 \leqq k < \omega$. By assumption there is some h satisfying $g = p^{k+1}h$ which yields the contradiction $v(g) = v(h) + k + 1$.

(iv) Let us first consider the case $\gamma = \omega(\beta+1) < c_0$. There is some h_0 such that $v(q^k g + q^s h_0) \geqq \omega(\beta+1)$. Thus $v_{q,s}(q^k g) = v_{q,s}(q^k g + g^s g_0) \geqq$ $\geqq v_{q,s}(g + q^{s-k}h_0) > v(g + q^{s-k}h_0) = v(q^k g + q^s h_0) \geqq \omega\cdot\beta + 1$. Since $v_{q,s}(g + q^{s-k}h_0)$ can only be a limit number $v_{q,s}(g + q^{s-k}h_0) = \gamma$ and we set $h = g + q^{s-k}h_0$. The argument for $\gamma = c_0$ in case there is a largest limit number $< c_0$ is analogous. If $v_{q,s}(q^k g) = c_1$ then $v(q^k g + q^s h_0) = v(g + q^{s-k}h_0) = c_0$ and we must have $v_{q,s}(g + q^{s-k}h_0) = c_1$ since $v_{q,s}(g + q^{s-k}h_0) = c_2$ is impossible. Thus we use $h = g + q^{s-k}h_0$. In case $v_{q,s}(q^k g) = c_2$ we may take any $h \in q^s G$.

(v) analogously to (iv). We observe that $v_{ps}(g) = c_1$ is only possible for $g \notin pG$.

If is of course possible to iterate the process that leads from the p-valuation v to its m-part v_m and consider mappings $v_{m;n}$. The next lemma records that nothing new is obtained in this way. Here we deal only with the q^s-parts for primes q different from p. The corresponding results for p^s-parts are contained in section 4, Lemma 4.2.

Lemma 1.5.

(i) $v_{k;m\cdot n}(g) = \min\{v_{k;m}(g), v_{k;n}(g)\}$ for $(m,n) = 1$.

(ii) $v_{m;n}(g) = c_3$ for $(m,n) = 1$.

(iii) $v_{q,s;q,t}(g) = v_{q,s}(g) + 1$ if $t \geqq s$.

In tamely p-valuated groups we have in addition

(iv) $v_{q,s;q,t}(g) = v_{q,s}(q^{s-t}g) + 1$ if $1 \leqq t \leqq s$.

Proofs:

(i) obvious.

(ii) It suffices to notice that by definition $v_{m;n}(g) = c_3$ if there exists an element h such that $g + nh \in mG$.

(iii) It suffices to observe that $v_{q,s}(g) > \gamma$ implies

$v_{q,s;q,t}(g) > \gamma+1$ which is a trivial consequence of the definition and second that $v_{q,s;q,t}(g) > \gamma$ implies $v_{q,s}(g) \geqq \gamma$ which follows from $t \geqq s$.

(iv) One easily verifies

(a) $v_{q,s;q,t}(g) > \gamma$ implies $v_{q,s}(q^{s-t}g) \geqq \gamma$, and

(b) $v_{q,s}(q^{s-t}g) > \gamma$ implies $v_{q,s;q,t}(g) > \gamma+1$.

To continue let $v_{q,s}(q^{s-t}g) = \gamma$. Then we get by (a) and (b) $\gamma \leqq v_{q,s;q,t}(g) \leqq \gamma+1$. Let us first assume $\gamma < c_o$, thus $\gamma = \omega(\beta+1)$ for some β. Let δ be some ordinal satisfying $\omega\cdot\beta < \delta < \gamma$. For some h we have $v_{q,s}(g+q^t h) > \delta$. Thus we must already have $v_{q,s}(g+q^t h) \geqq \gamma$ and therefore $v_{q,s;q,t}(g) = \gamma+1$. Similarly $v_{q,s}(q^{s-t}g) = c_o$ is handled while in case $v_{q,s}(q^{s-t}g) = c_1$ or c_2 the result follows directly from (a) and (b).

The dimensions $r_q(\gamma) = \dim(G(q,\gamma)/G(q,\gamma+1))$ will turn out to be important elementary invariants for tamely p-valuated groups. We extend the basic language L_V to L_V^* which will allow us to speak of these dimensions in quantifier-free formulas. L_V^* contains in addition to the symbols of L_V:

unary value predicates	$r_{q,n}$	for all primes q, $n \geqq 1$
unary function symbols	v_m	for $m \geqq 2$ denoting the m-part of v
a constant symbol	c_2 .	

We need only explain the interpretation of $r_{q,n}$:

$(G,v) \models r_{q,n}(\gamma)$ iff $r_q(\gamma) \geqq n$ in case $\gamma < c_1$

and $r_{q,n}(c_2)$ is definied to be false for all n.

Similarly $L_<^*$ is $L_<$ enriched by $r_{q,n}$ and c_2.

For a p-valuated group (G,v) we denote ist L_V^*-expansions by $(G,v)^*$ and its value-part by $Val^*(G,v)$.

<u>Lemma 1.6</u>. L_V^* is a definitional extension of L_V.

Proof: clear. □

We let TV_p^* denote the L_V^*-theory of tamely p-valuated groups.

Let $\{(G_i,\alpha_i,v_i): i \in I\}$ be a family of p-valuated groups and $\alpha = \sup\{\alpha_i : i \in I\}$. The direct sum $(G,\alpha,v) = \oplus\Sigma\{(G_i,\alpha_i,v_i): i \in I\}$ is obtained by taking G to be the direct sum of the family $(G_i : i \in I)$ of groups and setting for $g \in G$:

$$v(g) = \min\{v_i(g_i) : i \in I\} .$$

It is easily verified that

Lemma 1.7.

(i) The class of tamely p-valued groups is closed under direct sums.

(ii) For $(G,\alpha,v) = \oplus \sum \{(G_i,\alpha_i,v_i) : i \in I\}$, all primes q and values γ: $r_q^G(\gamma) = \sum \{r_q^{G_i}(\gamma) : i \in I\}$.

Lemma 1.8. For tamely p-valued groups (G,v):

(i) For all primes q different from p and $s \geqq 1$, $\gamma \leqq c_o$:

$$G(q^s,\gamma)\cap q^{s-1}G/G(q^s,\gamma+1)\cap q^{s-1}G \simeq G(q^{s+1},\gamma)\cap q^sG/G(q^{s+1},\gamma+1)\cap q^sG.$$

(ii) For all $s \geqq 1$ and $\gamma < c_o$:

$$G(p^s,\gamma)\cap p^{s-1}G/G(p^s,\gamma+1)\cap p^{s-1}G \simeq G(p^{s+1},\gamma+1)\cap p^sG/G(p^{s+1},\gamma+2)\cap p^sG$$

Proof: (i) By Lemma 1.2 (f) & (g) the required isomorphism is induced by multiplication with q. (ii) The required isomorphism is induced by multiplication with p. - Use Lemma 1.4 (ii). □

The following Lemma is an easy consequence of the definition of $v_{q,s}$:

Lemma 1.9. For tamely p-valued groups (G,v), all primes $q, s \geqq 1$ and $\gamma < c_o$

$$G(q^s,\gamma+1)\cap q^{s-1}G/G(q^s,\gamma+2)\cap q^{s-1}G \simeq G(\gamma)\cap q^{s-1}G/G(\gamma+1)+(q^sG\cap G(\gamma)).\square$$

The dimension of $G(\gamma)/G(\gamma+1)$ is certainly an elementary invariant. The next lemma shows how it can be expressed in terms of the invariants we chose.

Lemma 1.10. For tamely p-valued groups, $\gamma < c_o$:

(i) $G(\gamma+1)/G(\gamma+2) \simeq G(\gamma)/G(\gamma+1) \oplus G(p,\gamma+2)/G(p,\gamma+3)$.

(ii) If $\gamma = \lambda + k$ with λ a limit number, $0 \leqq k < \omega$ then

$$G(\gamma)/G(\gamma+1) \simeq \oplus \sum_{0\leqq i\leqq k} G(p,\lambda+i+1)/G(p,\lambda+i+2) .$$

Proof: (i) $G(\gamma+1)/G(\gamma+2)$ is a group of exponent p, thus $G(\gamma+1) \cap pG/G(\gamma+2) \cap pG$ is a direct summand. Using 1.9 its complementary summand is seen to be isomorphic to $G(p,\gamma+2)/G(p,\gamma+3)$. Finally $G(\gamma+1) \cap pG/G(\gamma+2) \cap pG$ is by axiom (T2) isomorphic to $G(\gamma)/G(\gamma+1)$.

(ii) Iterate (i) and notice that $G(\lambda) \cap pG/G(\lambda+1) = \{0\}$ for λ a limit number. □

The next lemma shows that also the Smielew-invariante G/qG are expressible in terms of the predicates $r_{q,n}$.

Lemma 1.11. For tamely p-valuated groups we have for all primes q

$$\dim(G/qG) = \sum_{\gamma \leqq c_1} r_q(\gamma) \pmod{\infty}$$

(i.e. either both cardinals are finite and then equal or both are infinite).

Proof: For every $\gamma \leqq c_1$ let $\{h_i^\gamma : i \in I_\gamma\}$ be representatives for a basis of $G(q,\gamma)/G(q,\gamma+1)$ considered as a vector space over the field F_q with q elements. We may assume that $v(h_i^\gamma) + 1 = \gamma$, if $q = p$ and $v(h_i^\gamma) + \omega = \gamma$ if $q \neq p$. We claim

(1) $H_o = \{h_i^\gamma : \gamma \leqq c_1, i \in I_\gamma\}$ is q-independent.

Assume this were not the case and we have $\sum_{i,\gamma} n_i^\gamma \cdot h_i^\gamma \in qG$ with not all coefficients n_i^γ divisible by q. Let γ be the smallest value such that for some $i \in I_\gamma$ q does not divide n_i^γ. Then $\sum_{i\in I_\gamma} n_i^\gamma \cdot h_i^\gamma \in G(\gamma) + pG$, i.e. $\sum_{i\in I_\gamma} n_i^\gamma \cdot h_i^\gamma \in G(q,\gamma+1)$ which is a contradiction. This establishes already (1) and the inequality $\geqq$ in the claim of the lemma. If H_o is infinite we are thus finished. If H_o is finite we claim

(2) $G = H + pG$ where H is the subgroup generated by H_o.

For $g \in G$ we find by repeated application of the definition of H_o elements $0 = h_o, h_1, \ldots, h_i \in H$ such that $\gamma_{i+1} = v_q(g-h_i) > \gamma_i$. Since we assumed H_o to be finite also the range of v_q is finite and we will have for some i $v_q(g-h_i) = c_2$ i.e. $g-h_i \in qG$.

From (2) we get by modularity $G/qG \simeq H/H \cap pG$. Since H is a q-pure subgroup of G we have in addition $H/H \cap pG \simeq H/qH$. Since H_o is a basis for H we are finished. □

Lemma 1.12. Let (G,v) be tamely p-valuated.

(i) For every g with $v(g) < c_o$ there is some n such that $v_{p,m}(g) = v(g) + 1$ for all $m \geqq n$.

(ii) If for some limit number $\lambda < c_o$, $G \models r_{q,1}(\lambda+\omega)$ for a prime $q \neq p$, then G satisfies $r_{p,1}(\gamma)$ for some γ, $\lambda < \gamma < \lambda+\omega$.

Proof: Because of $v_{p,n+1}(g) \leq v_{p,n}(g)$ there must be some n such that for all $m \geqq n$ $v_{p,m}(g) = v_{p,n}(g) = \delta$. Assume $v_{p,m}(g) > v(g) + 1$ for some $m \geqq n$. For some $m' \geqq m$ there is thus an element $h_{m'}$ such that $v(g + p^{m'} h_{m'}) = \delta > v(g) + 1$, in particular $v(g) = v(p^{m'} h_{m'})$. If $v(g) = \lambda + k$ with λ a limit number, $0 \leqq k < \omega$ this is for $m' > k$ a contradiction to Lemma 1.4 (v).

(ii) The assumption $r_{q,1}(\lambda+\omega)$ implies the existence of some element h such that $\lambda \leqq v(h) < \lambda + \omega$. By (i) we have $v_{p,n}(h) = v(h) + 1$ for some n. If $n = 1$ then we are done; otherwise we may assume $v_{p,n-1}(h) > v(h) + 1$ which yields the existence of an element h' satisfying $v(h + p^{n-1}h') > v(h)$. In particular $v(h) = v(p^{n-1}h') = v(h') + n-1$. If $v_{p,n}(h') > v(h') + 1$ then $v(h' + ph'') > v(h')$ for some h'' which implies $v(p^{n-1}h' + p^n h'') > v(h)$ and therefore the contradiction $v(h + p^n h'') > v(h)$. Thus $v_p(h') = v(h') + 1$ and G satisfies $r_{p,1}(v(h') + 1)$. □

SECTION 2: GROUPS WITH MORE THAN ONE VALUATION

The purpose of this section is to show that admitting more than one p-valuation, be it for different primes or the same, yields undecidability even for very restricted classes.

We first present a hereditarily undecidable auxiliary class AUX which is designed so that it can easily be interpreted in theories of valuated groups. The language of AUX contains three unary relation symbols X_1, X_2, X_3 and three binary relation symbols $<, R_1, R_2$. AUX will be the class of structures $(A, X_1, X_2, X_3, <, R_1, R_2)$ satisfying the following conditions:

(1) $(A,<)$ is a well-ordering.

(2) X_i is strictly less then X_{i+1} for $i = 1,2$
i.e. for all $a \in X_i$, $b \in X_{i+1}$ we have $a < b$.

(3) $X_1 \cup X_2 \cup X_3 = A$ and for every i $(X_i,<)$ is of order type $\leqq \omega$.

(4) For all a,b in A $R_1(a,b)$ implies $a \in X_1$, $b \in X_2$ and $R_2(a,b$ implies $a \in X_1$, $b \in X_3$.

(5) R_i is minimumclosed for $i = 1,2$, i.e. for all a,b,c,d $R_i(a,b)$ and $R_i(c,d)$ implies $R_i(\min(a,c),\min(b,d))$.

Lemma 2.1. AUX is hereditarily undecidable.

Proof: We aim to interpret the theory of two equivalence relations E_1, E_2 in AUX using the formulas

$$\psi_i(x,y) = \exists z(R_i(x,z) \& R_i(y,z) \& \forall u(R_i(x,u) \& R_i(y,u) \to u \leqq z) \vee x=y) \& \\ \& X_1(x) \& X_1(y) .$$

We notice that in very model A of the theory of AUX ψ_i defines an equivalence relation on $X_1(A)$. Let on the other hand two equivalence relations E_1, E_2 on the set B be given. We may assume w.l.o.g. that B is countable and $B = \{n : n < k\}$ for some $k \leqq \omega$. Let $\{C_{j,i} : i < k_j\}$ for some $j = 1,2$, $k_j \leqq \omega$ be enumerations without repetitions for all E_j equivalence classes. We construct a model A of AUX by taking $(X_1,<)$ to be $(k,<)$ and $(X_2,<)$ (resp. $(X_3,<)$) to be $(k_1,<)$ (resp. $(k_2,<)$). Thus the universe of A is the ordinal $k + k_1 + k_2$ and the order relation of A is the natural ordering. The relations R_j for $j = 1,2$ are defined on A by:

$R_1(n,m)$ iff $n < k$, $m = k+i$ for some $i < k_1$ and $i < l$ where $C_{1,l}$ is the E_1 equivalence class of n.

$R_2(n,m)$ iff $n < k$, $m = k + k_1 + i$ for some $i < k_2$ and $i < l$ where $C_{2,l}$ is the E_2 equivalence class of n.

It is easily seen that the structure A defined in this way satisfies requirements (1)-(5). Furthermore we have for $j = 1,2$ and $n_1, n_2 < k$

$$A \models \psi_j(n_1 \cdot n_2) \quad \text{iff} \quad E_j(n_1,n_2) \; . \qquad \square$$

<u>Lemma 2.2</u>. Let p_1, p_2 be primes, different or not. Let K be the class of all structures (G,v_1,v_2) where for $j = 1,2$ (G,v_j) is a tamely p_j-valuated group. Then K is hereditarily undecidable.

Proof: We will interpret AUX in the theory of K. We will use $\sigma_1,\sigma_2,\sigma_3$ to interpret X_1,X_2,X_3. Since some of these sets may be finite we first fix formulas σ_i' which will later be narrowed down to σ_j.

$$\sigma_1'(\gamma) = \text{"}\exists\beta(0 \leqq \beta < \omega \;\&\; \gamma = \omega\cdot\beta)\text{"}$$

$$\sigma_2'(\gamma) = \text{"}\exists\beta(\omega \leqq \beta < \omega\cdot 2 \;\&\; \gamma = \omega\cdot\beta)\text{"}$$

$$\sigma_3'(\gamma) = \text{"}\exists\beta(\omega\cdot 2 \leqq \beta < \omega\cdot 3 \;\&\; \gamma = \omega\cdot\beta)\text{"}$$

The ordering relation of AUX will be interpreted by the restriction to $\sigma_0 = \sigma_1 \vee \sigma_2 \vee \sigma_3$ of the ordering on the ordinals. Finally the relations R_j will be interpreted by φ_j as follows:

$$\varphi_1(\alpha,\beta) = \sigma_1'(\alpha) \;\&\; \sigma_2'(\beta) \;\&\; \exists x(v_1(x) = \alpha \;\&\; v_2(x) = \beta)$$

$$\varphi_2(\alpha,\beta) = \sigma_1'(\alpha) \;\&\; \sigma_3'(\beta) \;\&\; \exists x(v_1(x) = \alpha \;\&\; v_2(x) = \beta)$$

$$\sigma_1(\alpha) = \exists\beta(\varphi_1(\alpha,\beta) \vee \varphi_2(\alpha,\beta))$$

$$\sigma_2(\alpha) = \exists\beta\varphi_1(\beta,\alpha)$$

$$\sigma_3(\alpha) = \exists\beta\varphi_2(\beta,\alpha)$$

For a given countable structure A in AUX we may assume w.l.o.g. that for $j = 1,2,3$ $X_j \subseteq [\omega(j-1),\omega\cdot j)$. The construction of a valuated group G in K satisfying $A \simeq (\sigma_0(G),\sigma_1(G),\sigma_2(G),\sigma_3(G),<,\varphi_1(G),\varphi_2(G))$ is greatly facilitated by the fact that K is closed under direct sums and the following observation:

> Assume that for every i in some index set I there is a group G_i in K such that for all ordinals α,β $G_i \models \varphi_j(\alpha,\beta)$ implies $R_j(\alpha,\beta)$.
> If furthermore G is the direct sum of the family $(G_i : i \in I)$ then we still have $G \models \varphi_j(\alpha,\beta)$ implies $R_j(\alpha,\beta)$, since the relations R_1,R_2 are minimumclosed.

It thus suffices to find for every pair α,β of ordinals satisfying $R_1(\alpha,\beta)$ in A a valuated group G in K such that

(i) $G \models \varphi_1(\alpha,\beta)$

(ii) for all γ,δ and $j = 1,2$ $G \models \varphi_j(\gamma,\delta)$ implies $R_j(\gamma,\delta)$

and likewise for pairs satisfying $R_2(\alpha,\beta)$.

But this can easily be achieved by taking G to be the integers $\mathbb{Z}$ with $v_1(1) = \alpha$ and $v_2(1) = \beta$. □

SECTION 3: RELATIVE QUANTIFIER ELIMINATION FOR TAMELY p-VALUATED GROUPS

The main result of this section is the following relative quantifier elimination theorem.

Theorem 3.1. Every L_v^*-formula is in TV_p^* equivalent to a formula without group quantifiers.

Corollary 3.2. For every L_v^*-sentence φ there is a $L_<^*$-sentence ψ such that for every tamely p-valuated group (G,α,v)

$$(G,\alpha,v) \models \varphi \quad \text{iff} \quad Val^*(G,\alpha,v) \models \psi .$$

Proof of Corollary 3.2. By the theorem there is a L_v^*-sentence ψ' without group quantifiers which is in TV_p^* equivalent to φ. We only have to get rid of atomic formulas in ψ' involving the group constant O, which can easily be done; we replace, to give just two examples $V(O) = \eta$ by $\eta = c_1$ and $O + O = O$ by $c_0 = c_0$ and so forth. □

Corollary 3.3. For two tamely p-valuated groups (G_1,v_1), (G_2,v_2):

$$(G_1,v_1) \equiv (G_2,v_2) \quad \text{iff} \quad Val^*(G_1,v_1) \equiv Val^*(G_2,v_2) .$$

Proof: follows from 3.2 and Lemma 1.6. □

Lemma 3.4. In order to prove theorem 3.1 it suffices to find for every L_v^*-formula of the form

(3.5) $\exists x\varphi$

where φ is a conjunction of formulas of the form

$$v(t(x)) = \eta \quad , \quad v_m(t(x)) = \eta$$

for $t(x)$ group terms involving x and η a value-variable or one of the constants c_0, c_1, c_2 an TV_p^*-equivalent formula without group quantifiers.

Proof of Lemma 3.4. We prove theorem 3.1 by induction on the complexity of L_v^*-formulas. The only difficult step arises when we are given a formula of the form $\exists x\varphi_0$, where φ_0 does not contain group quantifiers and we are to eliminate $\exists x$. We may certainly assume that all equations between group terms occuring in φ_0 are of the form $t = 0$. These we replace by $v(t) = c_1$. Let S be the set of all group terms in φ_0 involving x and M the set of all natural numbers $m > 1$ such that v_m occurs in φ_0. Then $\exists x\varphi_0$ is equivalent to:

$$(\exists..\xi_t..)(\exists..\xi_t^m..)_{t\in S,m\in M}[\exists x(\bigwedge_{s\in S} v(t)=\xi_t \& \bigwedge_{m\in M}\bigwedge_{t\in S} v_m(t)=\xi_t^m)\&\varphi_0']$$

where φ_0' arises from φ_0 by replacing each occurence of a value term $v(t)$ by the variable ξ_t and likewise $v_m(t)$ by ξ_t^m. Notice that x does no longer occur in φ_0'. □

Lemma 3.6. In order to prove theorem 3.1 it suffices to find an TV_p^*-equivalent formula without group quantifiers for every L_v^*-formula of the form

(3.7) $\exists x\psi$

where ψ is a conjunction of formulas of the form

$$v_m(t(x)) = \eta$$

with varying $m \geq 2$, group terms $t(x)$ and η a value variable or one of the constants c_0, c_1, c_2.

Proof: We will in a series of reductions replace a formula of type (3.5) by a Boolean combination of formulas without group quantifiers or of type (3.7).

In order to keep notation down to managable size we will argue rather informally. In particular we will

- frequently usethe trick to replace $\exists x\varphi$ by the conjunction of $\exists x(\varphi\&\psi_i)$ for $1 \leq i \leq k$ with $\psi_1 \vee \ldots \vee \psi_k$ a consequence of the axioms

- not display assumptions which do not involve the quantified variable after they have been explained once
- switch to semantic verifications without extra warning.

Let us first fix in greater detail our point of departure:

$$(3.8) \qquad \exists x(\bigwedge_{i\in I_1} v(n_i x + t_i) = \eta_i \ \& \bigwedge_{i\in I_2} v_{m_i}(n_i x + t_i) = \eta_i)$$

where t_i are terms not involving x and η_i are value variables or one of the constants c_0, c_1, c_2 and $I_1 \neq \emptyset$.

Using torsionfreeness, Axiom (T2) and Lemmas 1.1 (b), 1.2 (b), 1.4 (i) we may assume that for all $i \in I_1 \cup I_2$: $n_i = n$. Replacing nx by x and adding $v_n(x) = c_2$ we may even assume for all $i \in I_1 \cup I_2$ $n_i=1$ (using Lemma 1.2 (f), (g)). At this point we see that we may assume $\eta_i < c_1$ for all $i \in I_1$. If $\eta_i = c_1$ occurs, we eliminate x by replacing all occurences of x by the term $-t_i$. If $\eta_i = c_2$ occurs the formula is contradictory and thus equivalent to, say $c_0 \neq c_0$. By distinguishing cases we may assume that some order between the η_i is fixed. Let $\eta = \max\{\eta_i : i \in I_1\}$. For $i \in I_1$ with $\eta_i < \eta$ we have by Lemma 1.1 (a)

$$v(x+t_i) = \eta_i \quad \text{iff} \quad v(x+t) = \eta \ \& \ v(t-t_i) = \eta_i \ .$$

Since $v(t-t_i) = \eta_i$ does no longer contain x we may assume w.l.o.g. that for all $i \in I_1$: $\eta_i = \eta$. Using Lemma 1.2 (b) and 1.4 (i) again we may assume furthermore that for all $i \in I_2$ $m_i = m$. Now the arguments of v_m are terms $s_i(x)$ which will in general not be of the form $x + t_i$, but this will no longer be of importance here. It seems appropriate to give an update of the formula (3.8) after all the above reductions have been performed:

$$(3.9) \qquad \exists x(\bigwedge_{i\in I_1} v(x+t_i) = \eta \ \& \bigwedge_{i\in I_2} v_m(s_i(x)) = \eta_i) \quad .$$

By a variable transformation $x \to x+t_i$ we may assume that for some $i \in I_1$ $t_i = 0$. Thus (3.9) is equivalent to the disjunction of the following two formulas:

$$(3.10) \quad \exists x(v(x)=\eta \ \& \bigwedge_{i\in I_1} v(x+t_i)=\eta \ \& \bigwedge_{i\in I_2} v_m(s_i(x))=\eta_i \& v_m(x)>\eta+1)$$

$$(3.11) \quad \exists x(v(x)=\eta \ \& \bigwedge_{i\in I_1} v(x+t_i)=\eta \ \& \bigwedge_{i\in I_2} v_m(s_i(x))=\eta_i \& v_m(x)=\eta+1) \ .$$

We will show that (3.10) is equivalent to the conjunction of the following two formulas:

$$(3.12)\quad \exists x\big(\bigwedge_{i\in I_1} v(x+t_i)=\eta \ \&\ v(x)=\eta \ \&\ v_m(x)>\eta+1\big)$$

$$(3.13)\quad \exists x\big(\bigwedge_{i\in I_2} v(s_i(x))=\eta_i \ \&\ v_m(x)>\eta+1\big)\ .$$

One of the implications of this equivalence is trivial. For the reverse implication we assume that g_0, resp. g_1 is a witness for the existential quantifier in (3.12) resp. in (3.13).

Since $v_m(g_0) > \eta+1$ is true we have $v(g_0-nh) \geqq \eta+1$ for some h. Thus

$$\bigwedge_{i\in I_1} v(nh+t_i)=\eta \ \&\ v(nh)=\eta\ .$$

Since also $v_m(g_1) > \eta+1$ we may by Lemma 1.2 (c) assume that $v(g_1) \geqq \eta+1$. Thus g_1+nh will be a witness for (3.10).

Let us next deal with (3.12).

We may assume that $\eta \leqq c_0$, since otherwise (3.12) is contradictory. Since for $i \neq j$ $v(x+t_i)=\eta \ \&\ v(t_i-t_j)>\eta$ implies $v(x+t_j)=\eta$ and $v(x)=\eta \ \&\ v(t_j)>\eta$ implies $v(x+t_j)=\eta$ we may assume

(3.14) for all $i,j \in I_1$ with $i \neq j$: $v(t_i-t_j)=\eta \ \&\ v(t_i)=\eta$.

For $i \in I_1$ $v(x)=\eta$ and $v(t_i)=\eta$ already imply $v(x+t_i)\geqq\eta$. Thus $v_m(t_i)=\eta+1$ and $v_m(x)>\eta+1$ would imply $v_m(x+t_i)=\eta+1$ and therefore $v(x+t_i)=\eta$. In this case we would drop i from I_1. Thus we may assume

(3.15) for all $i \in I_1$: $v_m(t_i) > \eta+1$.

This leads us to the conclusion that (3.12) is equivalent to

(3.16) $|G(\eta)\cap mG/G(\eta+1)\cap mG| > |I_1|$.

To prove this let g be a witness for (3.12), then we have $v(g-mg_0)\geqq\eta+1$ for some g_0 and by (3.15) also elements t_i^0 for $i \in I_1$ such that $v(t_i-mt_i^0)\geqq\eta+1$. Using the properties of g and (3.14) we see that mg_0, mt_i^0 for $i \in I_1$ represent different elements in $G(\eta)\cap mG/G(\eta+1)\cap mG$.

If on the other hand (3.16) is satisfied we are guaranteed to find an element mg_0 in $G(\eta)\cap mG$, $mg_0 \notin G(\eta+1)$ such that $mg_0+G(\eta+1) \neq -t_i+G(\eta+1)$ for all $i \in I_1$. Thus

$$v(mg_0)=\eta \ \&\ \bigwedge_{i\in I_1} v(mg_0+t_i)=\eta \ \&\ v_m(mg_0)=c_2>\eta+1.$$

By Lemma 1.10 (3.16) can be expressed by an L_v^*-formula without group quantifiers.

Now we take up the further reduction of (3.11).

We may again assume (3.14). But instead of (3.15) we have this time

(3.17) for all $i \in I_1$: $v_m(t_i) = \eta+1$.

For assume to the contrary that $v_m(t_i) > \eta+1$ for some $i \in I_1$, then there would be g_o with $v(t_i + mg_o) \geqq \eta+1$ and $v(mg_o) = v(t_i) = \eta$. Because of $v_m(x) = \eta+1$ we must have $v(x + ng_o) = \eta$ thus $v(x + t_i) = \eta$. We would consequently drop i from I_1. We finally claim that (3.11) is equivalent to

$$(3.18) \quad \exists x (\bigwedge_{i \in I_1} v_m(x+t_i) = \eta \ \& \ \bigwedge_{i \in I_2} v_m(s_i(x)) = \eta \ \& \ v_m(x) = \eta+1) .$$

One implication is easy using (3.17) and Lemma 1.2 (e). If on the other hand g is a witness for (3.18) then $v(g + mg_o) = \eta$ for some g_o. Since by Lemma 1.2 (c) $g + mg_o$ is also a witness of (3.18) we may have assumed $v(g) = \eta$ right away. Since $v(t_i) = \eta$, we have $v(g + t_i) \geqq \eta$ for all $i \in I_1$. But since $v(g + t_i) > \eta$ would contradict $v_m(g + t_i) = \eta$ we must have $v(g + t_i) = \eta$. □

Lemma 3.19. To prove theorem 3.1 it suffices to eliminate the group quantifier in L_v^*-formulas of the form

$$(3.20) \qquad \exists x \psi_q$$

where ψ_q is a conjunction of formulas of the form

$$v_{q,s}(ms + t) = \eta$$

for a fixed prime q, which may be equal to or different form p and varying $m,s \geqq 1$, group terms t without x and η a value variable or one of the constants c_o, c_1, c_2.

Proof: We start with a formula of type (3.7). Using the same transformations as in the first step in the proof of Lemma 3.6 and Lemma 1.2 (h) we may restrict attention to formulas

$$(3.21) \quad \exists x \bigwedge_{0 \leqq j \leqq k} \psi_{q_j}$$

where the ψ_{q_j} already meet the requirements of (3.20).

It remains to be shown that the existential quantifier distributes over the conjunction. Let s be the highest exponent for which $v_{q,s}$ appears for some prime q in (3.21). Let n be the product of all q_j^s, $j \leqq k$ and $n_j = n/q_j^s$. Then there are integers m_j with $1 = m_o n_o + \ldots + m_k n_k$. If g_j are witnesses for

$$\exists x \psi_{q_j} = \exists x \bigwedge_{i \in I_j} v_{q_j, s_i}(x + t_i) = \eta_i$$

let $g = m_o n_o g_o + \ldots + m_k n_k g_k$. We have by Lemma 1.2 (f), (g)

$$v_{q_j, s_i}(m_j n_j g_j + m_j n_j t_i) = \eta_i$$

for all j and all $i \in I_j$. Since n_{j_0} is divisible by q_j^s for every $j_0 \neq j$ we obtain by Lemma 1.2 (c) $v_{q_j,s_i}(g+t_i) = \eta_i$ for all j and $i \in I_j$. □

In the remainder of this section we will eliminate the group quantifier in formulas of type (3.20) for a prime q different from p. The equal prime case will be dealt with in the next section in slightly greater generality.

Starting from the formula

$$(3.22) \quad \exists x \bigwedge_{i\in I} v_{q,s_i}(m_i q^{r_i} x + t_i) = \eta_i$$
$$\text{with } (m_i,q) = 1 \text{ for all } i \in I$$

we use Lemma 1.2 (b), (f), (g) to obtain for all $i \in I : s_i = s$ and $m_i = m$. The x-free terms t_i will of course change in this procedure. Now we may w.l.o.g. replace m by 1. Indeed if some element b satisfies for all $i \in I$ $v_{q,s}(q^{r_i}b+t_i) = \eta_i$ then we have $1 = n_1 m + n_2 q^s$ for appropriate n_1, n_2 and hence $v_{q,s}(mq^{r_i}n_1 b + t_i) = \eta_i$ by Lemma 1.2 (c).

We will use in the following for notational simplicity w to denote $v_{q,s}$.

Thus it suffices to consider formulas of the form

$$(3.23) \quad \exists x (\bigwedge_{0\leqq i<s} \bigwedge_{t\in M_i} w(q^i x + t) = \eta_{i,t}) .$$

By introducing possibly new existential value quantifiers as we did in the proof of Lemma 3.4 we may assume w.l.o.g.

$$0 \in M_0 \quad\text{and}\quad qM_i \subseteq M_{i+1} \quad \text{for all} \quad i,\ 0 \leqq i < s-1 .$$

By distinguishing cases we may assume that some linear order is fixed among the value-variables and constants $\eta_{i,t}$. With respect to this order we define $\eta_i = \max\{\eta_{i,t} : t \in M_i\}$. If $\eta_i = \eta_{i,t_0}$ and $\eta_{i,t_1} < \eta_i$ for $t_0, t_1 \in M_i$ then $w(q^i x + t_1) = \eta_{i,t_1}$ iff $w(q^i x + t_o) = \eta_{i,t_0}$ and $w(t_0 - t_1) = \eta_{i,t_1}$. Thus t_1 may be dropped from M_i.

$$(3.24) \quad \exists x (\bigwedge_{0\leqq i<s} \bigwedge_{t\in N_i} w(q^i x + t) = \eta_i)$$
$$\text{where } N_i = \{t \in M_i : \eta_{i,t} = \eta_i\} \text{ and } \eta_o \leqq \eta_1 \leqq \dots \leqq \eta_{s-1} .$$

To see why the given chain of inequalities may be assumed, take some $t \in N_i$ for some i, $0 \leqq i < s-1$. Because $qt \in M_{i+1}$ we get $\eta_i = \eta_{i,t} \leqq w(q^{i+1}xqt) = \eta_{i+1,qt} \leqq \eta_{i+1}$. If in the assumed ordering of the η_i we have $\eta_{i+1} < \eta_i$ then we are facing a contradictory formula and its group quantifier can trivially be eliminated. We also assume $\eta_i \neq c_o$ since otherwise (3.24) is again contradictory.

To prepare the most important reduction step we first collect those i and j together for which $\eta_i = \eta_j$. Formally we define:

$$k_o = 0$$

$$k_{j+1} = \begin{cases} \text{the least element in } \{i : k_j \leqq i \leqq s-1 \text{ and } \eta_{k_j} < \eta_i\} \text{ if this set} \\ \text{is not empty} \\ s \quad \text{otherwise} \end{cases}$$

$$r = \text{the least } j \text{ such that } k_{r+1} = s$$

$$K_j = [k_j, k_{j+1}) \subseteq \{0,\ldots,s-1\} .$$

Furthermore we set

(3.25) (a) $\varphi_j = \bigwedge_{i \in K_j} \bigwedge_{t \in N_i} w(q^i x + t) = \eta_j^* \;\&\; w(q^{l_j} x + \tau) > \eta_j^*$

for $0 \leq j < r$, $l_j = k_{j+1}+1$, for all $i \in K_j$, $\eta_i = \eta_j^*$ and τ an arbitrary element from N_{l_j} which will be kept fixed for the remainder of this section.

(b) $\varphi_r = \bigwedge_{i \in K_r} \bigwedge_{t \in N_i} w(q^i x + t) = \eta_j^*$

where for all $i \in K_r : \eta_i = \eta_j^*$.

Theorem 3.26. The formula (3.24) is equivalent to $\bigwedge_{0 \leqq j \leqq r} \exists x \varphi_j$.

Proof: One implication is obvious. For the other assume that $\exists x \varphi_j$ is satisfied for all j, $0 \leqq j \leqq r$. We will prove by induction on m, $0 \leqq m \leqq r$ that this implies

$$\exists x (\bigwedge_{k_{r-m} \leqq i < s} \bigwedge_{t \in N_i} w(q^i x + t) = \eta_i) .$$

For m = 0 this is $\exists x \varphi_r$ while for m = r this is just (3.24).

For the induction step from m to m+1 we assume that there are elements h,g with

(3.27) $\bigwedge_{k_{j+1} \leqq i < s} \bigwedge_{t \in N_i} w(p^i h + t) = \eta_i \quad \text{with} \quad j+1 = r-m$

and $\varphi_j(g)$ which we write in greater detail as:

(3.28) $\bigwedge_{k_j \leqq i < k_{j+1}} \bigwedge_{t \in N_i} w(q^i g + t) = \eta_i \;\&\; w(q^{l+1} g + \tau) > \eta_l$

where $l+1 = k_{j+1}$ and $\tau \in N_{l+1}$.

From $w(q^{l+1} g + \tau) > \eta_l$ and $w(q^{l+1} h + \tau) = \eta_{l+1} > \eta_l$ we get

$w(q^{l+1}(g-h)) > \eta_1$.

By Lemma 1.4 (iv) there is an element c such that

(i) $w(g-h+c) > \eta_1$

(ii) $q^{l+1}c = q^s c'$ for some c'.

Now (3.28) and (i) yields by Lemma 1.2 (e)

$$\bigwedge_{k_j \le i < k_{j+1}} \bigwedge_{t \in N_i} w(q^i(h-c)+t) = \eta_i .$$

On the other hand h occurs in (3.27) at least as a multiple of q^{l+1}, thus (ii) implies that (3.27) is also satisfied by h-c in place of h. Altogether this shows that h-c is a witness for

$$\exists x \bigwedge_{k_{r-m-1} \le i < s} \bigwedge_{t \in N_i} w(q^i x + t) = \eta_i .$$

We may rephrase what we have achieved by saying that it suffices to eliminate the group quantifier in formulas of the form:

(3.29) $$\exists x \bigwedge_{k \le i \le l} \bigwedge_{t \in N_i} w(q^i)x+t) = \eta \ \& \ w(q^{l+1}x+\tau) > \eta$$

where the last conjunct is missing exactly when $l = s-1$.

We may assume that:

(3.29) (a) for all $i,j,k \le i \le j \le l$ and $t \in N_i$, $t' \in N_j$

$$w(q^{j-i}t - t') \ge \eta \quad \text{and} \quad w(q^{l-i+1}t - \tau) \ge \eta$$

since otherwise $w(qx) \ge w(x)$ and Lemma 1.2 (d) would yield a contradiction. □

The further treatment of (3.29) is split into two cases.

Case I: for all $i,k \le i \le l$ and $t \in N_i$ $w(q^{s-i}t) \le \eta$

Case II: complement of Case I.

We may assume for every i and every $t \in N_i$ $w_{q,i}(t) \ge \eta+1$ which by Lemma 1.5 (iv) is equivalent to $w(q^{s-i}t) \ge \eta$. Thus the assumption of Case I actually gives us $w(q^{s-i}t) = \eta$ for all relevant i and t.

Claim (3.30): Under the assumption of Case I formula (3.29) is equivalent to

(a) $\exists x(w(q^{l+1}x+\tau) > \eta)$ if $l+1 < s$

(b) $\exists x(w(q^l x + t_o) \ge \eta)$ if $l+1 = s$, with t_0 an arbitrary element from N_l.

Proof of Claim 3.30. Only one implication of the claimed equivalence is non-trivial. Let g be an element satisfying $w(q^{l+1}g+\tau) > \eta$. Let t_1 be an arbitrary element of N_k. From 3.29 (a) we have $w(q^{l-k+1}t_1-\tau) \geq \eta$ which yields $w(q^{l+1}g+q^{l-k+1}t_1) \geq \eta$. Lemma 1.4 (iv) provides us with an element g_0 such that $w(g_0) \geq \eta$ and $q^{l-k+1}(g_0-t_1) \equiv q^{l+1}$ (mod q^s). By this last congruence we find an element u such that $q^k u \equiv (g_0-t_1)$ (mod q^s) and also $q^{l+1}u \equiv q^{l+1}g$ (mod q^s). Therefore we have $w(q^{l+1}u+\tau) > \eta$. But now we have in addition $w(q^k u+t_1) = w(g_0) \geq \eta$. By 3.29 (a) this gives $w(q^i u+t) \geq \eta$ for all $i, k \leq i \leq l$ and all $t \in N_i$. But $w(q^i u+t) > \eta$ would give $w_{q,i}(t) > \eta+1$ which is by Lemma 1.5 equivalent to $w(q^i t) > \eta$, contradicting the assumptions of Case I. Thus u is a witness for (3.29). □

By definition $\exists x(w(q^{l+1}x+\tau) > \eta)$ is equivalent to $w_{q,l+1}(\tau) > \eta+1$ which is by Lemma 1.5 (iv) equivalent to $w(q^{s-l+1}\tau) > \eta+1$, an L_V^*-formulas without group quantifiers. Likewise $\exists x(w(q^l x+t_0) \geq \eta$ can be replaced by $w(q^{s-l}t_0) \geq \eta$.

We may thus assume from now on Case II.

Let k_0 be the least number $i, k \leq i \leq l$ such that for some $t \in N_{k_0}$ $w(q^{s-k_0}t) > \eta$. We fix some $t_0 \in N_{k_0}$ with this property. We claim that (3.29) is equivalent to

(3.31) $\exists x(\bigwedge_{k_0 \leq i \leq l} \bigwedge_{t \in N_i^*} w(q^i x+t) = \eta \;\&\; w(q^{l+1}x+\tau^*) > \eta)$

where of course the last conjunct is omitted for $l+1 = s$ and $N_i^* = \{t-q^{i-k_0}t_0 : t \in N_i\}$, $\tau^* = \tau - q^{l-k_0+1}t_0$.

To verify this claim we first note that by the choice of t_0 and Lemma 1.5 (iv) there is some t_0' such that $w(t_0 - q^{k_0}t_0') > \eta$ which implies $w(q^{i-k_0}t_0 - q^i t_0') > \eta$ for all $i, k_0 \leq i \leq l$ and $w(q^{l-k_0+1}t_0 - q^{l+1}t_0') > \eta$. From this we see that $h = g+t_0'$ is a witness for (3.31). If on the other hand we are given h satisfying (3.31) then $g = h-t_0'$ certainly satisfies

(3.32) $\bigwedge_{k_0 \leq i \leq l} \bigwedge_{t \in N_i} w(q^i g+t) = \eta \;\&\; w(q^{l+1}g+\tau) > \eta$.

But what about the conjuncts for $i < k_0$. Let t_1 be some arbitrary element in N_k. As in the verification of claim (3.30) we find using Lemma 1.4 (iv) some element u with $q^{k_0}u \equiv q^{k_0}g$ (mod q^s) and $w(q^k u+t_1) \geq \eta$. Thus u satisfies (3.32) in place of g and as before we find

$$\bigwedge_{k_0 \leq i \leq k} \bigwedge_{t \in N_i} w(q^i u+t) = \eta \;. \qquad \square$$

Considering the way we obtained N_i from M_i (see (3.24)) we notice that for all $i,k \leq i < l$ $qN_i \subseteq N_{i+1}$. This yields

(3.33) (i) $qN_i^* \subseteq N_{i+1}^*$ for all $i,k_0 \leq i < l$

(ii) $0 \in N_{k_0}^*$

(iii) for all $i,j,k_0 \leq i \leq j \leq l$, $t \in N_i^*$, $t' \in N_j^*$
$w(q^{j-i}t - t') \geq \eta$.

This allows us to reduce (3.31) further to

(3.34) $$\exists x(\bigwedge_{t \in N} w(q^l x + t) = \eta \;\&\; w(q^{l+1}x + \tau^*) > \eta)$$
where we have used N to abbreviate N_l^* .

Indeed assume that g is a witness for (3.34). Since $0 \in N$ we have in particular $w(q^l g) = \eta$. Using Lemma 1.4 (iv) we may assume w.l.o.g. $w(g) = \eta$, which gives by (3.33) (iii) at least $w(q^i g + t) \geq \eta$ for all $i,k_0 \leq i < l$ and all $t \in N_i$. But $w(q^i g + t) > \eta$ would imply $w(q^l g + q^{l-i}t) > \eta$ which contradicts $q^{l-i}t \in N$. □

We may further assume

(3.33) (iv) $w(t) = \eta$ for all $t \in N$, $t \neq 0$

since we have in any case $w(t) \geq \eta$ and those $t \in N$, $t \neq 0$ for which $w(t) > \eta$ is true may be dropped since $w(q^l x + t) = \eta$ iff $w(q^l x) = \eta \;\&\; w(t) > \eta$.

If $\eta = c_2$ then we may assume that the last conjunct in (3.34) is missing and the remaining formula is equivalent to the purely group theoretic formula

$$\exists x(\bigwedge_{t \in N} \exists y(q^s y = q^l x + t)) .$$

By the results in [1] this formula is equivalent to a quantifier free formula ψ_0 in a language with divisibility predicates and predicates for the Szmielew invariant $\beta(G) = \dim G/q'G$ for all primes q'. But ψ_0 is in TV_p^* equivalent to a quantifier-free formula by Lemma 1.2 (b) and Lemma 1.11.

In case $\eta \leq c_1$ (3.34) is equivalent to the formula

(3.35) $$\exists X(\bigwedge_{t \in N} q^l X \neq -\bar{t} \;\&\; q^{l+1}X = \overline{-\tau^*}$$
in the group $H(\eta) = G(q^s,\eta)/G(q^s,\eta+1)$ with the bar $^-$ denoting the canonical homomorphismus from $G(q^s,\eta)$ in $H(\eta)$.

Again using [1] we know that (3.35) is equivalent in the theory of groups which are direct sums of copies of $\mathbb{Z}(q^s)$ to a quantifier-free formula ψ_1 in the language containing predicates for the relevant Szmielew invariants, which in this case is just the dimension of $H(\eta)[q]$. But ψ_1 is in TV_p^* equivalent to a quantifier free formula since $r_q(\eta) \geq n$ iff $\dim H(\eta)[q] \geq n$ by Lemma 1.8 (i) for $\eta \leq c_0$ and simply checked for $\eta = c_1$. □

SECTION 4: VALUATED p^s-GROUPS

Let p be a prime, $s \geq 1$ an integer. A p-valuated group (G,w) is called a valuated p^s-group if

(K1) G (as a group) is a direct sum of copies of $Z(p^s)$

(K2) For every m, $0 < m < s$ and every value $\beta \leq c_0$, $g \in G$
if $w(p^m g) = \beta$
then there is an element $h \in G$ such that $\beta = w(h) + m$
and $p^m h = p^m g$.

One reason for studying this class of p-valuated groups is given by:

<u>Lemma 4.1</u>. If (G,v) is a tamely p-valuated group then $(G/q^s G, v_{p,s})$ is a valuated p^s-group.

Remark: In defining the structure $(H,w) = (G/p^s G, v_{p,s})$ a renaming of the constants of L_v is needed: c_0^G is omitted, it never is a value for $v_{p,s}$, $c_1^G = c_0^H$ and $c_2^G = c_1^H$.

Proof: Since G is torsionfree (K1) is true. (K2) follows from Lemma 1.4 (v). That $v_{p,s}$ is a p-valuation follows from Lemma 1.4 (ii). □

Since we do not require that in valuated p^s-groups $w(g)$ is never a limit there are valuated p^s-groups which do not derive from tamely p-valuated groups in the way given by lemma 4.1.

<u>Examples</u>.

(E1) Valuated p-groups are valuated vector spaces over the field with p elements in the sense of [2]. L. Fuchs' definition is more general by allowing arbitrary complete linear orderings as set of values.

(E2) Let g be a generator for $Z(p^s)$. We define w by $w(p^i g) = \alpha + i$ for $0 \leq i < s$ and $w(0) = c_1$. Then $(Z(p^s),w)$ is a valuated p^s-group.

(E3) Direct sums of valuated p^s-groups are again valuated p^s-groups.

In studying valuated p^s-groups we will use the extended language

$$L_v^{**} = L_v \cup \{c_2\} \cup \{d_n(\ ) : n \geq 1\}$$

where the interpretation of d_n in the valuated p^s-group (G,w) is given by:

$$G \models d_n(\eta) \quad \text{iff} \quad \dim(G(\eta) \cap p^{s-1}G/G(\eta+1) \cap p^{s-1}G) \quad \text{for} \quad \eta \leq c_o$$

where $G(\eta)$ is the subgroup $\{g \in G : w(g) \geq \eta\}$.

$TV(p,s)$ denotes the L_v-theory of valuated p^s-groups, $TV^*(p,s)$ the L_v^{**}-theory. We need not include in L_v^{**} names for the mappings w_m since they can be expressed in terms of w without using group quantifiers as the next lemma shows.

Lemma 4.2. For valuated p^s-groups (G,w) the following are true for all $g \in G$ and m, $0 < m < s$:

(i) $w_{p,n}(g) = w(g) + 1$ for $n \geq s$

(ii) $w_n(g) = c_2$ for all n prime to p

(iii) if $w_{p,m}(g)$ is a limit number, c_o, c_1 or c_2 then $w_{p,m}(g) = w_{p,m+1}(pg)$

(iv) if $w_{p,m}(g)$ is a successor number $< c_o$ then $w_{p,m+1}(pg) = w_{p,m}(g) + 1$

(v) $w_{p,m}(g) + s - m - 1 = w(p^{s-m}g)$.

Proof:

(i) trivial since $p^m h = 0$ for all $h \in G$.

(ii) trivial since g is divisible by any n prime to p.

Before proving (iii) we note: If $w_{p,m}(g) < w_{p,m+1}(pg)$ then $w(pg + p^{m+1}h) \geq w_{p,m}(g) + 1$ for some h. By axiom (K2) we have for some $g_o \in G$ $w(g_o) \geq w_{p,m}(g) + 1$ and $pg_o = pg + p^{m+1}h$. By (K1) $g_o - g + p^m h \in p^{s-1}G$. Thus $(g_o - g) \in p^m G$ and we obtain the contradiction $w_{p,m}(g) = w_{p,m}(g_o) > w(g_o) \geq w_{p,m}(g) + 1$. Since $w_{p,m}(g) \leq w_{p,m+1}(pg)$ is true for any p-valuation we thus have

$$w_{p,m}(g) \leq w_{p,m+1}(pg) \leq w_{p,m}(g) + 1 \ .$$

(iii) If $w_{p,m}(g) = \beta$ is a limit number or c_o then there can be by (K2) no h such that $w(pg + p^{m+1}h) = \beta$. Thus $w_{p,m+1}(pg) = w_{p,m}(g)$. If $w_{p,m+1}(pg) = c_2$ then $pg \in p^{m+1}G$ which implies by (K1) $g \in p^m G$, i.e. $w_{p,m}(g) = c_2$.

(iv) Let $w_{p,m}(g)$ be $\beta+1$. Then there is some h such that $w(g+p^m h) = \beta$ and thus $w(pg + p^{m+1}h) \geq \beta+1$, i.e. $w_{p,m+1}(pg) \geq \beta+1$.

(v) Follows from (iii) & (iv). □

Lemma 4.3. For all $\eta < c_o$:

(i) $G(\eta)/G(\eta+1)$ is a direct sum of copies of $Z(p)$

(ii) $G(\eta)/G(\eta+1) \simeq G(\eta+l) \cap p^l G/G(\eta+l+1) \cap p^l G$

(iii) If $\eta = \beta\cdot\omega + r$, $0 \leq r < l$ then $G(\eta) \cap p^l G/G(\eta+1) \cap p^l G = \{0\}$

(iv) $G(c_o) \cap p^l G/G(c_1) \cap p^l G = \{0\}$ for $l > 0$.

Proof:

(i) clear, since $w(pg) > w(g)$ for $w(g) < c_o$.

(ii) Since $p^l G(\eta) \subseteq G(\eta+l) \cap p^l G$ the mapping $g \mapsto p^l g$ is a homomorphismus from $p^l G(\eta)$ into $G(\eta+l) \cap p^l G/G(\eta+l+1) \cap p^l G$. By (K2) it is both injective and surjective.

(iii) By (K2) $w(p^l g) \geq \eta$ implies already $w(p^l g) \geq \eta + (l-r)$.

(iv) By (K2) $w(pg) \geq c_o$ implies already $w(pg) = c_1$, i.e. $pg = 0$. □

To apply the results of this section to complete the proof of Theorem 3.1 we need the following observations connecting the invariants $r_{p,n}$ and d_n.

Lemma 4.4. Let (H,v) be a tamely p-valuated group, $(G,w) = (H/p^s H, v_{p,s})$. Then

(i) in (G,w) $d_n(\eta)$ is false for all $n \geq 1$ and η of the form $\eta = \beta\cdot\omega + r$, $0 \leq r < s$ or $\eta = c_1$;

(ii) for $l \geq s$: $(G,w) \models d_n(\eta + l)$ iff $(H,v) \models r_{p,n}(\eta + l - s + 1)$;

(iii) in (G,w) $d_n(c_o)$ is false if $s > 1$ and $(G,w) \models d_n(c_o)$ is equivalent to $(H,v) \models r_{p,n}(c_1)$ if $s = 1$.

Proof: (i) By (K2), (ii) by Lemma 1.8 (ii), (iii) by definition. □

Theorem 4.5. Every L_v^{**}-formula is in $TV^*(p,s)$ equivalent to a formula without group quantifiers.

Lemma 4.6. In order to prove theorem 4.5 it suffices to find for every L_v^{**}-formula of the form

(4.7) $\exists x\varphi$

where φ is a conjunction of formulas of the form $w(t) = \eta$ for group terms t involving x and η a value variable or one of the constants c_o, c_1

an $TV^*(p,s)$-equivalent formula without group quantifiers.

Proof: same as for lemma 3.4. □

The elimination of the group quantifier in formulas of the form (4.7) greatly parallels the reductions performed in section 3 from (3.23) onward where the role of the prime q is now taken over by p. The main difference lies in the fact that we have now $w(pg) > w(g)$ while in section 3 $w(qg) = w(g)$ was possible. For this reason we will give a very sketchy proof of theorem 4.5 indicating only the major steps in the reduction and refering to the corresponding parts of section 3 for the trick to be used.

Starting from (4.7) we obtain as a first reduction

(4.8) $\exists x(\bigwedge_{0 \le i < s} \bigwedge_{t \in M_i} w(p^i x + t) = \eta_{i,t})$

with t group terms not containing x and $0 \in M_o$, $pM_i \subseteq M_{i+1}$ for $0 \le i < s-1$.

This uses only the fact that w is a p-valuation and the possible introduction of new value quantifiers.

Set $\eta_i = \max\{\eta_{i,t} : t \in M_i\}$. Then (4.8) is equivalent to

(4.9) $\exists x(\bigwedge_{0 \le i < s} \bigwedge_{t \in N_i} w(p^i x + t) = \eta_i)$

with $N_i = \{t \in M_i : \eta_{i,t} = \eta_i\}$ and $\eta_o < \eta_1 < \ldots < \eta_{s-1}$ (see (3.24)).

Set

$k_o = 0$

$k_{j+1} = \begin{cases} \text{the least element in } \{i : k_j \le i \le s-1 \ \& \ \eta_{k_j} + i-k_j < \eta_i\} \\ \quad \text{if this set is not empty} \\ s \quad \text{otherwise} \end{cases}$

r = the least j such that $k_{r+1} = s$

$K_j = [k_j, k_{j+1})$.

By this definitions we have for $i \in K_j$

if also $i+1 \in K_j$ then $\eta_{i+1} = \eta_i + 1$

if $i+1 \notin K_{j+1}$ then $\eta_{i+1} > \eta_i + 1$.

Furthermore we set

(a) $\varphi_j = \bigwedge_{i \in K_j} \bigwedge_{t \in N_i} w(p^i x + t) = \eta_j \ \& \ w(p^{l_j} + \tau) > \eta_j + 1$

where $0 \le j < r$, $l_j = k_{j+1} + 1$ and τ is an arbitrary element from N_{l_j} which will be kept fixed for the remainder of this section.

(b) $\varphi_r = \bigwedge_{i \in K_r} \bigwedge_{t \in N_i} w(p^i x + t) = \eta_j$.

As in (3.26) we show (4.9) equivalent to $\bigwedge_{0 \leq j \leq r} \exists x \varphi_j$.

This uses (K2). From now on it suffices to deal with formulas of the form

(4.10) $\exists x (\bigwedge_{k \leq i \leq l} \bigwedge_{t \in N_i} w(p^i x + t) = \eta_i \;\&\; w(p^{l+1} x + \tau) > \eta_l + 1)$
with $\eta_{i+1} = \eta_i + 1$ for all $i, k \leq i < l$. The least conjunct is missing exactly for $l = s-1$ (compare (3.39)).

At this point the further procedure is split into the cases

Case I: for all $i, k \leq i \leq l$ and all $t \in N_i$ $w(p^{s-i} t) \leq \eta_i + s-i$

Case II: complement of case I.

We remark that by lemma 4.2 the inequality specifying case I could equivalently be written as $w_{p,i}(t) \leq \eta_i + 1$.

If case I applies then (4.10) is equivalent to

(4.11) $\exists x (w(p^{l+1} x + \tau) > \eta_l + 1$ if $l < s-1$

$w(pt_0) \geq \eta_l + 1$ if $l = s-1$

where t_0 is an arbitrary element from N_l (corresponds to claim 3.30).

Lemma 4.2 tells us that also (4.11) is equivalent to a formula without group quantifiers.

From now on we are working in Case II.

By the same type of argument as used in verifying (3.31) and (3.32) we equivalently replace (4.11) by a formula of the following form

(4.12) $\exists x (w(p^l x) = \eta \;\&\; \bigwedge_{t \in N} w(p^l x + t) = \eta \;\&\; w(p^{l+1} x + \tau) > \eta + 1)$
with the last conjunct missing just for $l+1 = s$ and for all $t_0, t_1 \in N$, $t_0 \neq t_1$

(4.12) (a) $w(t) = w(t_0 - t_1) = \eta$

(4.12) (b) $w_{p,l}(t) > \eta + 1$.

For $l = s-1$ (4.12) is equivalent to $d_n(\eta)$ with n the smallest number such that $p^n \geq |N| + 1$.

Let us assume $l < s-1$. We may assume $\eta < c_0$ since otherwise (4.12) would be contradictory. If $w(\tau) > \eta + 1$ then (4.12) is equivalent to

(4.13) $\exists x (\bigwedge_{t \in N} w(p^l x + t) = \eta \;\&\; w(p^{l+1} x) > \eta + 1)$.

This is certainly a consequence of

(4.14) $\exists x(\bigwedge_{t\in N} w(p^{s-1}x+t) = \eta)$.

For assume g is a witness for (4.14) then $p^{s-l-1}g$ would be a witness for (4.13). If on the other hand g is a witness for (4.13) then $w(p^{l+1}g) > \eta+1$ implies by (4.2) $w_{p,s-1}(p^l g) > \eta+1$, thus $w(p^l g + p^{s-1}h) \geq \eta+1$ for some h and h would be a witness for (4.14).

Thus (4.13) and (4.14) are actually equivalent and we are reduced to the case of (4.12) with $l = s-1$.

We assume therefore from now on

(4.12) (c) $w(\tau) = \eta+1$.

We also note that

(4.12) (d) $w_{p,l+1}(\tau) > \eta+2$

is implied by (4.12).

We define $N_o = \{t \in N : w(pt-\tau) > \eta+1\}$.

Claim 4.15. (4.12) is equivalent to $d_n(\eta)$ with n the smallest number such that $p^n \geq |N_o|+1$.

Proof: Assume first that g is a witness for (4.12) in some valuated p^s-group (G,w). For all $t \in N_o$ we get $w(p(p^l g+t)) > \eta+1$. Thus there are by axiom (K2) elements c_t' such that $pc_t' = p(p^l g+t)$ and $w(c_t') = w(p(p^l g+t))-1 > \eta$. Setting $c_t = c_t' - p^l g + t$ we get $pc_t = 0$ and $w(c_t) = \eta$. By axiom (K1) we must have $c_t \in p^{s-1}G$. For $t_o, t_1 \in N_o$ with $t_o \neq t_1$ we have $w(c_{t_o} - c_{t_1}) = \eta$, since $w(c_{t_o} - c_{t_1}) > \eta$ would imply $w(c_{t_o}' - c_{t_1}' + t_o - t_1) \geq \eta+1$ and therefore $w(t_o - t_1) > \eta+1$ contradicting (4.12) (a). Thus $|G(\eta) \cap p^{s-1}G/G(\eta+1) \cap p^{s-1}G| \geq |N_o|+1$ which implies $d_n(\eta)$.

If on the other hand $c_1,\ldots,c_m$ are $m = |N_o|+1$ representatives of different cosets in $G(\eta) \cap p^{s-1}G$ modulo $H = G(\eta+1) \cap p^{s-1}G$. Let $c_i' \in G$ be such that $p^l c_i' = c_i$. By (4.12) (d) we find τ_o such that $w(p^{l+1}\tau_o + \tau) \geq \eta+2$ which gives $w(p^{l+1}\tau_o) = \eta+1$ by (4.12) (c). Axiom (K2) allows us to assume $w(p^l\tau_o) = \eta$. Since $p^l\tau_o + H$, $p^l(\tau_o + c_i') + H$ $1 \leq i \leq m$ are (m+1) different cosets, we find some c in $\{c_i' : 1 \leq i \leq m\} \cup \{0\}$ such that for all $t \in N_o$ $w(p^l(\tau_o + c) + t) = \eta$ while $p^{l+1}(\tau_o + c) = p^{l+1}\tau_o$ still guarantees $w(p^{l+1}\tau_o + \tau) > \eta+1$.

Finally consider $t \in N \setminus N_o$ for which we certainly have

$w(p^l(\tau_o + c) + t) \geq \eta$. But strict inequality would imply $w(p^{l+1}\tau_o + pt) > \eta + 1$ contrary to $t \notin N_0$. Thus $\tau_0 + c$ is a witness for (4.12).

As easy consequence of theorem 4.1 we obtain

Corollary 4.16. For every L_V^{**}-sentence φ there is some $L_<^{**}$-sentence ψ such that for all valuated p^s-groups (G,w)

$$(G,w) \models \varphi \quad \text{iff} \quad Val^{**}(G,w) \models \psi .$$

Corollary 4.17. For any two valuated p^s-groups (G_1,w_1), (G_2,w_2):

$$(G_1,w_1) \equiv (G_2,w_2) \quad \text{iff} \quad Val^{**}(G_1,w_1) \equiv Val^{**}(G_2,w_2) .$$

SECTION 5: DECIDABILITY RESULTS

The major problem in proving decidability of the class of all tamely p-valuated groups, namely which $L_<^*$-structures do occur as the value part of a tamely p-valuated group, is solved in the following lemma.

Lemma 5.1. Let α be a well-ordered set, c_0, c_1, c_2 element such that $\alpha < c_0 < c_1 < c_2$ and $r_{q,n}$ subsets of $\alpha \cup \{c_0, c_1, c_2\}$ then the following conditions are equivalent:

(I) there is a tamely p-valuated group (G,α,v) such that $(\alpha \cup \{c_0,c_1,c_2\},<,r_{q,n}) \simeq Val^*(G,v)$;

(II) (i) for all q and $n \geq 1$ and all γ: $r_{q,n+1}(\gamma)$ implies $r_{q,n}(\gamma)$

(ii) for all γ: $r_{p,1}(\gamma)$ implies $\gamma \neq c_2$ and γ is not a limit number

(iii) for all primes $q \neq p$ and all γ: $r_{q,1}(\gamma)$ implies that γ is a limit value or c_1 but not a limit of limits

(iv) if for $q \neq p$ $r_{q,1}(\lambda + \omega)$ then $r_{p,1}(\lambda + i)$ for some i, $1 \leq i < \omega$.

Proof: Necessity of II (i) is obvious, follows for II (ii), (iii) from axiom (T3) and Lemma 1.3. Necessity of II (iv) derives from Lemma 1.12 (ii).

To prove sufficiency we define for every prime q and value γ:

$$f_q(\gamma) = \begin{cases} 0 & \text{if } \gamma \notin r_{q,1} \\ n & \text{if } \gamma \in r_{q,n} \setminus r_{q,n+1} \\ \omega & \text{if } \gamma \in r_{q,n} \text{ for all } n \end{cases}$$

By II (ii) it suffices to find a tamely p-valuated group (G,α,v) such that for all values γ and all primes q: $r_q^G(\gamma) = f_q(\gamma)$.

By Lemma 1.7 it suffices to find (G_1,α,v_1) and $(G_\lambda,\alpha,v_\lambda)$ for $\lambda = 0$ and every limit number $\lambda < c_0$ such that

$$\text{range}(v_1) = \{c_0,c_1\} \text{ and } \text{range}(v_\lambda) = \{\lambda + i : 0 \leq i < \omega\} \cup \{c_1\}$$

and

(a) $r_q^{G_1}(c_1) = f_q(c_1)$

(b) $r_q^{G_\lambda}(\gamma) = f_q(\gamma)$ for all γ, $\lambda < \gamma \leq \lambda + \omega$.

If a largest limit number λ exists in α we adopt the convention $\lambda + \omega = c_0$. For G_1 we have $r_q^{G_1}(c_1) = \dim G_1/qG_1$; so we pick a group G_1 with $\dim G_1/qG_1 = f_q(c_1)$ and valuate it trivially by $v(g) = c_0$ if $g \neq 0$ and $v(g) = c_2$ for $g = 0$. For G_λ and $q \neq p$ we have $f_q(\gamma) = 0$ for all $\gamma \neq \lambda + \omega$ by II (iii) and $r_q^{G_\lambda}(\lambda+\omega) = \dim G_\lambda/qG_\lambda$ by 1.11. Furthermore by II (ii) $f_p(\lambda+\omega) = 0$. Let γ_0 be $\lambda < \gamma_0 < \lambda + \omega$ such that $f_p(\gamma_0) > 0$. If no such γ_0 exists then by II (iv) $f_q(\lambda+\omega) = 0$. Otherwise we define

$$G_{\gamma,i} \simeq Z_p = \{r/s : r,s \in Z,\ (p,s) = 1\}$$

for $\gamma,\lambda < \gamma < \lambda+\omega$, $0 \leq i < f_p(\gamma)$ for $\gamma \neq \gamma_0$ and $0 < i < f_p(\gamma_0)$

with the valuation $v_{\gamma,i}$ given by $v_{\gamma,i}(1) = \gamma$.

Let (G_0,α,v_0) be the direct sum of all these groups. We have $\dim G_0/qG_0 = 0$. It remains to choose $(G_{\gamma_0,0},\alpha,v_{\gamma_0,0})$, which we call (G_2,α,v_2) for short, to satisfy (b). Let $(\overline{Z}_p,\alpha,w)$ be the additive group of the p-adic numbers valuated by $w(\sum_{0<i<\infty} a_i p^i) = \gamma_0 + i$ where i is the least integer with $a_i \neq 0$. (G_2,α,v_2) will be the valuated subgroup of $(\overline{Z}_p,\alpha,w)$ generated by

$$Z \cup \{e_{q,i,m} : q \neq p,\ 0 \leq i < f_q(\lambda+\omega)\ \&\ (m,q) = 1\} \cup \{e_m : (m,p) = 1\}$$

where Z is the canonically embedded copy of the integers in $\overline{Z}_p$ with generating element say e and $\{e\} \cup \{e_{q,i,1} : q \neq p,\ 0 \leq i < f_q(\lambda+\omega)\}$ are linearly independent, $m\cdot e_{q,i,m} = e_{q,i,1}$ and $m\cdot e_m = e$.

Now $(G_\lambda,\alpha,v_\lambda) = (G_0,\alpha,v_0) + (G_2,\alpha,v_2)$ satisfies (b). □

Theorem 5.2. The theory W_ω of the class of all well-orderings with countably many additional unary predicates is decidable.

Proof: By replacing the given predicates P_n by the collection of all finite intersections of P_n's and complements of P_n's it is easily seen that we may w.l.o.g. assume that the predicates P_n are mutually disjoint. Call the resulting theory W^d. Let N_ω be the structure $\langle T_\omega, \leq, \preceq, s_n \rangle$ with T_ω the set of all finite sequences of natural numbers, $\leq$ the partial ordering of T_ω by initial segments, $\preceq$ the lexicographic ordering of T_ω and s_n defined for $a \in T_\omega$ by $s_n(a) = a^\frown\langle n \rangle$. It was proved in [4] that the second order theory of N_ω, knows as SωS, is decidable. Let B be the subset of T_ω consisting of all elements b such that

(i) b has at least length 4

(ii) the last three elements of b are 101 and there is no occurrence of a subsequence 101 in b except the one at the end

(iii) all entries in b except the first one, b(0), are in $\{0,1\}$.

For an element a of T_ω different from the empty sequence, we denote by e(a) the tail of a, obtained by omitting the first entry, a(0).

We claim

(A) For every countable well-ordering $(X, \leq, P_n)$ with countably many mutually disjoint predicates P_n there is a subset $C \subseteq B$ such that

   (i) for $a, b \in C$ $e(a) = e(b)$ implies $a = b$

   (ii) there is an isomorphism $f : (e(C), \preceq) \to (X, \leq)$

   (iii) for $c \in C$

   $$c(0) = 0 \quad \text{iff} \quad f(e(c)) \notin P_m \quad \text{for all } m$$
   $$c(0) = n+1 \quad \text{iff} \quad f(e(c)) \in P_n \ .$$

That $C_0 \subseteq e(B)$ can be found to satisfy (ii) is already proved in [4]. From C_0 we easily find $C \subseteq B$ such that (i)-(iii) hold true.

From the decidability of SωS we get in particular that the first-order theory of the countable structures $(X, \leq, P_n)$ considered in (A) is decidable. But by the Löwenheim-Skolem Theorem for first-order logic this theory coincides with W^d_ω. □

Remark: The above proof shows that even the weak second order theory of the class of all well-orderings with countably many disjoint unary predicates is decidable.

Theorem 5.3. The theory of tamely p-valuated groups is decidable.

Proof: We describe a recursive procedure that determines for every L_V-sentence φ in finitely many steps whether φ is a theorem of TV_p

or not. First we determine the sentence ψ in $L^*_<$ that is equivalent to φ in the sense of corollary 3.2. Let Q_o be the finite set of primes q and N the largest natural number such that some predicate $r_{q,n}$, $n \leq N$ occurs in ψ. Let ψ^* be the conjunction of the finitely many sentences in condition (II) of Lemma 5.1 where $r_{q,n}$ for $q \in Q_o$ and $n \leq N$ occur. We claim

$$TV_p \vdash \varphi \quad \text{iff} \quad W_\omega \vdash \psi^* \rightarrow \psi .$$

We need only remark on the implication from right to left. Assume that $(X,\leq,r_{q,n})$ is a model of W_ω which satisfies ψ^* & $\neg\psi$. We may assume that $(X,\leq)$ is a well-ordering, since changing the predicates $r_{q,n}$ for $q \notin Q_o$ and $q \in Q_o$ with $n > N$ does not affect the truth of ψ^* & $\neg\psi$ we may assume that the hypothesis of Lemma 5.1 are satisfied and we get a tamely p-valuated group (G,α,v) such that $Val^*(G,v) \simeq (X,\alpha,r_{q,n})$. By choice of ψ we have $G \vdash \neg\varphi$ and thus not $TV_p \vdash \varphi$. The procedure now continues by an apeal to Theorem 5.2. □

In a very parallel way we obtain:

Theorem 5.4. The theory of valuated p^s-groups is decidable.

We conclude by

Theorem 5.5. A p-valuated group (G,α,v) is elementary equivalent to a direct sum of torsion-free rank-one p-valuated groups satisfying $v(pg) = v(g) + 1$ iff (G,α,v) is tamely p-valuated and satisfies for every prime $q \neq p$ and every limit number $\lambda < c_o$ (including $\lambda = 0$)

$$r_q(\lambda + \omega) \leq \sum_{1 \leq i < \infty} r_p(\lambda + i) \pmod{\infty} .$$

Proof: Necessity of the conditions is simply checked. For sufficiency we note that under the present hypothesis we may in the construction preformed in the proof of Lemma 5.1 dispence with the direct summand $G_{\gamma_o,o}$ which was the only not rank one summand. □

In [6] a free valuated Z_p-module was defined to be a direct sum of cyclic Z_p-modules G with a p-valuation v satisfying $v(pg) = v(g) + 1$ for every $g \neq 0$.

Theorem 5.6. A p-valuated group (G,α,v) is elementary equivalent to a free valuated Z_p-module iff (G,α,v) is tamely p-valuated and $qG = G$ for every prime $q \neq p$.

Proof: Necessity is simply checked. For sufficiency we see that in Lemma 5.1 only Z_p-modules are used. □

We conclude with the following open problem:

Give a criterion on the pair $G \subseteq H$ such that the restriction of the p-height function of H turns G into a tamely p-valuated group.

REFERENCES

[1] Eklof, P. and Fischer, E.: Elementary properties of Abelian groups. Annals of Math. Logic 4 (1972), 115-171.

[2] Fuchs, L.: Vector Spaces with Valuations. Journal of Algebra 35 (1975), 23-28.

[3] Hunter, R. and Walker, E.A.: Valuated p-groups; in: "Abelian Group Theory", Proc. Oberwolfach 1981, Springer Lecture Notes in Math. 874, 350-373.

[4] Rabin, M.O.: Decidability of second order theories and automata on infinite trees. Transaction AMS 141 (1969), 1-35.

[5] Richman, F.: A guide to valuated groups; in: "Abelian Group Theory", 2nd New Mexico State University Conference 1976, Springer Lecture Notes in Math. 616, 73-86.

[6] Richman, F. and Walker, E.A.: Valuated abelian groups. Journal of Algebra 56 (1979), 145-167.

[7] Schmitt, P.H.: Undecidable theories of valuated abelian groups. to appear in Memoirs de la societé mathematique de la France.

LOGIC COLLOQUIUM '84
J.B. Paris, A.J. Wilkie, and G.M. Wilmers (Editors)

COMPLETE UNIVERSAL LOCALLY FINITE GROUPS OF LARGE CARDINALITY

Simon Thomas

Mathematisches Institut
Albert-Ludwigs-Universitat
7800 Freiburg i. Br.
West Germany

Department of Mathematics
Yale University
New Haven, Connecticut
U.S.A.

1. INTRODUCTION

A group G is in the class ULF of universal locally finite groups if:

(i) G is locally finite.

(ii) Every finite group is isomorphic to a subgroup of G.

(iii) Any two isomorphic finite subgroups of G are conjugate in G.

Hall introduced the concept of a ULF group in [1] and showed that:

(iv) There is a countable ULF group C which is unique up to isomorphism.

(v) Every locally finite group of cardinality $\varkappa$ is contained in a ULF group of cardinality $\varkappa$. In particular, every countable locally finite group is embedded in C.

In [5], answering questions of Kegel and Wehrfritz [3], Macintyre and Shelah proved:

(vi) For every $\varkappa > \omega$, there are $2^{\varkappa}$ nonisomorphic ULF groups of cardinality $\varkappa$.

(vii) There exists a locally finite group H of cardinality ω_1 such that for every $\varkappa \geq \omega_1$ there exists a ULF group of cardinality $\varkappa$ in which H does not embed.

Macintyre and Shelah used Ehrenfeucht-Mostowski models to construct their nonisomorphic ULF groups. Consequently we do not have a very clear idea of the structure of these groups, and it remains an interesting problem to construct $2^{\varkappa}$ nonisomorphic ULF groups which are nonisomorphic for simple "group-theoretic" reasons. Hickin solved this problem for $\varkappa = \omega_1$ in [2], where he constructed 2^{ω_1} nonisomorphic complete ULF groups. He also showed that no locally finite group of cardinality ω_1 is inevitable. (A locally finite group H is said to be inevitable if it embeds in every ULF group of cardinality greater than or equal to $|H|$. Macintyre introduced this notion in [4] where, assuming $\diamondsuit$, he showed that there are no inevitable abelian groups of cardinality ω_1.) Similar results were obtained for $\varkappa = 2^{\omega}$ by Shelah [8].

In this paper, we shall partially extend Hickin's results to arbitrary successor cardinals. Our main result is:

THEOREM

Let $\lambda > \omega$. Then there exists a family $\{G_\xi \mid \xi < 2^{\lambda^+}\}$ of nonisomorphic ULF groups of cardinality λ^+ satisfying:

(a) if $S \subseteq G_\xi$ is a soluble subgroup, then $|S| \leq \lambda$.

(b) (G.C.H.) G_ξ is a complete group.

The methods of Hickin, Macintyre and Shelah rely heavily on the fact that the free product with amalgamation of two finite groups is residually finite. There is no analogue of this result for infinite locally finite groups, as Neumann [6] has shown that amalgamation fails in this category. This failure is the root of the difficulty of constructing large ULF groups with restrictions on their subgroups.

The work in this paper was carried out while I was studying for a Ph.D. at Bedford College, London, with financial assistance from the Science Research Council. I would like to thank Wilfrid Hodges for some very helpful discussions and suggestions. Thanks are also due to the Alexander von Humboldt Stiftung for support in the form of a research fellowship in 1983/84.

2. THE CONSTRICTED SYMMETRIC GROUP

In this section, we shall prove the technical lemmas which form the heart of the construction.

DEFINITION

Let G be a locally finite group and let Sym(G) be the full symmetric group on the set G. The constricted symmetric group, S(G), is the subgroup of

Sym(G) consisting of those σ which satisfy: there exists a finite subgroup F_σ of G such that $(x F_\sigma)^\sigma = x F_\sigma$ for all $x \in G$.

It is easily seen that S(G) is a locally finite group, and that G embeds in S(G) via the right regular representation $\rho: G \hookrightarrow S(G)$. (Throughout this paper, ρ will always denote the right regular representation.) A proof of the following lemma may be found in Kegel and Wehrfritz [3].

LEMMA 1

Any two finite isomorphic subgroups of G^ρ are conjugate in S(G). □

We begin with a simple observation.

LEMMA 2

Let $H \subseteq G$ be locally finite groups. Then the following diagram can be completed so that for each $h \in H$ and $\tau \in S(H)$, $h^{f(\tau)} = h^\tau$.

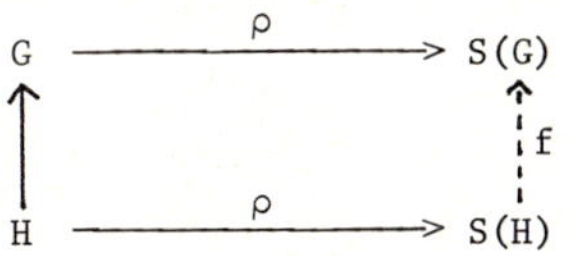

PROOF

Let I be a set of left coset representatives of H in G, chosen so that $1 \in I$. For $\tau \in S(H)$, $x \in I$ and $h \in H$, define

$$(x\ h)^{f(\tau)} = x\ h^\tau.$$

It is easily checked that f satisfies our requirements. (It is just a repeated representation.) □

Notice that there are many completion maps f, depending on the choice of I. We shall exploit this in the next two lemmas. It is understood that we always choose $1 \in I$.

LEMMA 3.

With the notation of lemma 2, suppose that:

(i) $a \in G \setminus H$.

(ii) $\tau \in S(H) \setminus H^\rho$.

(iii) Coset representatives have been chosen for H in $K \subset G$.

(iv) There exist $b, c \in G$ such that

$$< K, a > \subsetneq < K,a,b > \subsetneq < K,a,b,c > .$$

Then f can be chosen so that

$$f(\tau)^{-1}\rho(a)\, f(\tau) \notin G^{\rho}.$$

PROOF

Let $h \in H$ satisfy $1^{\tau} = h$. Since $\tau \notin H^{\rho}$, there exists $g \in H$ such that $g^{\tau} \neq gh$. We still have freedom in the choice of coset representatives for bH, baH, cH and caH. We select b, ba, c and cag^{-1} respectively. Then, regardless of the choice of the other coset representatives, we have

$$(bh)^{f(\tau)^{-1}\rho(a)f(\tau)} = bah$$

$$(ch)^{f(\tau)^{-1}\rho(a)f(\tau)} = cag^{-1}\, g^{\tau}. \qquad \square$$

This lemma will be used to restrict the size of soluble subgroups in the construction. The next lemma enables us to build many nonisomorphic ULF groups.

LEMMA 4.

With the notation of lemma 2, suppose that:

(i) $a \in G \setminus H$ is an involution.

(ii) $\tau \in S(H) \setminus H^{\rho}$.

(iii) there exists $h \in H$ such that $ah^{-1}h^{\tau}\, a \notin H$.

Then f can be chosen so that

$$[\rho(a)f(\tau)\rho(a),\ f(\tau)\ ] \neq 1.$$

PROOF.

By hypothesis, haH, $ha\, h^{-1}h^{\tau}a\, H$ and H are distinct cosets. We choose hah^{-1}, $hah^{-1}h^{\tau}a\, h^{-1}$ as coset representatives. Thus

$$(ha)^{f(\tau)\rho(a)f(\tau)\rho(a)} = (hah^{-1}h^{\tau}\ a)^{f(\tau)\rho(a)}$$
$$= hah^{-1}h^{\tau}\ a\ h^{-1}\ h^{\tau}\ a\ .$$

CASE 1. $hah^{-1}h^{\tau}a\ h^{-1}h^{\tau}a \notin h^{\tau}a\ H$.

We already know that

$$(ha)^{\rho(a)f(\tau)\rho(a)f(\tau)} = (h^{\tau}\ a)^{f(\tau)} \in h^{\tau}\ a\ H.$$

This holds regardless of whether we still have a choice of coset representative for $h^{\tau}\ a\ H$. Thus we have

$$(ha)^{\rho(a)f(\tau)\rho(a)f(\tau)} \neq (ha)^{f(\tau)\rho(a)f(\tau)\rho(a)}\ .$$

Case 2. $hah^{-1}h^{\tau}\ a\ h^{-1}\ h^{\tau}\ a \in h^{\tau}\ a\ H$.

Clearly $h^{\tau}\ a\ H \neq H$ and by hypothesis $h^{\tau}\ a\ H \neq haH$. Suppose that $h^{\tau}aH = ha\ h^{-1}h^{\tau}\ a\ H$. Then there exists $g \in H$ such that

$$hah^{-1}\ h^{\tau}\ a\ h^{-1}\ h^{\tau} = h^{\tau}\ a\ g.$$

The assumption in this case implies that

$$h^{\tau}\ aga \in h^{\tau}\ a\ H.$$

Thus $ga \in H$ and so $a \in H$, a contradiction. We conclude that $h^{\tau}\ a\ H \neq hah^{-1}\ h^{\tau}\ a\ H$. A second application of the assumption yields an element $w \in H$ such that

$$hah^{-1}\ h^{\tau}\ a\ h^{-1}\ h^{\tau}\ a = h^{\tau}\ aw.$$

Since $\tau \notin H^{\rho}$, there exists $z \in H$ such that $z^{\tau} \neq zw$. We choose $h^{\tau}\ a\ z^{-1}$ as coset representative for $h^{\tau}\ a\ H$. Thus

$$(ha)^{\rho(a)f(\tau)\rho(a)f(\tau)} = (h^{\tau}\ a)^{f(\tau)}$$
$$= h^{\tau}\ a\ z^{-1}z^{\tau}\ .$$

Once again, we have

$$(ha)^{\rho(a)f(\tau)\rho(a)f(\tau)} \neq (ha)^{f(\tau)\rho(a)f(\tau)\rho(a)} . \qquad \square$$

We now use lemma 3 to prove:

LEMMA 5

Let G be a locally finite group of cardinality $\lambda > \omega$. Let F_1, F_2 be finite isomorphic nonconjugate subgroups of G. Then there exists a locally finite group $< G,\tau > \supsetneq G$ such that:

(i) $F_1^{\tau} = F_2$.

(ii) If $\sigma \in < G,\tau > \setminus G$, then $|\{g \in G \mid g^{\sigma} \in G\}| < \lambda$.

PROOF.

G may be expressed as the union of a strictly increasing smooth chain, $G = \bigcup_{\xi<\lambda} G_\xi$, satisfying:

(a) $< F_1, F_2 > \subseteq G_0$.

(b) $|G_\xi| = |\xi| + \omega$ for all $\xi < \lambda$.

(c) If $|G_\xi| = \mu < \lambda$, then $G_{\xi+1} = \bigcup_{i<\mu} H_i^\xi$ is the union of a strictly increasing smooth chain with $H_0^\xi = G_\xi$.

(In later sections, this chain will be chosen to satisfy further conditions.) We shall show that there exists $\tau \in S(G)$ such that $< G^\rho, \tau >$ satisfies the requirements of the lemma. The action of τ on G_ξ will be defined inductively, by completing diagrams as in lemma 3.

$\underline{\xi = 0}$

By lemma 1, there exists $\tau_0 \in S(G_0)$ such that $\tau_0^{-1} F_1^\rho \tau_0 = F_2^\rho$.

$\underline{\xi = \eta + 1}$

Let $|G_\eta| = |G_{\eta+1}| = \mu$. We may assume that the chain, $G_{\eta+1} = \bigcup_{i<\mu} H_i^\eta$, has been constructed so that for each $i < \mu$ there exist b_i, c_i such that

$$H_i^\eta \subsetneq < H_i^\eta, b_i > \subsetneq < H_i^\eta, b_i, c_i > \subseteq H_{i+1}^\eta .$$

Let $X = (G_{\eta+1} \setminus G_\eta) \times (< G_\eta^\rho, \tau_\eta > \setminus G_\eta^\rho)$. Then $|X| = \mu$ and we may choose a well-ordering, $X = < x_i \mid 0 < i < \mu >$, so that if $x_i = < a,\sigma >$ then $a \in H_i^\eta$.

We shall use lemma 3 to complete the diagram

$$\begin{array}{ccc} G_{\eta+1} & \xrightarrow{\ \rho\ } & S(G_{\eta+1}) \\ \uparrow & & \vdots\ f \\ G_{\eta} & \xrightarrow{\ \rho\ } & S(G_{\eta}) \end{array}$$

Assume inductively that coset representatives for G_η in H_i^η have been chosen so that, regardless of further choices, we will have $f(\pi)^{-1}\rho(g)f(\pi) \notin G^\rho_{\eta+1}$ for each $x_j = < g, \pi >$ with $j < i$. By lemma 3, we can choose coset representatives for G_η in H^η_{i+1} so that $f(\sigma)^{-1}\rho(a)f(\sigma) \notin G^\rho_{\eta+1}$, where $x_i = < a,\sigma >$. We define $\tau_{\eta+1} = f(\tau_n)$.

<u>ξ is a limit ordinal</u>

Assume inductively that we have defined elements $\tau_i \in S(G_i)$ for all $i < \xi$ so that if $i < j < \xi$ then $g^{\tau_j} = g^{\tau_i}$ for all $g \in G_i$. Then $\tau_\xi \in S(G_\xi)$ is defined by $g^{\tau_\xi} = g^{\tau_i}$ for $g \in G_i$. Notice that we have taken the image of τ_i under the direct limit mapping

$$\begin{array}{ccc} G_{\xi} & \xrightarrow{\ \rho\ } & S(G_{\xi}) \\ \uparrow & & \vdots \\ G_{j} & \xrightarrow{\ \rho\ } & S(G_{j}) \\ \uparrow & & \uparrow \\ G_{i} & \xrightarrow{\ \rho\ } & S(G_{i}) \end{array}$$

and that every element of $< G^\rho_\xi, \tau_\xi >$ has a preimage in some $< G^\rho_i, \tau_i >$ for $i < \xi$.

Thus $< G^\rho, \tau >$ satisfies the requirements of the lemma, where $\tau = \tau_\lambda$. □

3. KILLING OUTER AUTOMORPHISMS

Suppose that π is an outer automorphism of the uncountable ULF group G. In this section, we shall show that there exists a locally finite group $H \supsetneq G$ such that π cannot be extended to an automorphism of H.

LEMMA 6

Let G be a simple group of cardinality $\lambda > \omega$. Suppose that $\pi \in \mathrm{Aut}\, G$ satisfies the condition:

there exists $H \subseteq G$ and $g \in G$ such that $|H| < \lambda$ and $x^{\pi} H = g\,x\,H$ for all $x \in G \backslash H$.

Then $x = g\,x\,g^{-1}$ for all x G.

PROOF

Choose any element $x \notin G_0 = \langle H, g \rangle$. Then $x^{\pi} = g\,x\,h$ for some $h \in H$. Suppose that $hg \neq 1$. Since G is simple, it is generated by the conjugacy class $(hg)^G$. Hence there exists $y \in G$ such that $y^{-1} hg\, y \notin G_1 = \langle G_0, x \rangle$. Then $y \notin G_1$ and so $yx \notin G_1$. Thus the equation

$$x^{\pi}\, y^{\pi} = (xy)^{\pi}$$

yields elements $h_0, h_1 \in H$ such that

$$g\,x\,h\,g\,y\,h_0 = g\,x\,y\,h_1.$$

Hence $y^{-1} hg\, y = h_1\, h_0^{-1} \in H \subseteq G_1$, a contradiction. So for all $x \notin G_0$, we have $x^{\pi} = g\,x\,g^{-1}$. Fix $x \notin G_0$. Then for all $z \in G_0$, it follows that $xz \notin G_0$. Hence

$$g\,x\,z\,g^{-1} = (xz)^{\pi} = g\,x\,g^{-1}\,z^{\pi}.$$

Thus $z^{\pi} = g\,z\,g^{-1}$, as required. □

From now on, $\mu : G \to \mathrm{Sym}(G)$ will denote the left regular representation, ie. $x^{\mu(g)} = g^{-1} x$ for all $x, g \in G$.

LEMMA 7

With the notation of lemma 2, suppose that $\pi \in \mathrm{Aut}\, G$ and $g \in H$ satisfy:

(i) $\pi[H] = H$.

(ii) $x \in G \backslash H$.

(iii) $\sigma, \tau \in S(H)\backslash H^{\rho}$

(iv) $x^{\pi} H \neq g^{\pi} x H$.

Then f can be chosen so that

$$(\mu(g)\pi)^{-1} f(\sigma) \mu(g)\pi \neq f(\tau).$$

PROOF

Since $\tau \neq \rho(1^{\sigma\pi})$, there exists $h \in H$ such that $h^{\tau} \neq h.1^{\sigma\pi}$. By assumption H, xH and $(g^{\pi})^{-1} x^{\pi} H$ are distinct cosets of H in G. We choose x and $(g^{\pi})^{-1} x^{\pi} h$ as coset representatives. Then

$$x^{f(\sigma)\mu(g)\pi} = (g^{-1} x 1^{\sigma})^{\pi} = (g^{\pi})^{-1} x^{\pi} 1^{\sigma\pi}$$

and

$$x^{\mu(g)\pi f(\tau)} = ((g^{\pi})^{-1} x^{\pi})^{f(\tau)} = (g^{\pi})^{-1} x^{\pi} h^{-1} h^{\tau}.$$

Thus $f(\sigma) \mu(g)\pi \neq \mu(g)\pi f(\tau)$. □

LEMMA 8.

Let G be a ULF group of cardinality $\lambda > \omega$ and let $\pi \in \mathrm{Aut}\, G$ be an outer automorphism. Then there exists $\sigma \in S(G)$ such that for all $\tau \in \langle G^{\rho}, \sigma\rangle \backslash G^{\rho}$:

(i) $|\{\rho(g) \in G^{\rho} \mid \tau^{-1}\rho(g)\tau \in G^{\rho}\}| < \lambda$.

(ii) for all $g \in G$, the set

$$\{x \in G \mid x^{(\mu(g)\pi)^{-1}\sigma\, \mu(g)\pi} \neq x^{\tau}\}$$

is infinite.

PROOF

Let $X = \langle g_{\xi} \mid \xi < \lambda\rangle$ be a list of the elements of G such that every element occurs λ times. Since G is simple, we may use lemma 6 to obtain an expression, $G = \bigcup_{\xi < \lambda} G_\xi$, satisfying conditions (a) to (c) of lemma 5, together with:

(d) $\pi [G_\xi] = G_\xi$.

(e) $g_\xi \in G_\xi$.

(f) There exists $x \in G_{\xi+1} \setminus G_\xi$ such that $x^\pi G_\xi \neq g_\xi^\pi x G_\xi$.

The result now follows easily from lemmas 5 and 7. □

DEFINITION.

If $\sigma \in Sym(G)$, then $sup(\sigma) = \{g \in G \mid g^\sigma \neq g\}$. $Sym(G,\omega) \subseteq Sym(G)$ is the subgroup consisting of those permutations satisfying $|sup(\sigma)| < \omega$. $Alt(G) \subseteq Sym(G,\omega)$ is the subgroup of finite even permutations.

Note that $Sym(G,\omega) \subseteq S(G)$. Hence if σ is the element given by lemma 8, then

$$H = < Sym(G,\omega), \sigma, G^\rho >$$

is a locally finite group of cardinality λ. It is easily checked that

$$H = Sym(G,\omega) \rtimes < \sigma , G^\rho >.$$

Thus if $x = \theta \tau \in H$ is any element, with $\theta \in Sym(G,\omega)$ and $\tau \in < \sigma, G^\rho >$, then

$$|\{g \in G \mid g^x \neq g^\tau\}| < \omega .$$

Consequently for all $g \in G$,

$$(\mu(g)\pi)^{-1} \sigma \mu(g)\pi \notin H.$$

Suppose that $\varphi \in Aut H$ extends the outer automorphism $\pi \in Aut G$. Since

$$Alt(G) \subseteq H \subseteq Sym(G)$$

theorem 11.4.6 of Scott [7] says that there exists $\Upsilon \in Sym(G)$ such that $\varphi(h) = \Upsilon^{-1} h \Upsilon$ for all $h \in H$. Thus for all $g \in G$,

$$\gamma^{-1}\ \rho(g)\gamma \;=\; \rho(g^{\pi}) \;=\; \pi^{-1}\ \rho(g)\pi\ .$$

By theorem 10.3.6 of Scott [7], there exists $g \in G$ such that $\gamma = \mu(g)\pi$. But this means that $(\mu(g)\pi)^{-1}\ \sigma\ \mu(g)\ \pi \in H$, a contradiction. So we have proved:

LEMMA 9.

The outer automorphism $\pi \in \mathrm{Aut}\ G$ cannot be extended to an automorphism of H. □

The following lemma will be useful during the main construction.

LEMMA 10.

Let $X \subseteq G$ be a subgroup of cardinality λ. Then $N_H(X) = N_G(X)$.

PROOF.

Suppose that $h \in H$ normalizes X^{ρ}. Then $h = \theta.\tau$ for some $\theta \in \mathrm{Sym}(G,\omega)$ and $\tau \in \langle G^{\rho}, \sigma\rangle$. Applying the canonical projection, $p: H \to \langle G^{\rho}, \sigma\rangle$, we see that $\tau \in N(X^{\rho})$ and hence $\tau \in G^{\rho}$. We also obtain $\theta \in N(X^{\rho})$. Clearly $\theta^{-1}\ \rho(g)\ \theta \in G^{\rho}$ if and only if $[\theta, \rho(g)] = 1$. Hence $[\theta, X^{\rho}] = 1$. Suppose that $\theta \neq 1$. Let $\sup(\theta) = \{g_1, \dots, g_n\}$, where $n > 1$. Then for $\rho(x) \in X^{\rho}$, we have

$$\sup(\rho(x)^{-1}\ \theta\ \rho(x)) \;=\; \{g_1x, \dots, g_nx\}\ .$$

Thus $[\theta, X^{\rho}] \neq 1$, a contradiction. □

4. THE CONSTRUCTION

To construct nonisomorphic ULF groups, we require a well known combinatorial fact.

LEMMA 11.

Let $\varkappa$ be a regular uncountable cardinal. Then there exists $A \subseteq {}^{\varkappa}2$ such that:

(i) $|A| = 2^{\varkappa}$

(ii) If $\eta \neq \tau \in A$, then $\{\xi < \kappa \mid \eta(\xi) \neq \tau(\xi)\}$ is stationary in κ. □

In particular, there exists $A \subseteq {}^{\lambda^+}2$ satisfying the above conditions. We shall build smooth strictly increasing chains

$$G^{\eta} = \bigcup_{\xi < \lambda^+} G^{\eta}_{\xi} \quad , \quad \eta \in A \quad ,$$

where each G^{η}_{ξ} is a ULF group of cardinality λ. Suppose that $\eta \neq \tau \in A$ and that $\pi : G^{\eta} \to G^{\tau}$ is an isomorphism. The regularity of λ^+ ensures that

$$\{\xi < \lambda^+ \mid \pi\,[G^{\eta}_{\xi}] = G^{\tau}_{\xi}\}$$

is a closed unbounded set in λ^+. By lemma 11, there exists $\xi < \lambda^+$ such that $\pi[G^{\eta}_{\xi}] = G^{\tau}_{\xi}$ and $\eta(\xi) \neq \tau(\xi)$. By putting in "obstructions to isomorphism", we shall eventually reach a contradiction.

For the next few pages, we will fix $\eta \in A$ and suppress index η, i.e. we write $G = \bigcup_{\xi < \lambda^+} G_{\xi}$. We shall now discuss our strategy for killing <u>all</u> of the outer automorphisms of G. The next lemma is an easy generalization of lemma 3.2 of Hickin [2].

LEMMA 12.

Let $\pi \in \mathrm{Aut}\, G$ be an outer automorphism. Then there exists a closed unbounded set $C \subseteq \lambda^+$ such that for all $\alpha \in C$:

(i) $\pi[G_{\alpha}] = G_{\alpha}$.

(ii) $\pi \restriction G_{\alpha}$ is an outer automorphism of G_{α}. □

Shelah [9] has shown that the G.C.H. implies that $\diamondsuit(\lambda^+)$ holds for all $\lambda > \omega$. Assuming that the underlying set of G_{α} is the limit ordinal $\lambda.\alpha$, a standard argument gives:

LEMMA 13 (G.C.H.)

There exists a sequence of functions $< f_{\alpha} : G_{\alpha} \to G_{\alpha} \mid \alpha < \lambda^+ >$ such that for any automorphism $\pi \in \mathrm{Aut}\, G$

$$\{\alpha < \lambda^+ \mid \pi \restriction G_\alpha = f_\alpha\}$$

is stationary in λ^+. □

COROLLARY (G.C.H.)

Let $\pi \in \text{Aut } G$ be an outer automorphism. Then there exists $\alpha < \lambda^+$ such that:

(i) $\pi[G_\alpha] = G_\alpha$

(ii) $\pi \restriction G_\alpha = f_\alpha$ is an outer automorphism of G_α. □

So we know in advance who the potential trouble makers are. When we have constructed G_α, we can check whether the function f_α is an outer automorphism. If it is, then lemma 9 provides an extension $H_\alpha \supsetneq G_\alpha$ such that f_α cannot be extended to an automorphism of H_α. The rest of the chain will then be constructed so that if $\pi \in \text{Aut } G$ satisfies $\pi \restriction G_\alpha = f_\alpha$, then π induces an automorphism $\varphi \in \text{Aut } H_\alpha$ with $\varphi \restriction G_\alpha = f_\alpha$, a contradiction.

We begin the construction (continuing to suppress the index η.)

STEP ONE $\xi = 0$.

Let $H = \bigoplus_{n \geq 2} A_n$, where each A_n is the direct sum of λ copies of the cyclic group C_n. Then G_0 is an arbitrary ULF group of cardinality λ which contains H.

LEMMA 15.

Let $g \in G_0$ be a nonidentity element. Then $|\{aga \mid a \in G_0$ is an involution$\}| = \lambda$.

PROOF.

Let g have order $n \geq 2$. G_0 contains a subgroup $[\langle z_0\rangle \oplus \langle z_1\rangle] \rtimes \langle z_2\rangle$ where $\langle z_0\rangle \approx \langle z_1\rangle \approx C_n$ and z_2 is an involution such that $z_2 z_0 z_2 = z_1$. Let $A_n = \bigoplus_{i<\lambda} \langle y_i\rangle$. There exists $h \in G_0$ such that $y_0^h = g$. Putting $g_i = y_i^h$ for $i > 0$,

$A_n^h = \langle g \rangle \oplus [\bigoplus_{i>0} \langle g_i \rangle]$. For each $i > 0$, there exists $u \in G_0$ such that $z_0^u = g$ and $z_1^u = g_i$. Then $a = z_2^u$ is an involution and $aga = g_i$. □

STEP TWO. ξ is a limit ordinal.

Define $G_\xi = \bigcup_{i<\xi} G_i$.

STEP THREE. $\xi = \alpha + 1$.

Assume inductively that G_α has been constructed.

CASE 1. f_α is an outer automorphism of G_α.

Let $\sigma_\alpha \in S(G_\alpha)$ be an element satisfying:

(i) f_α cannot be extended to an automorphism of $\langle \mathrm{Sym}(G_\alpha,\omega), G_\alpha^\rho, \sigma_\alpha \rangle$.

(ii) If $\tau \in \langle G_\alpha^\rho, \sigma_\alpha \rangle \setminus G_\alpha^\rho$, then

(a) $|\{\rho(g) \in G_\alpha^\rho \mid \tau^{-1}\rho(g)\tau \in G_\alpha^\rho\}| < \lambda$

(b) there exists $\rho(a) \in G_\alpha^\rho$ such that $[\rho(a)^{-1}\tau\rho(a), \tau] \neq 1$.

CASE 2 otherwise.

Let $\sigma_\alpha = S(G_\alpha)$ satisfy (ii) above. (We slightly delay the proof that clause (ii)(b) can always be satisfied.)

Define $H_\alpha = \langle \mathrm{Sym}(G_\alpha,\omega), G_\alpha^\rho, \sigma_\alpha \rangle$

$$G_\alpha^+ = C_{n_\alpha} \text{ wr } H_\alpha$$

where $n_\alpha = 2$ if $\eta(\alpha) = 0$.

$= 3$ if $\eta(\alpha) = 1$.

(Here "wr" denotes the restricted wreath product.) Let K_α be the base group of G_α^+. Then $|K_\alpha| = \lambda$, and it is easily checked that K_α is a maximal abelian subgroup of G_α^+.

We shall define $G_{\alpha+1}$ to be the union of a smooth strictly increasing chain, $G_{\alpha+1} = \bigcup_{i<\lambda} G_i^\alpha$, with $G_0^\alpha = G_\alpha^+$. The G_i^α are chosen, using lemmas 4 and 5, so that the following conditions are satisfied.

(1) $G_{\alpha+1}$ is a ULF group.

(2) If $Y \subseteq G_\alpha$ is a subset of cardinality λ, then $N_{G_i^\alpha}(Y) = N_{G_\alpha}(Y)$.

(3) If $x \in G_i^\alpha \setminus G_0^\alpha$, there exists $a \in G_0$ such that $[a^{-1}x\, a, x] \neq 1$.

Clearly (1) presents no difficulty. Suppose that $Y \subseteq G_\alpha$ has cardinality λ. By lemma 10, $N_{H_\alpha}(Y) = N_{G_\alpha}(Y)$ and, by Kegel and Wehrfritz [3] p. 74, $N_{G_0^\alpha}(Y) = N_{H_\alpha}(Y)$. Lemma 5 deals with (2) for $i > 0$.

Finally we show that condition (3) can always be satisfied. (The proof will allow us to choose σ_α satisfying clause (ii)(b).) Suppose inductively that G_i^α has been constructed. To ensure that (3) holds, we slightly modify the construction in lemma 5. We inductively construct a smooth chain $G_i^\alpha = \bigcup_{\xi<\lambda} H_\xi$ satisfying conditions (a) to (c) of lemma 5, and use this to define the action of a suitable $\tau \in S(G_i^\alpha)$. Suppose that H_ξ and $\tau_\xi \in S(H_\xi)$ have been defined. Then $|\langle H_\xi^\rho, \tau_\xi\rangle \setminus H_\xi^\rho| < \lambda$. So as we proceed through the construction, we may inductively define a list of the "new" elements. Let $\sigma \in \langle H_\xi^\rho, \tau_\xi\rangle \setminus H_\xi^\rho$ be the ξth element in this list. Then there exists $h \in H_\xi$ such that $h^\sigma \neq h$.

CLAIM

There is an involution $a \in G_0 \setminus H_\xi$ such that $a\, h^{-1} h^\sigma\, a \notin H_\xi$.

PROOF OF CLAIM

If $h^{-1} h^\sigma \in G_0$, the claim follows from lemma 15. Otherwise $h^{-1} h^\sigma \in H_\xi \setminus G_0$ first appeared as the result of one of the following types of extension:

(a) a right regular extension, as in lemma 5.

(b) a wreath product extension.

(c) a killing extension, i.e.

$$\langle G_\beta^\rho, \sigma_\beta\rangle \hookrightarrow \langle \mathrm{Sym}(G_\beta, \omega), G_\beta^\rho, \sigma_\beta\rangle$$

for some $\beta \leq \alpha$.

(Throughout the rest of this paper, an element $x \in G \setminus G_0$ will be said to

have type (a), (b) or (c) depending on where it first appeared.) Let $g = h^{-1} h^{\sigma}$, $A = A_2 \subseteq G_0$ and $B = C_A(g)$.
Then $g \in N(B)$ and so $|B| < \lambda$. It follows that $[A:C_A(g)] = \lambda$, and the claim holds. □

We choose $H_{\xi+1}$ so that it includes a, and then use lemma 4 to ensure that $[a\sigma a, a] \neq 1$. The construction is completed!

5. THE CONSTRUCTION WORKS

LEMMA 16

Let $\eta \in A$. If $S \subseteq G^{\eta}$ is a soluble subgroup, then $|S| \leq \lambda$.

PROOF

By construction, if $H \subseteq G^{\eta}$ has cardinality λ, then $|N(H)| = \lambda$. Suppose that $S \subseteq G^{\eta}$ is a soluble subgroup of cardinality λ^+. We shall prove inductively that for each $i \in \omega$ the ith derived subgroup $D^{(i)}(S)$ has cardinality λ^+, an obvious contradiction.

$i = 0$

By assumption.

$i = j + 1$

Assume that $K = D^{(j)}(S)$ has cardinality λ^+, and suppose that $|[K,K]| \neq \lambda^+$. Since $[K,K] \lhd K$, we must have $|[K,K]| < \lambda$. Hence $K/[K,K]$ is an abelian group of cardinality λ^+. Let $K_0/[K,K] \lhd K/[K,K]$ be an abelian group of cardinality λ. Then $|K_0| = \lambda$ and $K_0 \lhd K$, a contradiction. Hence $[K,K] = D^{(i)}(S)$ also has cardinality λ^+. □

The theorem will follow from the next lemma.

LEMMA 17

Suppose that $\eta, \tau \in A$ and $\pi : G^{\eta} \to G^{\tau}$ is an isomorphism. If $\pi[G^{\eta}_{\alpha}] = G^{\tau}_{\alpha}$, then:

(i) $\pi[(G^{\eta}_{\alpha})^+] = (G^{\tau}_{\alpha})^+$.

(ii) $\pi[K_\alpha^\eta] = K_\alpha^\tau$.

Before proving this lemma, we shall show how it completes the proof of the theorem.

LEMMA 18

If $\eta \neq \tau \in A$, then G^η and G^τ are nonisomorphic ULF groups.

PROOF.

Suppose that $\pi : G^\eta \to G^\tau$ is an isomorphism. Then there exists $\alpha \in \lambda^+$ such that:

(i) $\pi[G_\alpha^\eta] = G_\alpha^\tau$.

(ii) $\eta(\alpha) \neq \tau(\alpha)$.

By lemma 17, $\pi[K_\alpha^\eta] = K_\alpha^\tau$, a contradiction. □

LEMMA 19 (G.C.H.)

If $\eta \in A$, then G^η is a complete ULF group.

PROOF.

We suppress the index η. Suppose that $\pi \in \mathrm{Aut}\, G$ is an outer automorphism. By corollary 14, there exists $\alpha \in \lambda^+$ such that:

(i) $\pi[G_\alpha] = G_\alpha$.

(ii) $\pi \restriction G_\alpha = f_\alpha$ is an outer automorphism of G_α.

By construction, f_α cannot be extended to an automorphism of $H_\alpha = < \mathrm{Sym}(G_\alpha, \omega), G_\alpha^\rho, \sigma_\alpha >$. By lemma 17, we have

(iii) $\pi[G_\alpha^+] = G_\alpha^+ = K_\alpha \rtimes H_\alpha$.

(iv) $\pi[K_\alpha] = K_\alpha$.

Consider the automorphism, $\varphi = p^{-1}\theta p$, given by

$$H_\alpha \xrightarrow{p^{-1}} G_\alpha^+/K_\alpha \xrightarrow{\theta} G_\alpha^+/K_\alpha \xrightarrow{p} H$$

where p is the canonical projection, and $\theta(K_\alpha x) = K_\alpha \pi(x)$. Then $\varphi \in \mathrm{Aut}\, H_\alpha$ satisfies $\varphi \restriction G_\alpha = f_\alpha$, a contradiction. □

We return to the proof of lemma 17.

LEMMA 20

Let $\eta \in A$. Then K_α^η is a maximal abelian subgroup of G^η for each $\alpha < \lambda^+$.

PROOF.

We have already noted that K_α^η is a maximal abelian subgroup of $(G_\alpha^\eta)^+$. Suppose that $g \in G^\eta \setminus (G_\alpha^\eta)^+$. Since $|K_\alpha^\eta| = \lambda$, $g \notin N(k_\alpha^\eta)$. □

For the next few pages, we fix $\eta \in A$ and suppress the index η. Let $a_\alpha \in K_\alpha$ be the function defined by

$$a_\alpha(g) = z_\alpha \quad \text{if } g = 1$$
$$= 1 \quad \text{otherwise}$$

where z_α is a generator of C_{n_α}. Then

$$K_\alpha = \langle a_\alpha^h \mid h \in H_\alpha \rangle.$$

LEMMA 21

Suppose that $x \in G \setminus G_\alpha$. If $\langle x^g \mid g \in G_\alpha \rangle$ is abelian, then $x \in K_\beta$ or $x \in \mathrm{Sym}(G_\beta, \omega)$ for some $\beta \geq \alpha$.

PROOF.

If x is an element of type (a), then $\langle x^g \mid g \in G_\alpha \rangle$ is nonabelian. So we need only consider elements of types (b) and (c).

<u>TYPE (b)</u>.

Assume that x first appears as

$$x = f.b \in K_\beta \rtimes H_\beta$$

where $f \neq 1$ and $\beta \geq \alpha$. Suppose that $b \neq 1$. Considering separately the cases that $b \in G_\alpha$ and $b \in H_\beta \setminus G_\alpha$, we see that $[G_\alpha, b] \neq 1$. Since G_α is simple, $[G_\alpha : C_{G_\alpha}(b)]$ is infinite and hence $\{b^g \mid g \in G_\alpha\}$ is infinite. Let $\sup(f) = \{g_1, \ldots, g_n\}$ be the support of the function f. There are only finitely many elements of $\{b^g \mid g \in G_\alpha\}$ such that:

(i) $\{g_i \mid 1 \le i \le n\} \cap \{g_i b^g \mid 1 \le i \le n\} \ne \phi$.

Similarly there are only finitely many $g \in G_\alpha$ such that:

(ii) $\{g_i \mid 1 \le i \le n\} \cap \{g_i g \mid 1 \le i \le n\} \ne \phi$

(iii) $\{g_i g^{-1} b\, g \mid 1 \le i \le n\} \cap \{g_i g \mid 1 \le i \le n\} \ne \phi$.

Choose $g \in G_\alpha$ so that conditions (i) to (iii) fail. Suppose that $[x^g, x] = 1$. Write

$$x^g = f^g \,.\, b^g = h.c.$$

Then

$$x\,x^g = f^c h.bc$$
$$x^g x = h^b f.cb.$$

Hence $f^c h = h^b f$. We have

$$\sup(f^c h) \subseteq \sup(f^c) \cup \sup(h)$$
$$= \{g_i b^g \mid 1 \le i \le n\} \cup \{g_i g \mid 1 \le i \le n\} .$$

Since condition (iii) fails, we have an equality above. Hence $h^b f$ has a support of cardinality $2n$, and we must have

$$\sup(h^b f) = \sup(h^b) \cup \sup(f)$$
$$= \{g_i\, g\, b \mid 1 \le i \le n\} \cup \{g_i \mid 1 \le i \le n\} .$$

But, since conditions (i) and (ii) fail, we have

$$\sup(f^c h) \cap \{g_i \mid 1 \le i \le n\} = \phi \quad ,$$

a contradiction. Consequently $b = 1$ and $x \in K_\beta$.

<u>TYPE (c)</u>

Suppose that x first appears as

$$x = \theta.\tau \in \mathrm{Sym}(G_\beta,\omega) \rtimes < G_\beta^\rho, \sigma_\beta >$$

where $\theta \ne 1$ and $\beta \ge \alpha$. Applying the canonical projection, it follows that

$< \tau^g \mid g \in G_\alpha >$ is abelian. Hence $\tau = \rho(b)$ for some $b \in G_\beta$. Suppose that $b \neq 1$. Let $\sup(\theta) = \{g_1, \ldots, g_n\}$ be the support of the permutation θ. (We are using "sup" in two senses in this paper. This should not cause confusion!) For any $g \in G_\alpha$,

$$x^g = \theta^{\rho(g)} \rho(b^g)$$

and

$$\sup(\theta^{\rho(g)}) = \{g_1 g, \ldots, g_n g\} .$$

Arguing as in the previous case, we reach a contradiction. Thus $x = \theta \in \mathrm{Sym}(G_\beta, \omega)$. □

Suppose that $\pi : G^\eta \to G^\tau$ is an isomorphism such that $\pi[G^\eta_\alpha] = G^\tau_\alpha$. Let $a_\alpha \in K^\eta_\alpha$ be the element previously defined. Then if $x = \pi(a_\alpha)$, we have $x \in K^\tau_\beta$ or $x \in \mathrm{Sym}(G^\tau_\beta, \omega)$ for some $\beta \geq \alpha$. Suppose that $x \in K^\tau_\beta$. Then

$$\pi \, [< a^g_\alpha \mid g \in G^\eta_\alpha >] \subseteq K^\tau_\beta .$$

Since $|< a^g_\alpha \mid g \in G^\eta_\alpha >| = \lambda$, we must have

$$\pi[< a^g_\alpha \mid g \in H^\eta_\alpha >] \subseteq K^\tau_\beta \rtimes H^\tau_\beta .$$

Let $z \in \pi[K^\eta_\alpha] \setminus < x^g \mid g \in G^\tau_\alpha >$. Then $L = < z^g \mid g \in G^\tau_\alpha >$ is abelian of cardinality λ. By lemma 21, $z \in K^\tau_\beta$ or $z \in H^\tau_\beta$. Suppose that $z \in H^\tau_\beta$. Then $L \subseteq H^\tau_\beta$ and

$$< x^g \mid g \in G^\tau_\alpha > \subseteq N(L) ,$$

a contradiction. Thus $z \in K^\tau_\beta$. Since K^η_α is a maximal abelian subgroup of K^η_α, $\pi[K^\eta_\alpha] = K^\tau_\beta$. So we have shown:

LEMMA 22

Suppose that the isomorphism $\pi : G^\eta \to G^\tau$ satisfies:

(i) $\pi[G^\eta_\alpha] = G^\tau_\alpha$.

(ii) $\pi(a_\alpha) \in K^\tau_\beta$ for some $\beta \geq \alpha$.

Then $\pi[K^{\eta}_{\alpha}] = {}^{\tau}_{\beta}$. □

Suppose that $\beta > \alpha$. Then

$$\pi[(G^{\eta}_{\alpha})^{+}] = \pi[N(K^{\eta}_{\alpha})] = (G^{\tau}_{\beta})^{+}.$$

This induces an isomorphism $\varphi = p^{-1}\,\theta\,p'$

$$H^{\eta}_{\alpha} \xrightarrow{p^{-1}} (G^{\eta}_{\alpha})^{+}/K^{\eta}_{\alpha} \xrightarrow{\theta} (G^{\tau}_{\beta})^{+}/K^{\tau}_{\beta} \xrightarrow{p'} H^{\tau}_{\beta}$$

where p, p' are the canonical projections, and $\theta(K^{\eta}_{\alpha}x) = K^{\tau}_{\beta}\,\pi(x)$. Hence φ satisfies:

(i) $\varphi[G^{\eta}_{\alpha}] = G^{\tau}_{\alpha}$.

By theorem 11.4.1 of Scott [7], $Alt(G^{\eta}_{\alpha})$ is the unique minimal normal nontrivial subgroup of H^{η}_{α}. Thus:

(ii) $\varphi[Alt(G^{\eta}_{\alpha})] = Alt(G^{\tau}_{\beta})$.

Let X^{η}_{α}, X^{τ}_{β} denote the sets of 3-cycles of $Alt(G^{\eta}_{\alpha})$, $Alt(G^{\tau}_{\beta})$ respectively. By theorem 11.4.2 of Scott [7], we have:

(iii) $\varphi[X^{\eta}_{\alpha}] = X^{\tau}_{\beta}$.

Let $x = (a\ b\ c) \in X^{\eta}_{\alpha}$. Its orbit under conjugation by elements of G^{η}_{α} is

$$0_x = \{(ag\ bg\ cg) \mid g \in G^{\eta}_{\alpha}\}.$$

<u>CLAIM</u>

If $x, y \in X^{\eta}_{\alpha}$ and $y \notin 0_x \cup 0_{x^2}$, then there exists $g \in G^{\eta}_{\alpha}$ such that $[x, y^{\rho(g)}] \neq 1$.

<u>PROOF OF CLAIM</u>

Suppose that $x = (a\ b\ c)$ and $y = (\alpha\ \ \beta\ \ \gamma)$. If $g = \alpha^{-1} a$, then $y^{\rho(g)} = (a\ \ \beta\alpha^{-1}a\ \ \gamma\alpha^{-1}a)$ satisfies our requirements. □

Since $\beta > \alpha$, there exist elements $a,b,c \in G^{\tau}_{\beta}\backslash G^{\tau}_{\alpha}$. Let $x = (a\ b\ c) \in X^{\tau}_{\beta}$ and $y = (\alpha\ \ \beta\ \ \gamma)$, where $\alpha, \beta, \gamma \in G^{\tau}_{\alpha}$. Then for all

$g \in G_\alpha^\tau$, we have $y^{\rho(g)} \in \mathrm{Alt}(G_\alpha^\tau)$. Thus:

(a) $y \notin x^{G_\alpha^\tau} \cup (x^2)^{G_\alpha^\tau}$.

(b) for all $g \in G_\alpha^\tau$, $[x, y^{\rho(g)}] = 1$.

But this contradicts the claim and (i), (iii) above. We conclude:

LEMMA 23

Suppose that the hypothesis of lemma 22 holds. Then:

(i) $\pi[(G_\alpha^\eta)^+] = (G_\alpha^\tau)^+$.

(ii) $\pi[K_\alpha^\eta] = K_\alpha^\tau$. □

Thus to prove lemma 17, it is enough to show that $\pi(a_\alpha) \notin \mathrm{Sym}(G_\beta^\tau,\omega)$ for $\beta \geq \alpha$. For the sake of contradiction, assume that $\pi(a_\alpha) \in \mathrm{Sym}(G_\beta^\tau,\omega)$. Then

$$\pi[< a_\alpha^g \mid g \in G_\alpha^\eta >] \subseteq \mathrm{Sym}(G_\beta^\tau,\omega) .$$

Suppose that $z \in \pi[K_\alpha^\eta] \setminus < x^g \mid g \in G_\alpha^\tau >$. Then, arguing as in the previous case, $z \in \mathrm{Sym}(G_\beta^\tau,\omega)$. Hence

$$\pi\ [K_\alpha^\eta] \subseteq \mathrm{Sym}(G_\beta^\tau,\omega).$$

Since $N(\pi[K_\alpha^\eta]) \subseteq H_\beta^\tau$, we must have

$$\pi[K_\alpha^\eta \rtimes H_\alpha^\eta] \subseteq H_\beta^\tau.$$

Consider any element $z \in H_\alpha^\eta$. Then for all $g \in G_\alpha^\eta$, $[z, a^g] \neq 1$. Suppose that $\pi(z) \in \mathrm{Sym}(G_\beta^\tau,\omega)$. Let

$$\sup(\pi(a_\alpha)) = \{a_1, \ldots, a_n\}$$
$$\sup(\pi(z)) = \{b_1, \ldots, b_m\} .$$

There exists $g \in G_\alpha^\tau$ such that

$$\{a_1 g, \ldots, a_n g\} \cap \{b_1, \ldots, b_m\} = \phi.$$

Hence, letting $\pi(h) = g$ for some $h \in G_\alpha^\eta$ we obtain

$$[\pi(z), \pi(a_\alpha)^{\pi(h)}] = 1 \quad ,$$

a contradiction. Hence for all $1 \neq z \in H_\alpha^\eta$, $\pi(z) \notin \mathrm{Sym}(G_\beta^\tau,\omega)$. We conclude that:

(*) $$\pi[K_\alpha^\eta \rtimes H_\alpha^\eta] \cap \mathrm{Sym}(G_\beta^\tau,\omega) = \pi[K_\alpha^\eta].$$

For the sake of concreteness, suppose that $a_\alpha^2 = 1$. (The other case is handled similarly.) Express $x = \pi(a_\alpha)$ as a product of disjoint transpositions

$$x = \prod_{i=1}^{s} (c_i \; d_i) \in \mathrm{Sym}(G_\beta^\tau,\omega).$$

Then for each $g \in G_\alpha^\tau = \pi[G_\alpha^\eta]$,

$$x^g = \prod_{i=1}^{s} (c_i g \;\; d_i g).$$

We may inductively define a sequence of elements of G_α^τ, $Z = \langle g_\xi | \; \xi < \lambda \rangle$, such that:

(i) $g_0 = 1$.

(ii) if $i \neq j < \lambda$, then $\sup(x^{g_i}) \cap \sup(x^{g_j}) = \phi$.

Then $\langle x^g | g \in Z \rangle$ is an abelian group of cardinality λ. Consider the element

$$\theta = \prod_{i=1}^{s} (c_i \; c_i g_1)(d_i \; d_i g_1).$$

Then $\theta \in \mathrm{Sym}(G_\beta^\tau,\omega)$ and

(iii) $x^\theta = x^{g_1}$.

(iv) $(x^{g_1})^\theta = x$.

(v) $(x^{g_i})^\theta = x^{g_i}$ for $1 < i < \lambda$.

Hence θ normalizes $\langle x^g | g \in Z \rangle$ and $[\theta, \pi[K_\alpha^\eta]] \neq 1$. Consider

$$B = \pi^{-1}[\langle x^g | g \in Z \rangle] \subseteq K_\alpha^\eta .$$

Since $|B| = \lambda$, $N(B) \subseteq K_\alpha^\eta \rtimes H_\alpha^\eta$. Hence there exists $z \in (K_\alpha^\eta \rtimes H_\alpha^\eta) \setminus K_\alpha^\eta$ such that $\pi(z) = \theta$, contradicting (*). This completes the proof of lemma 17.

REFERENCES

[1] P. Hall, Some constructions for locally finite groups, J. London Math. Soc. 34 (1959), 305-319.

[2] K. Hickin, Complete universal locally finite groups, Trans. Amer. Math. Soc. 239 (1978), 213-227.

[3] O. H. Kegel and B.A.F. Wehrfritz, Locally finite groups, North-Holland, Amsterdam, 1973.

[4] A. Macintyre, Existentially closed structures and Jensen's Principle $\diamondsuit$, Israel J. Math. 25 (1976), 202-210.

[5] A. Macintyre and S. Shelah, Uncountable universal locally finite groups, J. Algebra 43 (1976), 168-175.

[6] B. H. Neumann, On amalgams of periodic groups, Proc. Roy. Soc. London, Ser. A. 255 (1960), 477-489.

[7] W. R. Scott, Group theory, Prentice-Hall, Englewood Cliffs, N.J., 1964.

[8] S. Shelah, Existentially closed models in continuum, preprint, 1977.

[9] S. Shelah, Models with second order properties III. Omitting types for L(Q), Arch. Math. Logik 21 (1980), 1-11.

LOGIC COLLOQUIUM '84
J.B. Paris, A.J. Wilkie, and G.M. Wilmers (Editors)

p-$\aleph_0$-CATEGORICAL STRUCTURES

Carlo Toffalori

Istituto Matematico "U.Dini"
Università degli Studi di Firenze
Florence, Italy

Let T be a countable, complete 1st order theory with no finite models; for every $M \models T$, let $B(M)$ be the Boolean algebra of parametrically definable subsets of M, notice that $B(M)$ is atomic and $|B(M)| = |M|$. For every infinite cardinal λ, define λ-Boolean spectrum of T the set of the isomorphism types of the algebras $B(M)$ for $M \models T$, $|M| = \lambda$. A problem closely linking model theory and Boolean (meta) algebra is to classify 1st order theories by looking at their Boolean spectra; a large discussion of this problem already appears in [MT2], we only recall that it looks more approachable if $\lambda = \aleph_0$ (in view of the Ketonen classification of isomorphism types of countable atomic Boolean algebras [K]), in particular if we limit ourselves to ω-stable theories (as a theory T is ω-stable if and only if, for every countable $M \models T$, $B(M)$ is a superatomic algebra, and a system of very simple invariants for superatomic algebras is provided by the Cantor-Bendixson analysis). A first step towards the classification of 1st order theories by their $\aleph_0$-Boolean spectra is the introducing of p-$\aleph_0$-categoricity [MT1]: a theory T is said to be p-$\aleph_0$-categorical if its $\aleph_0$-Boolean spectrum contains only one isomorphism type. p-$\aleph_0$- categorical theories were studied in a satisfactory way in [MT1], the aim of this paper is to develop those results, emphasizing an algebraic direction, namely the study of p-$\aleph_0$-categorical structures (a structure M is said to be p-$\aleph_0$-categorical if Th(M) is). One could notice that a lot of examples of p-$\aleph_0$- categorical structures are just provided by the general theory, for instance $\aleph_0$-categorical structures, $\aleph_1$-categorical structures, ω-stable structures of Morley rank 1 are p-$\aleph_0$-categorical. Some more results, mainly concerning rings, were proved in [MT1], it may be useful to recall that algebraically closed fields, real closed fields, differentially closed fields of characteristic 0 are p-$\aleph_0$-categorical. We want here to deal with this problem for groups. We will proceed step by step, starting of course from

Abelian groups, or, more generally, from modules (over a countable ring); by combining some classical model theoretic results about modules (see [Z]) and Ketonen's analysis of countable atomic Boolean algebras [K], we will show that every module (hence every Abelian group) is p-$\aleph_0$-categorical.
The general problem for groups seems to be very difficult; even if we limit ourselves to the ω-stable case, the situation keeps confusing, because the problem of classifying ω-stable groups is still open and looks essentially intractable. No benefit is got by restricting ourselves to some comparatively slight generalization of Abelian groups, like nilpotent groups of class 2; in fact, ω-stable nil-2 groups are a real enigma, as well as all ω-stable groups (see [BCM] and [Me]); this difficulty hinders the understanding of p-$\aleph_0$-categorical groups, but lets us give an example of an ω-stable nil-2 group which is not p-$\aleph_0$-categorical.
Most of our notation comes from [MT1] and [MT2]; we will use the multiplicative notation $(\cdot, ^{-1}, u)$ for groups, but we will prefer the additive one $(+, -, 0)$ while dealing with Abelian groups.
Thanks to G. Cherlin and M. Ziegler for their valuable suggestions.

§ 1 - All modules are p-$\aleph_0$-categorical

1. We consider here (left) modules over a countable ring R with identity. R-modules are represented in a 1st order language L(R) containing +, -, 0, and a 1-ary functional symbol for every $r \in R$. We recall some basic facts about the model theory of modules (the main ones concerning the definable subsets of a module).
A formula (of L(R)) $\gamma(\bar{v})$ is said to be an equation if $\gamma(\bar{v})$ is of the following kind

$$r_0 v_0 + r_1 v_1 + \dots + r_n v_n = 0$$

where $r_0, r_1, \dots, r_n \in R$. A formula $\phi(\bar{v})$ is said to be a pp-formula (positive primitive formula) if $\phi(\bar{v})$ is of the following kind

$$\exists \bar{z} \, (\bigwedge_{i \leq m} \gamma_i(\bar{v}, \bar{z}))$$

where $\gamma_i(\bar{v}, \bar{z})$ is an equation for every $i \leq m$. Notice that, if M is a module and $\phi(v, \bar{w})$ is a pp-formula, then $\phi(M, \bar{0})$ is a subgroup of M (called a pp-definable subgroup) and, for every $\bar{a} \in M$, either $\phi(M, \bar{a}) = \emptyset$ or $\phi(M, \bar{a})$ is a coset of $\phi(M, \bar{0})$.

Baur's quantifier elimination procedure implies that every parametrically definable subset X of M can be represented in the following way

$$X = \bigcup_{i<m} ((\phi_i(M) + a_i) - \bigcup_{j<n_i} (\phi_{ij}(M) + a_{ij}))$$

where $m, n_i < \omega$ and, for every $i < m$, $j < n_i$, $\phi_i(M)$, $\phi_{ij}(M)$ are pp-definable subgroups of M (without loss of generality $\phi_{ij}(M)$ is a subgroup of $\phi_i(M)$) and $a_i, a_{ij} \in M$. We can also say that, if $X \neq \emptyset$, then

$$X = \dot{\bigcup_{i<m}} ((\phi_i(M) + a_i) - \bigcup_{j<n_i} (\phi_{ij}(M) + a_{ij}))$$

where $0 < m < \omega$, $n_i < \omega$ and, for every i, j, $\phi_i(M)$, $\phi_{ij}(M)$, a_i, a_{ij} are defined as above, moreover $[\phi_i(M) : \phi_{ij}(M)]$ is infinite.

A very important result for our purposes is the Neumann Lemma.

<u>Lemma 1.1 (B. H. Neumann)</u> - If G is an Abelian group, $H_0, H_1, \ldots, H_n, H$ are subgroups of G, $a_0, a_1, \ldots, a_n, a \in G$, $H + a \subseteq \bigcup_{i\leq n} (H_i + a_i)$ and there is $k \leq n$ such that $H / H \cap H_i$ is finite if and only if $i \leq k$, then

$$H + a \subseteq \bigcup_{i\leq k} (H_i + a_i) .$$

In fact, the following proposition is a direct consequence of Lemma 1.1.

<u>Lemma 1.2</u> - Let M be a (countable) module,

$$X = (\phi(M) + a) - \bigcup_{k<m} (\phi_k(M) + a_k)$$

where $\phi(M)$, $\phi_k(M)$ are pp-definable subgroups of M, $a, a_k \in M$ and, for every $k < m$, $\phi_k(M)$ is a subgroup of $\phi(M)$ and $[\phi(M) : \phi_k(M)]$ is infinite. Then, for every subgroup $\phi'(M)$ of $\phi(M)$ satisfying $[\phi(M) : \phi'(M)]$ infinite, there are infinitely many elements $(b_i : i \in \omega)$ in X such that

* if $i \neq j$, then $b_i - b_j \notin \phi'(M)$;
* for every $k < m$, either $[\phi'(M) : \phi'(M) \cap \phi_k(M)]$ is infinite or, for all $i \in \omega$, $(\phi'(M) + b_i) \cap (\phi_k(M) + a_k) = \emptyset$.

<u>Proof</u> - We proceed by induction on m.

$m = 0$: remember that $[\phi(M) : \phi'(M)]$ is infinite.

$m \to m+1$: let $X_0 = (\phi(M) + a) - \bigcup_{k<m} (\phi_k(M + a_k)$, then there exist infinitely many elements b_i $(i \in \omega)$ in X_0 such that, if $i \neq j$, then $b_i - b_j \notin \phi'(M)$ and, for every $k < m$, either $[\phi'(M) : \phi'(M) \cap \phi_k(M)]$ is infinite or $(\phi'(M) + b_i) \cap (\phi_k(M) + a_k) = \emptyset$ for all $i \in \omega$. Choose a maximal set $\{b_i : i \in \omega\}$ satisfying these properties, consider $\phi_m(M) + a_m$ and set $I = \{i \in \omega : (\phi'(M) + b_i) \cap (\phi_m(M) + a_m) = \emptyset\}$.

1st case: $[\phi'(M) : \phi'(M) \cap \phi_m(M)]$ is infinite. Therefore $\{b_i : i \in \omega\}$ still works.

2nd case: $[\phi'(M) : \phi'(M) \cap \phi_m(M)]$ is finite and I is coinfinite. $\{b_i : i \notin I\}$ works.

3rd case: $[\phi'(M) : \phi'(M) \cap \phi_m(M)]$ is finite and I is cofinite, namely for almost all $i \in \omega$ $(\phi'(M) + b_i) \cap (\phi_m(M) + a_m) \neq \emptyset$. Let $n = [\phi'(M) : \phi'(M) \cap \phi_m(M)]$, then there exist $a_{m,0}$ $(= a_m)$, ..., $a_{m,n-1}$ such that

$$\bigcup_{i \in I} (\phi'(M) + b_i) \subseteq \bigcup_{h<n} (\phi_m(M) + a_{m,h}).$$

For every $k < m$ such that $[\phi'(M) : \phi'(M) \cap \phi_k(M)]$ is finite ($n(k)$, say), let $I_k = \{b \in X_0 : (\phi'(M) + b) \cap (\phi_k(M) + a_k) \neq \emptyset\}$.
Again, there exist $a_{k,0}$ $(= a_k)$, ..., $a_{k,n(k)-1}$ satisfying

$$\bigcup_{b \in I_k} (\phi'(M) + b) \subseteq \bigcup_{h<n(k)} (\phi_k(M) + a_{k,h}).$$

Therefore

$$\phi(M) + a = \bigcup_{i \in \omega,\ i \notin I} (\phi'(M) + b_i) \cup \bigcup_{h<n} (\phi_m(M) + a_{m,h}) \cup$$
$$\cup \bigcup_{k<m} \bigcup_{h<n(k)} (\phi_k(M) + a_{k,h}).$$

This contradicts Lemma 1.1, since $\phi'(M)$, $\phi_0(M)$, ..., $\phi_m(M)$ have infinite index in $\phi(M)$.-

Some last remarks about modules: every module is stable, furthermore, if M is a module and $p \in S(M)$ (the dual space of $B(M)$), define

$$p^+ = \{ \phi(v, \bar{a}) \in p : \phi(v, \bar{w}) \text{ pp-formula} \}$$

$$p^- = \{ \neg\phi(v, \bar{a}) \in p : \phi(v, \bar{w}) \text{ pp-formula} \}.$$

Apparently p is axiomatized by $p^+ \cup p^-$ and determined by p^+.

2. Let us turn our attention now to countable atomic Boolean algebras. It may be useful to recall the Ketonen results [K] about the classification of isomorphism types of these algebras. Let B ($\neq (0)$) be a countable atomic Boolean algebra then we can define an increasing sequence $(I_\nu(B) : \nu \text{ ordinal})$ of ideals of B in the following way:

* $I_0(B) = (0)$,
* $I_1(B)$ = ideal generated by At B,
* $I_\lambda(B) = \bigcup_{\nu<\lambda} I_\nu(B)$ if λ is a limit ordinal,
* $I_{\nu+1}(B)$ = preimage of $I_1(B/I_\nu(B))$ in the natural homomorphism of B onto $B/I_\nu(B)$.

Let $b \in B$; if $b \notin I_\nu(B)$ for every ordinal ν, then we define

$$\text{CB-rank } b = \infty$$

(CB = Cantor-Bendixson); otherwise, it is easy to see that the least ordinal ν such that $b \in I_\nu(B)$ is a successor ordinal, and we set in this case

$$\text{CB-rank } b = \min \{ \nu : b \in I_{\nu+1}(B) \}.$$

Apparently, if CB-rank $b = \nu$, then $b \,|\, I_\nu(B)$ is a finite element in $B/I_\nu(B)$, and we define

$$\text{CB-degree } b = |\{a \in At(B/I_\nu(B)) : a \leq b|I_\nu(B)\}| .$$

In general, an obvious cardinality argument implies that there is $\alpha < \omega_1$ such that $I_\alpha(B) = I_{\alpha+1}(B)$. The least ordinal with this property is called the CB-rank of B, hence

$$\text{CB-rank } B = \sup \{ \text{CB-rank } b + 1 : b \in B, \text{ CB-rank } b < \infty \} .$$

If CB-rank $B = \alpha$, then either $B/I_\alpha(B) \simeq (0)$ or $B/I_\alpha(B) \simeq \Lambda$ (the free countable Boolean algebra).
Finally, if $p \in S(B)$ (the dual space of B), we define CB-rank $p =$ $= \min \{ \text{CB-rank } b : b \in p \}$.

Then, the isomorphism type of B can be determined in the following way:

a. B is superatomic (namely, $B/I_\alpha(B) \simeq (0)$): in this case, α is a successor ordinal and, if $\alpha = \beta + 1$, then $B/I_\beta(B)$ is finite. The isomorphism type of B is given by the ordered pair (α, d) where

$$\alpha = \text{CB-rank } B = \text{CB-rank } 1 + 1$$

$$d = \text{CB-degree } 1 = |At(B/I_\beta(B))|$$

(d is called the CB-degree of **B**, the ordered pair (α, d) is called the CB-type of B: there is a slight difference between the definition of CB-rank B we give here, and the one of [MT1] in the superatomic case).

b. B is uniform, namely, for every $b \in B$ such that CB-rank $b < \infty$, CB-rank $B[b] \leq$ CB-rank $B[b^*]$ (we denote here by $B[b]$ the Boolean algebra of sub-elements of b in B). Of course, if B is uniform and $B \not\simeq (0)$, then $B/I_\alpha(B) \simeq \Lambda$. In this case, we define:

* for every $b \in B$, $U(b) = \{ p \in S(B) : b \in p, \text{ CB-rank } p = \infty \}$ (notice that, if $b \in I_\alpha(B)$, then $U(b) = \emptyset$; and, if $b|I_\alpha(B) = b'|I_\alpha(B)$, then $U(b) = U(b')$);
* for every $p \in S(B)$ (such that CB-rank $p = \infty$), $r(p) = \min \{ \text{CB-rank } B[x] : x \in p \}$.

Then we set, for every $b \in B$, $f_B(b|I_\alpha(B)) = \sup \{ r(p) : p \in U(b) \}$.
f_B is a strictly additive function of $B/I_\alpha(B)$ ($\simeq \Lambda$) in ω_1 (if we put, for every $\nu, \mu \in \omega_1$, $\nu + \mu = \max \{\nu, \mu\}$). Moreover, we define the derived functions f_B^β ($\beta < \omega_1$) of f_B in the following way: for every $b \in B$,

$$f_B^0(b \mid \alpha) = f_B(b \mid \alpha),$$

$$f_B^\beta(b \mid \alpha) = \left\{ ((f_B^\gamma(b_0 \mid \alpha))_{\gamma<\beta}, \ldots, (f_B^\gamma(b_n \mid \alpha))_{\gamma<\beta}) : b \mid \alpha = \dot{\bigvee}_{i \le n} b_i \mid \alpha \right\}$$

when $\beta > 0$ (here, $b \mid \alpha$ abridges $b|I_\alpha(B)$). Therefore, the isomorphism type of B is given by α = CB-rank B and by the sequence $(f_B^\beta(1 \mid \alpha) : \beta < \omega_1)$.

c. In general, every countable atomic Boolean algebra B ($\not\simeq (0)$) can be decomposed in one and only one way (up to isomorphisms) as a direct product $B \simeq B_1 \times B_2$ where B_1 is superatomic, B_2 is uniform and either $B_1 = (0)$ or CB-rank $B_1 >$ $>$ CB-rank B_2.

3. (The superatomic case) The idea we will follow to prove that every (countable) module M is p-$\aleph_0$-categorical is to show that the Ketonen isomorphism invariants of $B(M)$ can be obtained from the lattice of pp-definable subgroups of $Th(M)$, and hence are preserved by elementary equivalence. Of course, the first step concerns the CB-analysis of $B(M)$. Let $X \in B(M)$, $X \neq \emptyset$; by recalling the general decomposition of X (see 1), it is easy to see that, in order to calculate CB-rank X and (in case) CB-degree X, we may limit ourselves to the case

$$X = (\phi(M) + a) - \bigcup_{j<n} (\phi_j(M) + a_j)$$

where $\phi(M)$ and $\phi_j(M)$ are pp-definable subgroups of M, and, for every $j < n$, $\phi_j(M)$ is a subgroup of $\phi(M)$ and $[\phi(M) : \phi_j(M)]$ is infinite.

<u>Lemma 1.3</u> - For every $a \in M$, CB-rank X = CB-rank $(X + a)$.

<u>Proof</u> - It suffices to show (by induction on ν) that, for every ordinal ν, CB-rank $X \ge \nu$ implies CB-rank $(X + a) \ge \nu$.-

<u>Lemma 1.4</u> - (i) CB-rank X = CB- rank $\phi(M)$.

(ii) If CB-rank X < ∞, CB-degree X = CB-degree $\phi(M)$.

<u>Proof</u> - (i) Let us show that, for every ordinal ν, CB-rank X $\geq \nu$ if and only if CB-rank $\phi(M) \geq \nu$.

($\rightarrow$) is obvious (see Lemma 1.3). ($\leftarrow$) We proceed by induction on ν; the cases $\nu = 0$, ν limit are trivial. Let $\nu = \mu + 1$, CB-rank $\phi(M) \geq \nu$; therefore CB-rank $(\phi(M) + a) \geq \mu + 1$. If CB-rank $(\bigcup_{j<n} (\phi_j(M) + a_j)) < \mu$, the claim is obvious. Otherwise, there is $j < n$ (for instance, $j = 0$) such that CB-rank $(\phi_j(M) + a_j) \geq \mu$. By Lemma 1.2, there exist infinitely many elements $\{b_i : i \in \omega\}$ in X such that, if $i \neq j$, then $b_i - b_j \notin \phi_0(M)$ and, for any $j < n$, if $[\phi_0(M) : \phi_0(M) \cap \phi_j(M)]$ is finite, then $(\phi_0(M) + b_i) \cap (\phi_j(M) + a_j) = \emptyset$ for all $i \in \omega$. As CB-rank $(\phi_0(M) + b_i) \geq \mu$ for all $i \in \omega$, the induction hypothesis implies

$$\text{CB-rank } ((\phi_0(M) + b_i) - \bigcup_{j<n} (\phi_j(M) + a_j)) \geq \mu$$

for all $i \in \omega$, hence CB-rank X $\geq \mu + 1$.

(ii) Suppose CB-rank X = CB-rank $\phi(M) = \nu < \infty$, therefore CB-rank $(\bigcup_{j<n} (\phi_j(M) + a_j)) < \nu$ (see (i)).-

<u>Lemma 1.5</u> - CB-rank $\phi(M) < \infty$ if and only if there is no infinite descending sequence of pp-definable subgroups $(\phi_n(M) : n \in \omega)$ such that $\phi_0(M) = \phi(M)$.

<u>Proof</u> - ($\rightarrow$) Otherwise, we can define an infinite descending sequence of ordinals.

($\leftarrow$) Let CB-rank $\phi(M) = \infty$; if α = CB-rank B(M) and $\beta \geq \alpha$, then CB-rank $\phi(M) \geq \beta + 1$, hence there exist $X_1, X_2 \in B(M)$ such that $X_1, X_2 \subseteq \phi(M)$, $X_1 \cap X_2 = \emptyset$, CB-rank $X_1 \geq \beta$, CB-rank $X_2 \geq \beta$ (namely CB-rank X_1 = = CB-rank $X_2 = \infty$); without loss of generality, we can assume, for $i = 1, 2$,

$$X_i = (\phi_j(M) + a_i) - \bigcup_{j<n_i} (\phi_{ij}(M) + a_{ij})$$

where $\phi_i(M)$, $\phi_{ij}(M)$ satisfy the usual conditions (and $\phi_i(M)$ is a subgroup of

$\phi(M))$. Suppose $\phi_1(M) = \phi(M)$, then

$$\phi_2(M) + a_2 \subseteq \bigcup_{i=1,2} \bigcup_{j<n_i} (\phi_{ij}(M) + a_{ij})$$

and, by Lemma 1.1, $\phi_2(M)$ has infinite index in $\phi(M)$; in particular $\phi_2(M) \subsetneq \phi(M)$.-

Notice that, if CB-rank $\phi(M) < \infty$, then $\phi(M)$ admits at most finitely many pp-definable subgroups of finite index, otherwise we can define a strictly decreasing infinite sequence of pp-definable subgroups $(\phi_n(M) : n \in \omega)$ in the following way

* $\phi_0(M) = \phi(M)$

* $\phi_{n+1}(M) = \phi_n(M) \cap \psi(M)$ where $\psi(M)$ is any pp-definable subgroup of $\phi(M)$ such that $[\phi(M) : \psi(M)]$ is finite and $\phi_n(M) \not\subseteq \psi(M)$.

In particular, the intersection of all pp-definable subgroups of $\phi(M)$ whose index in $\phi(M)$ is finite is a pp-definable subgroup of $\phi(M)$, has finite index in $\phi(M)$ and admits no proper pp-definable subgroup of finite index. Let us denote by $\phi^\circ(M)$ this subgroup; notice that, for every $N \equiv M$, $\phi^\circ(N)$ is a pp-definable subgroup of $\phi(N)$, $[\phi(N) : \phi^\circ(N)] = [\phi(M) : \phi^\circ(M)]$ and $\phi^\circ(N)$ has no proper pp-definable subgroup of finite index.

<u>Lemma 1.6</u> - If CB-rank $\phi(M) < \infty$, then, for every $\nu < \omega_1$, CB-rank $\phi(M) \geq \nu + 1$ if and only if there is a pp-definable subgroup $\phi'(M)$ of $\phi(M)$ such that $\phi'(M)$ has infinite index in $\phi(M)$ and CB-rank $\phi'(M) \geq \nu$.

<u>Proof</u> - ($\leftarrow$) is trivial.

($\rightarrow$) By replacing (if necessary) $\phi(M)$ with $\phi^\circ(M)$ (apparently, CB-rank $\phi(M)$ = = CB-rank $\phi^\circ(M)$), we may assume that $\phi(M)$ has no pp-definable proper subgroup of finite index. As in the proof of Lemma 1.5, we have that there exist $X_1, X_2 \in$ $\in B(M)$ such that $X_1, X_2 \subseteq \phi(M)$, $X_1 \cap X_2 = \emptyset$, CB-rank $X_1 \geq \nu$, CB-rank $X_2 \geq \nu$. Without loss of generality, for every $i = 1, 2$,

$$X_i = (\phi_i(M) + a_i) - \bigcup_{j<n_i} (\phi_{ij}(M) + a_{ij})$$

where $\phi_i(M)$, $\phi_{ij}(M)$ satisfy the usual conditions and $\phi_i(M)$ is a subgroup of $\phi(M)$. Lemma 1.1 implies $\phi_1(M) \subsetneq \phi(M)$ or $\phi_2(M) \subsetneq \phi(M)$.-

Lemma 1.7 - If CB-rank $\phi(M) < \infty$, then CB-degree $\phi(M) = [\phi(M) : \phi^\circ(M)]$.

Proof - It suffices to show that CB-degree $\phi^\circ(M) = 1$. The proof is similar to the one of 1.6.-

Theorem 1.8 - If M is a countable module and $B(M)$ is superatomic, then M is p-$\aleph_0$-categorical.

Proof - Let $N \equiv M$, $|N| = \aleph_0$. We have

* $B(N)$ is superatomic; for, CB-rank $M < \infty$, and, by Lemma 1.5, CB-rank $N < \infty$, too;
* CB-rank $B(M)$ = CB-rank $B(N)$; recall that CB-rank $B(M)$ = CB-rank $M + 1$ and similarly CB-rank $B(N)$ = CB-rank $N + 1$, hence we have to show that CB-rank M = = CB-rank N; more generally, we will prove that, for every pp-formula $\phi(v)$ and for every ordinal ν,

$$\text{CB-rank } \phi(M) \geq \nu \quad \text{if and only if} \quad \text{CB-rank } \phi(N) \geq \nu.$$

  The cases $\nu = 0$, ν limit are trivial. If $\nu = \mu + 1$, Lemma 1.6 implies that CB-rank $\phi(M) \geq \nu$ if and only if $\phi(M)$ admits a pp-definable subgroup $\phi'(M)$ satisfying $[\phi(M) : \phi'(M)]$ infinite and CB-rank $\phi'(M) \geq \mu$. Therefore, $\phi'(N)$ is a subgroup of $\phi(N)$, $[\phi(N) : \phi'(N)]$ is infinite and CB-rank $\phi'(N) \geq \mu$. Hence CB-rank $\phi(N) \geq \nu$.
* CB-degree $B(M)$ = CB-degree $B(N)$: see Lemma 1.7.-

4. (The uniform case) First we will show:

Theorem 1.9 - For every countable module M, either $B(M)$ is superatomic or $B(M)$ is uniform.

Proof - Suppose that $B(M)$ is not superatomic, and choose $X \in B(M)$ such that CB-rank $X = \nu < \infty$. As usually, we set

$$X = \dot{\bigcup_{i<m}} ((\phi_i(M) + a_i) - \bigcup_{j<m_i} (\phi_{ij}(M) + a_{ij}))$$

where a_i, $a_{ij} \in M$, $\phi_i(M)$, $\phi_{ij}(M)$ are pp-definable subgroups of M and, for every $i < m$ and $j < m_i$, $\phi_{ij}(M)$ is a subgroup of $\phi_i(M)$ and $[\phi_i(M) : \phi_{ij}(M)]$ is infinite. Therefore CB-rank $B[X] = \nu + 1$; on the other side, there is $i < m$ ($i = 0$, for simplicity) such that

$$\text{CB-rank } \phi_0(M) = \text{CB-rank } ((\phi_0(M) + a_0) - \bigcup_{j<m_0} (\phi_{0j}(M) + a_{0j})) = \nu ;$$

since CB-rank $M = \infty$, $[M : \phi_i(M)]$ is infinite for every $i < n$, hence, by Lemma 1.2, there exist infinitely many elements $(b_h : h \in \omega)$ in M satisfying

* $b_h - b_k \notin \phi_0(M)$ if $h \neq k$,
* for every $i < m$, if $[\phi_0(M) : \phi_0(M) \cap \phi_i(M)]$ is finite, then for all $h \in \omega$ $(\phi_0(M) + b_h) \cap (\phi_i(M) + a_i) = \emptyset$.

In particular, for all $h \in \omega$, CB-rank $(\phi_0(M) + b_h)$ = CB-rank $((\phi_0(M) + b_h) - X) = \nu$. Consequently, CB-rank $B[M - X] \geq \nu + 1$. It follows that $B(M)$ is uniform.-

Let now M be a countable module such that $B(M)$ is uniform. Put α = CB-rank $B(M)$, in order to get the isomorphism type of $B(M)$, we have to consider the strictly additive function $f_M = f_{B(M)}$ of $B(M) / I_\alpha(B(M))$ ($\simeq \Lambda$) in ω_1 which is defined in the following way (see 2):

* for any $X \in B(M)$, let $U(X) = \{ p \in S(M) : X \in p$, CB-rank $p = \infty \}$.
* for any $p \in S(M)$ (satisfying CB-rank $p = \infty$), let $r(p) = \min \{$ CB-rank $B[Y] : Y \in p \}$,
* finally, for any $X \in B(M)$, set $f_M(X|\alpha) = \sup \{ r(p) : p \in U(X) \}$ where $X|\alpha$ abbreviates $X|I_\alpha(B(M))$.

We define as in 2 the derived functions f_M^β $(\beta < \omega_1)$ of f_M, then the isomorphism type of $B(M)$ is determined by α and $(f_M^\beta(M|\alpha) : \beta < \omega_1)$.

<u>Lemma 1.10</u> - For every $X \in B(M)$, $a \in M$, $\beta < \omega_1$, $f_M^\beta(X|\alpha) = f_M^\beta((X + a)|\alpha)$.

<u>Proof</u> - If $\beta = 0$, recall that, for every $X \in B(M)$, $a \in M$, CB-rank X = = CB-rank $(X + a)$ and notice that $p \in U(X)$ if and only if $p + a$ =

$= \{ Y + a : Y \in p \} \in U(X + a)$, and $r(p) = r(p + a)$. Let now $\beta > 0$, we assume our claim true for every $\gamma < \beta$ and we show it for β. In fact, we have $X|\alpha = \dot{\bigcup}_{j<m} X_j|\alpha$ if and only if $(X + a)|\alpha = \dot{\bigcup}_{j<m} (X_j + a)|\alpha$, where, for every $\gamma < \beta$ and $j < m$, $f^\gamma_M(X_j|\alpha) = f^\gamma_M((X_j + a)|\alpha)$.-

Let now $X \in B(M)$, $X = (\phi(M) + a) - \bigcup_{j<n} (\phi_j(M) + a_j)$ (with the usual conditions on $\phi(M)$ and $\phi_j(M)$).

<u>Lemma 1.11</u> - For every $\beta < \omega_1$, $f^\beta_M(X|\alpha) = f^\beta_M(\phi(M)|\alpha)$.

<u>Proof</u> - First suppose $\beta = 0$, we proceed by induction on n. $n = 0$: the claim is trivial.

$n \to n + 1$: if $X' = (\phi(M) + a) - \bigcup_{j<n} (\phi_j(M) + a_j)$, then $f_M(X'|\alpha) = f_M(\phi(M)|\alpha)$, we have to consider now $\phi_n(M) + a_n$. If $f_M(\phi_n(M)|\alpha) < f_M(X'|\alpha)$, our claim follows from the additivity of f_M. In any case, Lemma 1.2 implies that there are infinitely many elements $\{ b_h : h \in \omega \}$ in X such that, if $h \neq k$, then $b_h - b_k \notin \phi_n(M)$ and, for every $j \leq n$, if $[\phi_n(M) : \phi_n(M) \cap \phi_j(M)]$ is finite, then $(\phi_n(M) + b_h) \cap (\phi_j(M) + a_j) = \emptyset$ for all $h \in \omega$. Put

$$Y_h = (\phi_n(M) + b_h) - \bigcup_{j\leq n} (\phi_j(M) + a_j) = (\phi_n(M) + b_h) - \bigcup_{j<n} (\phi_j(M) + a_j) .$$

Therefore $f_M(Y_h|\alpha) = f_M(\phi_n(M)|\alpha) = f_M(X'|\alpha)$. Since $Y_h \subseteq X \subseteq X'$, it follows $f_M(X|\alpha) = f_M(X'|\alpha) = f_M(\phi(M)|\alpha)$.

Let now $\beta > 0$; as above, we suppose our claim true for every $\gamma < \beta$ and we prove it for β. The problem lies in showing that, if $X = (\phi(M)+a) - \bigcup_{j\leq n} (\phi_j(M)+a_j)$ and $X' = (\phi(M) + a) - \bigcup_{j<n} (\phi_j(M) + a_j)$, then, for every decomposition $X|\alpha = \dot{\bigcup}_{i<m} X_i|\alpha$, we can define a decomposition $X'|\alpha = \dot{\bigcup}_{i<m} X'_i|\alpha$ such that, for all $\gamma < \beta$ and $i < m$, $f^\gamma_M(X_i|\alpha) = f^\gamma_M(X'_i|\alpha)$, and conversely.

First suppose $X|\alpha = \dot{\bigcup}_{i<m} X_i|\alpha$; with no loss of generality, we can assume $X = \dot{\bigcup}_{i<m} X_i$; moreover, for every $i < m$, if $X_i \neq \emptyset$,

$$X_i = \dot{\bigcup}_{h < t_i} X_{ih}$$

where $t_i > 0$ and for every $h < t_i$ X_{ih} is contained in a suitable coset

$\psi_{ih}(M) + a_{ih}$ ($\psi_{ih}(M)$ a pp-definable subgroup of $\phi(M)$) and $(\psi_{ih}(M) + a_{ih}) - X_{ih}$ is the union of finitely many smaller cosets corresponding to pp-definable subgroups of infinite index in $\psi_{ih}(M)$. Notice that Lemma 1.1 implies that $\phi(M) + a$ equals the union of those cosets $\psi_{ih}(M) + a_{ih}$ such that $[\phi(M) : \psi_{ih}(M)]$ is finite. Of course we can assume $(\phi_n(M) + a_n) - \bigcup_{j<n} (\phi_j(M) + a_j) \neq \emptyset$ and, furthermore, $[\phi_n(M) : \phi_n(M) \cap \phi_j(M)]$ infinite for every $j < n$ (otherwise we replace $\phi_n(M)$ with the intersection of $\phi_n(M)$ and those subgroups $\phi_j(M)$ such that $[\phi_n(M) : \phi_n(M) \cap \phi_j(M)]$ is finite, and we repeat several times the following procedure). As $[\phi(M) : \phi_n(M)]$ is infinite, there exist $i < m$, $h < t_i$ such that $[\phi(M) : \psi_{ih}(M)]$ is finite and $(\phi_n(M) + a_n) \cap (\psi_{ih}(M) + a_{ih}) \neq$ $\neq \emptyset$ (hence $[\phi_n(M) : \phi_n(M) \cap \psi_{ih}(M)] = q$ is finite, while $[\psi_{ih}(M) : \phi_n(M) \cap \psi_{ih}(M)]$ is infinite; let $b_0, \ldots, b_{q-1} \in \psi_{ih}(M) + a_{ih}$ such that, if $u < v < q$, then $b_u - b_v \notin \phi_n(M) \cap \psi_{ih}(M)$). By using the induction hypothesis, the additivity of f_M^γ and Lemma 1.10, we get, for every $\gamma < \beta$,

$$f_M^\gamma\Big(\bigcup_{u<q} (\phi_n(M) \cap \psi_{ih}(M) + b_u) \,|\, \alpha\Big) = f_M^\gamma((\phi_n(M) + a_n) \,|\, \alpha) =$$

$$= f_M^\gamma\Big((\phi_n(M) + a_n) - \bigcup_{j<n} (\phi_j(M) + a_j) \,|\, \alpha\Big),$$

hence

$$f_M^\gamma(X_{ih} \,|\, \alpha) = f_M^\gamma((\psi_{ih}(M) + a_{ih}) \,|\, \alpha) =$$

$$= f_M^\gamma\Big(((\psi_{ih}(M) + a_{ih}) - \bigcup_{u<q} (\phi_n(M) \cap \psi_{ih}(M) + b_u)) \,|\, \alpha\Big) +$$

$$+ f_M^\gamma\Big(\bigcup_{u<q} (\phi_n(M) \cap \psi_{ih}(M) + b_u) \,|\, \alpha\Big) =$$

$$= f_M^\gamma(X_{ih} \,|\, \alpha) + f_M^\gamma\Big((\phi_n(M) + a_n) - \bigcup_{j<n} (\phi_j(M) + a_j) \,|\, \alpha\Big) =$$

$$= f_M^\gamma\Big(X_{ih} \cup ((\phi_n(M) + a_n) - \bigcup_{j<n} (\phi_j(M) + a_j)) \,|\, \alpha\Big).$$

Define

$$X_i' = X_i \cup ((\phi_n(M) + a_n) - \bigcup_{j<n} (\phi_j(M) + a_j)),$$

$$X_l' = X_l \text{ if } l < n,\ l \neq i \text{ (hence } X_l' = \emptyset \text{ if } X_l = \emptyset),$$

then $X' = \bigcup_{l<n} X'_l$ (where $X'_l \cap X'_s = \emptyset$ if $l < s < n$), and $f^{\gamma}_M(X'_l|\alpha) = f^{\gamma}_M(X_l|\alpha)$ for every $l < n$ and $\gamma < \beta$. Consequently, $f^{\beta}_M(X|\alpha) \subseteq f^{\beta}_M(X'|\alpha)$.

Let now $X'|\alpha = \dot{\bigcup}_{i<m} X'_i|\alpha$, we claim that there is a decomposition $X|\alpha = \dot{\bigcup}_{i<m} X_i|\alpha$ such that, for every $i < m$, $\gamma < \beta$, $f^{\gamma}_M(X_i|\alpha) = f^{\gamma}_M(X'_i|\alpha)$ (then, $f^{\beta}_M(X|\alpha) \supseteq f^{\beta}_M(X'|\alpha)$). As above, we can assume $X' = \dot{\bigcup}_{i<m} X'_i$ and, for every $i < m$, if $X'_i \neq \emptyset$, $X'_i = \dot{\bigcup}_{h<t_i} X'_{ih}$ $(t_i > 0)$, where each X'_{ih} is contained in a suitable coset $\psi_{ih}(M) + a_{ih}$ ($\psi_{ih}(M)$ a pp-definable subgroup of $\phi(M)$) and $(\psi_{ih}(M) + a_{ih}) - X'_{ih}$ is the union of finitely many smaller cosets corresponding to pp-definable subgroups of infinite index in $\psi_{ih}(M)$. For every $i < m$ (such that $X'_i \neq \emptyset$) and $h < t_i$, consider

$$X''_{ih} = X'_{ih} - (\phi_n(M) + a_n).$$

1st case: $X''_{ih} = X'_{ih}$; then $f^{\gamma}_M(X''_{ih}|\alpha) = f^{\gamma}_M(X'_{ih}|\alpha)$ for all $\gamma < \beta$;

2nd case: $X''_{ih} \neq X'_{ih}$, but $[\psi_{ih}(M) : \phi_n(M) \cap \psi_{ih}(M)]$ is infinite; again $f^{\gamma}_M(X''_{ih}|\alpha) = f^{\gamma}_M(X'_{ih}|\alpha)$ for all $\gamma < \beta$ (use the induction hypothesis);

3rd case: $X''_{ih} \neq X'_{ih}$ and $[\psi_{ih}(M) : \phi_n(M) \cap \psi_{ih}(M)]$ is finite (hence, $[\phi(M) : \psi_{ih}(M)]$ is infinite); by Lemma 1.1 $\phi(M) + a$ equals the union of those cosets $\psi_{lk}(M) + a_{lk}$ such that $[\phi(M) : \psi_{lk}(M)]$ is finite; therefore there exist $l < n$, $k < t_l$ such that $[\phi(M) : \psi_{lk}(M)]$ is finite and

$$(\psi_{lk}(M) + a_{lk}) \cap (\psi_{ih}(M) + a_{ih}) \neq \emptyset .$$

Notice that:

* $\phi_n(M) \cap \psi_{ih}(M) \cap \psi_{lk}(M)$ has infinite index in $\psi_{lk}(M)$, but finite index, say q, in $\phi_n(M) \cap \psi_{ih}(M)$;
* $[\psi_{lk}(M) : \phi_n(M) \cap \psi_{lk}(M)]$ is infinite.

It follows from Lemma 1.2 that there are infinitely many elements $\{b_u : u \in \omega\}$ in X''_{lk} such that $b_u - b_v \notin \phi_n(M) \cap \psi_{ih}(M) \cap \psi_{lk}(M)$ if $u \neq v$, and, for every $u \in \omega$, $(\phi_n(M) \cap \psi_{ih}(M) \cap \psi_{lk}(M) + b_u) \cap X''_{lk}$ excludes only finitely many cosets corresponding to some pp-definable subgroups of infinite index in

$\phi_n(M) \cap \psi_{ih}(M) \cap \psi_{lk}(M)$ (in particular, it is disjoint from $\psi_{ih}(M) + a_{ih}$ and, of course, from $\phi_n(M) + a_n$). We replace

$$X''_{ih} \text{ with } X''_{ih} \cup \bigcup_{u<q} ((\phi_n(M) \cap \psi_{ih}(M) \cap \psi_{lk}(M) + b_u) \cap X''_{lk})$$

$$X''_{lk} \text{ with } X''_{lk} - \bigcup_{u<q} (\phi_n(M) \cap \psi_{ih}(M) \cap \psi_{lk}(M) + b_u),$$

notice that, for all $\gamma < \beta$ (and for a suitable element c in $\psi_{ih}(M) + a_{ih}$),

$$f^{\gamma}_M(X'_{ih}|\alpha) = f^{\gamma}_M(X''_{ih}|\alpha) + f^{\gamma}_M((X'_{ih} \cap (\phi_n(M) + a_n))|\alpha) =$$

$$= f^{\gamma}_M(X''_{ih}|\alpha) + f^{\gamma}_M((\psi_{ih}(M) \cap \phi_n(M) + c)|\alpha) =$$

$$= f^{\gamma}_M(X''_{ih}|\alpha) + f^{\gamma}_M(\bigcup_{u<q} ((\phi_n(M) \cap \psi_{ih}(M) \cap \psi_{lk}(M) + b_u) \cap X''_{lk})|\alpha) =$$

$$= f^{\gamma}_M(X''_{ih} \cup \bigcup_{u<q} ((\phi_n(M) \cap \psi_{ih}(M) \cap \psi_{lk}(M) + b_u) \cap X''_{lk})|\alpha)$$

and

$$f^{\gamma}_M(X'_{lk}|\alpha) = f^{\gamma}_M(X''_{lk}|\alpha) = f^{\gamma}_M((X''_{lk} - \bigcup_{u<q} (\phi_n(M) \cap \psi_{ih}(M) \cap \psi_{lk}(M) + b_u))|\alpha).$$

For every $i < m$ (satisfying $X'_i \neq \emptyset$) and $h < t_i$, let X_{ih} be the set which comes out eventually in this way. Put

$$X_i = \bigcup_{h<t_i} X_{ih} \text{ if } i<m,\ X'_i \neq \emptyset \ (X_{ih} \cap X_{ik} = \emptyset \text{ if } h<k<t_i)$$

$$X_i = \emptyset \text{ if } X'_i = \emptyset .$$

Apparently $X = \bigcup_{i<m} X_i$ (and $X_i \cap X_j = \emptyset$ if $i<j<m$), $f^{\gamma}_M(X_i|\alpha) = f^{\gamma}_M(X'_i|\alpha)$ for every $i < m$ and $\gamma < \beta$.-

<u>Lemma 1.12</u> - If $p \in S(M)$ (and CB-rank $p = \infty$), then $r(p) =$ $= \min \{ \text{CB-rank } B[\phi(M)] : \exists a \in M \text{ such that } \phi(M) + a \in p \}$.

<u>Proof</u> - $\leq$ is obvious. Conversely, let $Y \in p$; as above, set $Y = \dot{\bigcup}_{i<m} Y_i$, where, for every $i < m$,

$$Y_i = (\phi_i(M) + a_i) - \bigcup_{j<n_i} (\phi_{ij}(M) + a_{ij})$$

and $\phi_i(M)$, $\phi_{ij}(M)$ satisfy the usual conditions for every $i < m$, $j < n_i$.
Therefore, for some $i < m$, $\phi_i(M) + a_i \in p$ and, for every $i < m$,
CB-rank B $[Y] \geq$ CB-rank B $[Y_i]$ = CB-rank B $[\phi_i(M) + a_i]$ = CB-rank B $[\phi_i(M)]$.-

<u>Theorem 1.13</u> - If M is a countable module and $B(M)$ is uniform, then M is p-$\aleph_0$-categorical.

<u>Proof</u> - Let $N \equiv M$, $|N| = \aleph_0$; $B(N)$ is uniform because of 1.8 and 1.9.
Moreover CB-rank B(N) = CB-rank B(M) since CB-rank B(M) =
= sup { CB-rank $\phi(M) + 1$: $\phi(v)$ pp-formula, CB-rank $\phi(M) < \infty$ } (and similarly for CB-rank B(N)) and, for every pp-formula $\phi(v)$, CB-rank $\phi(M)$ = CB-rank $\phi(N)$.
Let α = CB-rank B(M), then it suffices to show that, for every pp-formula $\phi(v)$ and $\beta < \omega_1$, $f_M^\beta(\phi(M)|\alpha) = f_N^\beta(\phi(N)|\alpha)$.
Suppose $\beta = 0$; without loss of generality we can assume $N > M$.

* $f_M(\phi(M)|\alpha) \leq f_N(\phi(N)|\alpha)$. Let $p \in S(M)$ such that $\phi(M) \in p$ and CB-rank $p = \infty$, and $p|N$ be the heir of p in $S(N)$ (see [LP], recall that Th(M) is stable); therefore $\phi(N) \in p|N$, CB-rank $p|N = \infty$, and, finally, $r(p) = r(p|N)$ (see Lemma 1.12, recall that, for every pp-formula $\psi(v)$, CB-rank B $[\psi(M)]$ = = CB-rank B $[\psi(N)]$). Then, for every $p \in U(\phi(M))$, there exists $q = p|N \in U(\phi(N))$ such that $r(p) = r(q)$.

* $f_M(\phi(M)|\alpha) \geq f_N(\phi(N)|\alpha)$. It suffices to show that, for every $p \in U(\phi(N))$, there is $q \in U(\phi(M))$ such that $r(p) = r(q)$. For any $p \in U(\phi(N))$, define $p_0 \in U(\phi(N))$ in the following way

$$p_0^+ = \{ \psi(N) : \exists\, b \in N \text{ such that } \psi(N) + b \in p \}.$$

$p_0 \in S(N)$; in fact, let $\psi_i(N) \in p_0^+$ $(i < n)$, $\phi_j(N) \notin p_0^+$, $a_j \in N$ $(j < m)$, in particular, for every $i < n$, there is $b_i \in N$ such that $\psi_i(N) + b_i \in p$; suppose towards a contradiction that

$$\bigcap_{i<n} \psi_i(N) - \bigcup_{j<m} (\phi_j(N) + a_j) = \emptyset;$$

if $b \in \bigcap_{i<n} (\psi_i(N) + b_i)$, then $\bigcap_{i<n} (\psi_i(N) + b_i) - \bigcup_{j<m} (\phi_j(N) + a_j + b) = \emptyset$, whereas $\psi_i(N) + b_i \in p$, $\phi_j(N) + a_j + b \notin p$ for every $i < n$, $j < m$. Furthermore $p_0 \in U(\phi(N))$ and $r(p) = r(p_0)$. Set $q = p_0 \upharpoonright_M$, then $q \in U(\phi(M))$, $r(q) = r(p_0) = r(p)$ (for, $q^+ = \{ \phi(M) : \phi(N) \in p_0^+ \}$).

Let now $\beta > 0$, we assume our claim true for all $\gamma < \beta$ and we prove it for β. Let $\phi(M)|\alpha = \dot{\bigcup}_{j<m} X_j|\alpha$ where, for every $j < m$, X_j admits the usual representation

$$X_j = \dot{\bigcup}_{h<t_j} ((\psi_{jh}(M) + a_{jh}) - \bigcup_{s<u(i,\ h)} (\psi_{jhs}(M) + a_{jhs})) .$$

By using the elementary equivalence of N and M, we get a similar decomposition $\phi(N)|\alpha = \dot{\bigcup}_{j<m} Y_j|\alpha$ where, because of Lemma 1.11 and the induction hypothesis, for every $\gamma < \beta$ and $j < m$,

$$f_N^\gamma(Y_j|\alpha) = \sum_{h<t_j} f_N^\gamma(\psi_{jh}(N)|\alpha) = \sum_{h<t_j} f_M^\gamma(\psi_{jh}(M)|\alpha) = f_M^\gamma(X_j|\alpha) .$$

Therefore, $f_M^\beta(\phi(M)|\alpha) \subseteq f_N^\beta(\phi(N)|\alpha)$, and conversely, of course.-

Theorems 1.8, 1.9, 1.13 imply the main result of this section:

<u>Theorem 1.14</u> - Every module is p-$\aleph_0$-categorical.

§ 2 - <u>A nil-2 ω-stable non-p-$\aleph_0$-categorical group</u>

Nilpotent groups of class 2 (shortly nil-2 groups) provide a comparatively slight generalization of the class of Abelian groups; in fact a group G is said to be nil-2 if the commutator subgroup of G is contained in the centre $Z(G)$ of G. However, while Abelian groups are quite understood from a model theoretic point of view and, in particular, there is a complete classification of ω-stable Abelian groups, the situation for nil-2 groups, even in the ω-stable case, is very confusing; Mekler's paper [Me] suggests that the stability classification of nil-2 groups is the key point towards the stability classification for groups, rings and, more generally, all structures with finitely many relations and operations; in particular, the hardness of classifying ω-stable groups transfers to ω-stable nil-2 groups, however the existence of so many ω-stable nil-2 groups just suggests that there may be ω-stable nil-2 groups which are not p-$\aleph_0$-categorical; we propose the following example, using Mekler's arguments [Me].

Let $\Gamma_0 = \{ a \in \omega^{<\omega} : a \neq \emptyset, a(i) \leq i$ for every $i < l(a) \}$, and for any a, $b \in \Gamma_0$, define

$$A(a, b) \leftrightarrow \begin{cases} l(a) = l(b) + 1,\ a(i) = b(i) \text{ for every } i < l(b) \\ \text{or} \\ l(b) = l(a) + 1,\ a(i) = b(i) \text{ for every } i < l(a) . \end{cases}$$

It is easy to see that Γ_0 is an ω-stable nice graph, hence, for every countable $\Gamma \equiv \Gamma_0$, $B(\Gamma)$ is a superatomic algebra, but Γ_0 is not p-$\aleph_0$-categorical, since CB-type $B(\Gamma_0) = (2, 1)$, while, if Γ is a countable model of $Th(\Gamma_0)$ and $\Gamma \not\cong \Gamma_0$, then CB-type $B(\Gamma) = (\omega + 1, 1)$, as Γ contains at least one star, possibly $\aleph_0$ many stars, whose elements admit infinitely many adjacent nodes.

Consider now the nil-2 group of exponent p $G_0 = G(\Gamma_0)$, for completeness' sake we remember that

$$G_0 = \underset{\nu \in \Gamma_0}{2} (\nu) / N$$

where, for every $\nu \in \Gamma_0$, $(\nu) \simeq Z/p$, 2 denotes the free nil-2 product and $N = ([\nu, \nu'] : \nu, \nu' \in \Gamma_0, A(\nu, \nu'))$. G_0 is ω-stable since Γ_0 is; give Γ_0 any well ordering $<$, then:

* $Z_0 = Z(G_0)$ is a vector space over $\mathbb{F}_p$, and $\{ [\nu, \nu'] : \nu, \nu' \in \Gamma_0, \neg A(\nu, \nu'), \nu < \nu' \}$ is a basis for Z_0;
* G_0/Z_0 is a vector space over $\mathbb{F}_p$, and $\{ Z_0 \nu : \nu \in \Gamma_0 \}$ is a basis for G_0/Z_0;
* Γ_0 is interpretable in G_0; in fact, set in G_0/Z_0

$$Z_0 x \sim Z_0 y \text{ if and only if } C_{G_0}(x) = C_{G_0}(y)$$

and in the quotient $(G_0/Z_0)/\sim$, excluding $Z_0|\sim = \{Z_0\}$,

$$A(Z_0x|\sim, Z_0y|\sim) \text{ if and only if } Z_0x \not\sim Z_0y \text{ and } [x, y] = u .$$

Let $\Gamma_1(G_0)$ be the corresponding graph, we have:

i. if $\nu \in \Gamma_0$, then $Z_0\nu|\sim = \{ Z_0\nu^h : 0 < h < p \}$, $Z_0\nu|\sim$ admits some adjacent nodes;

ii. if $\nu_1, \nu_2 \in \Gamma_0$ and $\Gamma_0 \models A(\nu_1, \nu_2)$, then $Z_0 \nu_1 \nu_2|\sim = \left\{ Z_0\nu_1^{h_1} \nu_2^{h_2} : 0 < h_1, h_2 < p \right\}$, and the only adjacent nodes are $Z_0 \nu_1|\sim$, $Z_0 \nu_2|\sim$;

iii. if $k \geq 2$, $\nu_1, \ldots, \nu_k \in \Gamma_0$ are adjacent to a suitable $\nu \in \Gamma_0$ and $0 < h_1, \ldots, h_k < p$, then

$$Z_0 \nu_1^{h_1} \ldots \nu_k^{h_k}|\sim = \left\{ Z_0(\nu_1^{h_1} \ldots \nu_k^{h_k})^t \nu^s : 0 < t < p, 0 \leq s < p \right\}$$

and the only adjacent node is $Z_0 \nu|\sim$;

iv. otherwise, $Z_0 \nu_1^{h_1} \ldots \nu_k^{h_k}|\sim = \left\{ Z_0(\nu_1^{h_1} \ldots \nu_k^{h_k})^t : 0 < t < p \right\}$ has no adjacent node.

Therefore, $\Gamma_2(G_0) = \{ Z_0x|\sim \in \Gamma_1(G_0) : |Z_0x|\sim| = p-1, Z_0x|\sim$ has some adjacent nodes $\}$ with the graph structure induced by $\Gamma_1(G_0)$ is isomorphic to Γ_0.

In order to determine the isomorphism type of $B(G_0)$, hence the CB-type of $B(G_0)$, we analyse the dual space $S(G_0)$, by considering $G > G_0$, G ω_1-saturated, $|G| = \aleph_1$, and, for every $x \in G$, $tp(x|G_0)$. Let us abbreviate $Z(G)$ by Z, first we recall some results from [Me] about the structure of G: define a subset Γ_2 of G by choosing an element γ for every $\sim$-class of $\Gamma_2(G)$ (choose ν in $Z\nu|\sim$ when $\nu \in \Gamma_0$); enlarge Γ_2 to a subset Γ_3 of G such that $\{Z\gamma : \gamma \in \Gamma_3\}$ is $\mathbb{F}_p$-independent and generates $\{ Zx : Zx|\sim \in \Gamma_1(G), \exists\, Z\gamma|\sim \in \Gamma_2(G)$ satisfying $A(Zx|\sim, Z\gamma|\sim) \}$ over $\mathbb{F}_p$ (see iii.); define now $B \subsetneq G$ such that $B \cap \Gamma_3 = \emptyset$ and $\{ Z\gamma : \gamma \in B \cup \Gamma_3 \}$ is a basis of G/Z over $\mathbb{F}_p$ (see iv.); notice that $B \cup \Gamma_3$ -as well as Γ_3- has a graph structure induced by $\Gamma_1(G)$; after noting that $[(B \cup \Gamma_3), (B \cup \Gamma_3)]$ is a direct addendum of Z, let X be a basis over $\mathbb{F}_p$ of its coaddendum (which is an Abelian group of exponent p, hence a vector space over $\mathbb{F}_p$). Then

$$G \simeq G(\Gamma_3)\ 2_\beta\ 2_B\ (\beta) \oplus \bigoplus_{\eta \in X} (\eta)\ ;$$

in other words, G has $\Gamma_3 \cup B \cup X$ as a set of generators, with the suitable relations. Enlarge now the well ordering of Γ_0 to a well ordering for $\Gamma_3 \cup B \cup X$ such that $\Gamma_0 < \Gamma_2 - \Gamma_0 < \Gamma_3 - \Gamma_2 < B < X$, then

* Z is a vector space over $\mathbb{F}_p$ and $X \cup \{ [\gamma_1, \gamma_2] : \gamma_1, \gamma_2 \in \Gamma_3 \cup B, \gamma_1 < \gamma_2, \neg A(\gamma_1, \gamma_2)$ (if $\gamma_1, \gamma_2 \in \Gamma_3$) $\}$ is a basis of Z;

* G/Z is a vector space over $\mathbb{F}_p$, too (and $\{ Z\gamma : \gamma \in \Gamma_3 \cup B \}$ is a basis of G/Z).

If $\Gamma'_3 \cup B' \cup X'$ is another set of generators of G (defined in the same way), then there exist a graph isomorphism ϕ_Γ of Γ_3 onto Γ'_3 fixing Γ_0, a bijection ϕ_B of B onto B', and a bijection ϕ_X of X onto X' (remember $|G| = \aleph_1$, G ω_1-saturated); hence an automorphism ϕ of G fixing G_0 can be defined by setting

$$\phi(x) = \begin{cases} \phi_\Gamma(x) \text{ if } x \in \Gamma_3 \\ \phi_B(x) \text{ if } x \in B \\ \phi_X(x) \text{ if } x \in X . \end{cases}$$

We can classify now $\{ tp(x | G_0) : x \in G \}$. We set for simplicity

$$\Delta = \{ Z\gamma : Z\gamma|\sim \in \Gamma_2(G) \}$$

$$\Delta' = \Delta \cup \{ Z\gamma : \exists\ Z\gamma_0|\sim \in \Gamma_2(G) \text{ satisfying } A(Z\gamma|\sim, Z\gamma_0|\sim) \}.$$

1. $x \in G_0$; in this case, CB-rank $tp(x| G_0) = 0$.

2. $x \notin G_0$, but there is $g \in G_0$ such that $Zx = Zg$; first suppose that $x \cdot g^{-1}$ can be expressed as a product of commutators of elements of G, the problem is what elements of $G - G_0$ are needed in this decomposition.

* Let $x = g\,[\nu, \gamma_0] \cdots [\nu, \gamma_n]$ where $\nu \in \Gamma_0$, $n \in \omega$, $Z\gamma_0|\sim, \ldots, Z\gamma_n|\sim \in$

$\Gamma_2(G) - \Gamma_2(G_0)$ are placed in the same star. Let $p = tp(Z\gamma_0|\sim, \ldots, Z\gamma_n|\sim|\Gamma_2(G_0))$, notice that, if

$$x' = g\,[\nu, \gamma'_0] \ldots [\nu, \gamma'_n]$$

where $Z\gamma'_0|\sim, \ldots, Z\gamma'_n|\sim \in \Gamma_2(G)$ and $(Z\gamma'_0|\sim, \ldots, Z\gamma'_n|\sim) \models p$, then there is an automorphims ϕ of G fixing G_0 and mapping x into x' (enlarge $\Gamma_0 \cup \{\gamma_0, \ldots, \gamma_n\}$ to Γ_3 and $\Gamma_0 \cup \{\gamma'_0, \ldots, \gamma'_n\}$ to Γ'_3; extend the isomorphism ϕ_Γ of Γ_3 onto Γ'_3 fixing Γ_0 and mapping γ_i into γ'_i for all $i \leq n$ to get ϕ). It is easy to deduce that CB-rank $tp(x|G_0) = 1$. More generally, if the decomposition of x only needs elements of G whose classes in $\Gamma_2(G) - \Gamma_2(G_0)$ are placed in the same star, then CB-rank $tp(x|G_0) = 1$.

* Let now

$$x = g \prod_{i \leq n} \prod_{j \leq n_i} [\nu, \gamma_{ij}]$$

where $\nu \in \Gamma_0$, $Z\gamma_{ij}|\sim \in \Gamma_2(G) - \Gamma_2(G_0)$ for every $i \leq n$ and $j \leq n_i$, and

$Z\gamma_{i0}|\sim, \ldots, Z\gamma_{in_i}|\sim$ lie on the same star for all $i \leq n$,

$Z\gamma_{i0}|\sim, Z\gamma_{l0}|\sim$, lie on different stars if $i < l \leq n$.

Let $p = tp((Z\gamma_{ij}|\sim)_{i \leq n,\, j \leq n_i} \mid \Gamma_2(G_0))$, we have again that, if

$$x' = g \prod_{i \leq n} \prod_{j \leq n_i} [\nu, \gamma'_{ij}]$$

for $Z\gamma'_{ij}|\sim \in \Gamma_2(G)$, $(Z\gamma'_{ij}|\sim)_{i,j} \models p$, then there is an automorphims ϕ of G fixing G_0 and mapping x into x' (proceed as above). Moreover, CB-rank $tp(x|G_0) = n+1$, in fact notice that, if

$$[\nu, \gamma_0]^{h_0} \ldots [\nu, \gamma_n]^{h_n} = [\nu, \gamma'_0]^{h_0} \ldots [\nu, \gamma'_n]^{h_n}$$

where $\nu \in \Gamma_0$, $0 < h_0 \ldots, h_n < p$, $Z\gamma_0|\sim, \ldots, Z\gamma_n|\sim$ (and similarly $Z\gamma'_0|\sim, \ldots, Z\gamma'_n|\sim$) belong to $\Gamma_2(G)$, are pairwise distinct, and neither equal nor adjacent to $Z\nu|\sim$, then there exists $\sigma \in S_{n+1}$ such that, for every $i \leq n$, $Z\gamma_i = Z\gamma'_{\sigma(i)}$ (without loss of generality, $Z\gamma_i = Z\gamma'_i$).

In general, if the decomposition of x needs elements of G whose classes in $\Gamma_2(G) - \Gamma_2(G_0)$ are placed in n different stars, then CB-rank $tp(x|G_0) \leq n$.

More generally, if the decomposition of x also needs some classes $Z\gamma'|\sim$ such that $Z\gamma'|\sim$ is adjacent but does not belong to $\Gamma_2(G)$ and $Z\gamma'$, Δ are $\mathbb{F}_p$-independent, and n_1 is the number of the stars involved in $\Gamma_2(G) - \Gamma_2(G_0)$, n_2 is the power of a maximal set Y of elements $Z\gamma'$ such that $(Z\gamma' : Z\gamma' \in Y)$, Δ are $\mathbb{F}_p$-independent, then CB-rank $tp(x|G_0) \leq n_1 + n_2$.

* Let now $x = g[\nu, \beta]$ where $\nu \in \Gamma_0$, $Z\beta|_\sim$ has no adjacent classes in $\Gamma_1(G)$, and $Z\beta$, Δ' are $\mathbb{F}_p$-independent. Notice that, if $x' = g[\nu, \beta']$ (β' like β), then there exists an automorphims ϕ of G fixing G_0 and mapping β into β', hence x into x'.

Furthermore, CB-rank $tp(x|G_0) = \omega$: the key remark to show this claim is that, if $[\nu,\beta]=[\nu,\beta']$, then $Z\beta$, Δ' are $\mathbb{F}_p$-independent if and only if $Z\beta'$, Δ' are.

* Let $x = g \prod_{h \leq m} [\nu_h, \beta_h] \prod_{i \leq n} \prod_{j \leq n_i} [\nu_{m+1}, \gamma_{ij}]$, where $\nu_0, \ldots, \nu_{m+1} \in \Gamma_0$ and

$$\nu_0 < \ldots < \nu_m < \nu_{m+1}$$

is an initial segment of the well ordering fixed in Γ_0; $Z\beta_0|\sim, \ldots, Z\beta_m|\sim$ have no adjacent nodes in $\Gamma_1(G)$, $Z\beta_0, \ldots, Z\beta_m$, Δ' are $\mathbb{F}_p$-independent; $Z\gamma_{ij}$ $(i \leq n,\ j \leq n_i)$ are as above. In this case, CB-rank $tp(x|G_0) = (m+1)\omega + n + 1$, because:

- if $x' = g \prod_{h \leq m} [\nu_h, \beta'_h] \prod_{i \leq n} \prod_{j \leq n_i} [\nu_{m+1}, \gamma'_{ij}]$ has a similar decomposition, then there is an automorphism ϕ of G fixing G_0 and mapping β_h, γ_{ij} into β'_h, γ'_{ij}, and hence x into x':

- the decomposition of x is essentially unique, since, if

$$\prod_{h \leq m} [\nu_h, \beta_h] \prod_{i \leq n} \prod_{j \leq n_i} [\nu_{m+1}, \gamma_{ij}] =$$

$$= \prod_{h \leq m} [\nu_h, \beta'_h] \prod_{i \leq n} \prod_{j \leq n_i} [\nu_{m+1}, \gamma'_{ij}]$$

and $Z\beta_h|_\sim \in \Gamma_1(G)$ has no adjacent nodes $(h \leq m)$, $Z\gamma_{ij}|\sim (i \leq n,\ j \leq n_i)$ belong to $\Gamma_2(G)$, are pairwise distinct and neither equal nor adjacent to $Z\nu_h|\sim$ for every $h \leq m+1$ -and similarly for $Z\beta'_h|\sim$, $Z\gamma'_{ij}|\sim$-, then, without loss of generality, $Z\beta_h = Z\beta'_h$ (provided some factors $Z\nu$ with $\nu \in \Gamma_0$, ν equal or adjacent to $\nu_0, \ldots, \nu_m$ are neglected) and $Z\gamma_{ij} = Z\gamma'_{ij}$.

In general, if the decomposition of x needs

- some classes $Z\beta|\sim$ such that $Z\beta|\sim$ has no adjacent nodes and $Z\beta$, Δ' are $\mathbb{F}_p$-independent,
- some classes $Z\gamma'|\sim$ such that $Z\gamma'|\sim$ is adjacent but does not belong to $\Gamma_2(G)$, and $Z\gamma'$, Δ are $\mathbb{F}_p$-independent,

in addition to some classes in $\Gamma_2(G)$, and

- m is the power of a maximal set Y of elements $Z\beta$ such that $(Z\beta : Z\beta \in Y)$, Δ' are $\mathbb{F}_p$-independent,
- n is the sum of the number of involved stars in $\Gamma_2(G) - \Gamma_2(G_0)$ and the power of a maximal set Y of elements $Z\gamma'$ such that $(Z\gamma' : Z\gamma' \in Y)$, Δ are $\mathbb{F}_p$-independent,

then CB-rank $tp(x \mid G_0) \leq m\,\omega + n$.

* Finally, if $Zx = Zg$ for some $g \in G_0$, but $x \cdot g^{-1}$ cannot be expressed as a product of $\mathbb{F}_p$-independent commutators, then these conditions isolate $tp(x|G_0)$ and it is easy to deduce from the previous remarks that CB-rank $tp(x \mid G_0) = \omega^2$.

3. For every $g \in G_0$, $Zx \neq Zg$; the problem is now to measure the complexity of the class Zx.

* There is $g_0 \in G_0$ such that $Z(x\,g_0^{-1})|\sim \in \Gamma_2(G)$ [$-\Gamma_2(G_0)$, otherwise there exist $\nu \in \Gamma_0$, $h \in \{1, \ldots, p-1\}$ such that $Zx = Z\nu^h g_0$]: then, CB-rank $tp(x \mid G_0) = \omega^2 + 1$, as well as in the following more general case.

* There is $g_0 \in G_0$ such that $Zx = Z\,g_0\,\gamma_0^{h_0} \ldots \gamma_n^{h_n}$ where $n \in \omega$, $0 < h_0, \ldots, h_n < p$, and $Z\gamma_0|\sim, \ldots, Z\gamma_n|\sim \in \Gamma_2(G)$ [$-\Gamma_2(G_0)$] are placed in the same star.

* More generally, suppose that there is $g_0 \in G_0$ such that

$$Zx = Z g_0 \prod_{i \le n} \prod_{j \le n_i} \gamma_{ij}^{h_{ij}}$$

where $n, n_i \in \omega$, $0 < h_{ij} < p$, $Z\gamma_{ij}|\sim \in \Gamma_2(G) - \Gamma_2(G_0)$ for every i, j, $Z\gamma_{i0}|\sim, \ldots, Z\gamma_{in_i}|\sim$ are placed in the same star for every $i \le n$, but $Z\gamma_{i0}|\sim, Z\gamma_{l0}|\sim$ belong to different stars if $i < l \le n$. Therefore

- if $x' \in G$ and Zx' has a similar decomposition, then there is an automorphism ϕ of G fixing G_0 and mapping x into x';

- a standard compactness argument implies that CB-rank $tp(x | G_0) = \omega^2 + n + 1$ (recall that, if $Z\gamma_0|\sim, \ldots, Z\gamma_m|\sim$ -and similarly $Z\gamma'_0|\sim, \ldots, Z\gamma'_m|\sim$ - belong to $\Gamma_2(G)$, are pairwise distinct, and $0 < h_0, \ldots, h_m < p$, then

$$Z\gamma_0^{h_0} \ldots \gamma_m^{h_m} = Z\gamma_0'^{h_0} \ldots \gamma_m'^{h_m}$$

if and only if there is $\sigma \in S_{m+1}$ such that, for all $i \le m$, $Z\gamma_i = Z\gamma'_{\sigma(i)}$).

A similar argument works if the decomposition of Zx also involves some classes $Z\gamma'|\sim$ such that $Z\gamma'|\sim$ is adjacent but does not belong to $\Gamma_2(G)$ and $Z\gamma'$, Δ are $\mathbb{F}_p$-independent.

* Let finally $x \in G$ such that Zx, Δ' are $\mathbb{F}_p$-independent, notice that this condition determines $tp(x | G_0)$, and it is easy to deduce from the previous remarks that CB-rank $tp(x | G_0) = \omega^2 + \omega$.

Therefore CB-type $B(G_0) = (\omega^2 + \omega + 1, 1)$, while, if Γ is a countable model of $Th(\Gamma_0)$ and $\Gamma \not\simeq \Gamma_0$ (so that CB-type $B(\Gamma) = (\omega + 1, 1)$), then $G(\Gamma)$ is elementarily equivalent to G_0, but, by proceeding as above, we can see that CB-rank $B(G(\Gamma)) > \omega^2 + \omega + 1$. Hence, $Th(G_0)$ is ω-stable but not p-$\aleph_0$-categorical.

<u>Theorem 2.1</u> - There are (ω-stable) non p-$\aleph_0$-categorical nil-2 groups.

REFERENCES

[BCM] W. Baur-G. Cherlin-A. Macintyre, Totally categorical groups and rings, J. Algebra 57 (1979) 407-440

[K] J. Ketonen, The structure of countable Boolean algebras, Ann. Math. 108 (1978) 41-89

[LP] D. Lascar-B. Poizat, An introduction to forking, J. Symbolic Logic 44 (1979) 330-350

[MT1] A. Marcja-C. Toffalori, On pseudo-$\aleph_0$-categorical theories, Zeitschr. f. Math. Logik, to appear

[MT2] A. Marcja-C. Toffalori, On Cantor-Bendixson spectra containing (1, 1), I - Logic Colloquium 83, II - J. Symbolic Logic, to appear

[Me] A. Mekler, Stability of nilpotent groups of class 2 and prime exponent, J. Symbolic Logic 46 (1981) 781-788

[Z] M. Ziegler, Model theory of modules, Annals of Pure and Applied Logic 26 (1984), 149-213

LOGIC COLLOQUIUM '84
J.B. Paris, A.J. Wilkie, and G.M. Wilmers (Editors)

ON SENTENCES INTERPRETABLE IN SYSTEMS OF ARITHMETIC

A.J. Wilkie

Department of Mathematics,
The University,
Manchester, M13 9PL,
UK.

We give a characterization of the Π_1 sentences that are interpretable in Π_2 extensions of bounded arithmetic.

1. INTRODUCTION

Let $L = \{0,1,+,.,<\}$ be the usual language of arithmetic. Recall that a formula of L is called bounded (or Δ_0 or Σ_0 or Π_0) if all its quantifiers occur in context $\exists\vec{x} < t\ldots$, $\forall\vec{x} < t\ldots$, where t is a term of L not containing the variables $\vec{x}$. The hierarchies of Σ_n and Π_n formulas are defined as usual. We denote by $I\Delta_0$ the scheme of induction for bounded formulas (with parameters) together with a suitable base theory (see e.g. [3]). Note that $I\Delta_0$ is a Π_1 thoery and hence preserved to initial segments (closed under $\cdot$) of models of $I\Delta_0$. If σ is a Π_2 sentence of L true in the standard model $\mathbb{N}$, say σ is $\forall\vec{u}\,\exists\vec{v}\sigma'(\vec{u},\vec{v})$ where σ' is Δ_0, then T_σ denotes the theory $I\Delta_0 + \forall x\exists y\,\forall\vec{u}<x\,\exists\vec{v} < y\,\sigma'(\vec{u},\vec{v})$. Our aim in this paper is to define a natural map $\sigma \to \bar{\sigma}$, taking true Π_2 sentences to true Π_2 sentences, having the property that for any Π_1 sentence A, if $T_{\bar{\sigma}} \vdash A$ then $T_\sigma + A$ is interpretable in T_σ and, providing σ satisfies a certain smoothness condition, conversely. Interpretability is being used here in the strong sense, namely if S,T are any theories of L and $T \vdash I\Delta_0$ then we say that S is interpreted in T by Φ, where Φ is a formula of L containing just one free variable, if

1.1 $T \vdash \Phi(0)$,

1.2 $T \vdash \forall x,y\ ((\Phi(y) \wedge x < y) \to \Phi(x))$,

1.3 $T \vdash \forall x,y\ ((\Phi(x) \wedge \Phi(y)) \to (\Phi(x+1)\wedge\Phi(x+y)\wedge\Phi(x.y)))$, and

1.4 $T \vdash A^\Phi$, for each sentence $A \in S$ where A^Φ denotes the sentence obtained be relativizing all quantifiers in A to Φ.

Thus 'S is interpreted in T by Φ' means that in any model M of T Φ defines an initial segment of M, which is also a substructure of M, satisfying S (and clearly also $I\Delta_0$).

We shall also investigate the closure properties of definable initial segments of models of T_σ in terms of the natural 'growth function' associated with σ. Concerning this we should mention the paper [2] where the following result was proved in answer to a question of R. Solovay:–

Proposition 1 (Paris–Dimitracopoulos)
Let M be a nonstandard countable model of Peano arithmetic (PA). Then there is an initial segment I of M (closed under .) such that if J is any initial segment of I definable in the structure (M,I) (i.e. a unary predicate symbol interpreting I is added to L) then J is not closed under exponentiation.

This result shows that $I\Delta_0$ + exp is not interpretable in $I\Delta_0$, where exp is the Π_2 sentence $\forall x,y\ \exists z\ z = x^y$, and '$z = x^y$' is the natural Δ_0 formula defining the graph of exponentiation (see [1]). It turns out, however, that if we take σ to be $\forall x\ \exists y\ y = x + 1$ then $\bar{\sigma}$ is (equivalent in $I\Delta_0$ to) exp, so a corollary to our main theorem is

Theorem 2
For any Π_1 sentence A, $I\Delta_0 + A$ is interpretable in $I\Delta_0$ if and only if $I\Delta_0 + \exp \vdash A$.

(Theorem 2 was stated without proof in [3] where it was used to obtain consistency results for certain systems of arithmetic. It is also referred to in [4] where similar problems are investigated by proof-theoretic methods. The techniques used in this paper are model-theoretic.)

To return to our main result notice that since a Π_2 sentence is asserting that a certain function (with Δ_0 graph) is total, the problem of determining the map $\sigma \dashrightarrow \bar{\sigma}$ is one of finding an operation, with suitable properties, which maps total functions to total functions. The next section investigates such operations.

2. UNIFORM FUNCTIONS AND OPERATIONS ON THEM

We call a function $f: \mathbb{N} \dashrightarrow \mathbb{N}$ uniform if it satisfies for all $x, y \in \mathbb{N}$,

2.1 $x < y \dashrightarrow f(x) < f(y)$, and

2.2 $f(0) = 1$

Clearly 2.1 and 2.2 imply, for all $x \in \mathbb{N}$,

2.3 $f(x) > x$.

We now define the operations $\hat{}$ (iteration) and * (iterated inverse) on uniform functions by the recursions:-

2.4 (i) $\hat{f}(0,x) = x$, (ii) $\hat{f}(y+1,x) = f(\hat{f}(y,x))$.

2.5 (i) $f^*(0) = 1$, (ii) $f(y) \leqslant x < f(y+1) \dashrightarrow f^*(x) = 1+f^*(y)$.

Then $f^*: \mathbb{N} \dashrightarrow \mathbb{N}$ is clearly non-decreasing (and well defined by 2.1, 2.2 and 2.3) and $\hat{f}: \mathbb{N}^2 \dashrightarrow \mathbb{N}$ is strictly increasing in both arguments. We now define $f_1: \mathbb{N}^2 \dashrightarrow \mathbb{N}$ by:-

2.6 $f_1(y,x) = \hat{f}(f^*(y),x)$.

Clearly the diagonal of f_1, $x \mapsto f_1(x,x)$, is uniform so we may repeat the construction on this function. To make the notation more concise, however, we first make the convention that if $h: \mathbb{N}^2 \dashrightarrow \mathbb{N}$ is any binary function then $h(x)$ denotes $h(x,x)$. Thus if the diagonal of h satisfies 2.1 and 2.2 then $\hat{h}$ and h^* are well defined and, for example, $\hat{h}$ satisfies $\hat{h}(y+1,x) = h(\hat{h}(y,x)) = h(\hat{h}(y,x), \hat{h}(y,x))$.

The binary functions $f_1, f_2, \ldots, f_n, \ldots$ are now defined by 2.6 and for $n \geqslant 1$,

2.7 $f_{n+1}(y,x) = \hat{f}_n(f_n^*(y), x)$.

These functions have some pleasant properties which are crucial for the sequel and which we now establish.

Lemma 3.
Setting $f_0 = f$, the following hold for all $n,x,y,z \in \mathbb{N}$:-

2.8 $f_n^*(f_n(x)) = 1 + f_n^*(x)$.

2.9 $f_n^*(\hat{f}_n(y,x)) = y + f_n^*(x)$.

2.10 $\hat{f}_n(y+z,x) = \hat{f}_n(y, \hat{f}_n(z,x))$.

2.11 $n \geqslant 1 \dashrightarrow f_n(f_n(y,x),x) = f_n(y, f_n(x,x))$.

2.12 (i) $f_{n+1}(x) \geqslant f_n(x)$, (ii) $f_n(x) \geqslant f_{n+1}(x,0)$.

2.13 $\hat{f}_{n+1}(y,x) = \hat{f}_n((2^y-1)\cdot f_n^*(x),x)$.

Proof

2.8 is immediate from the definition of f_n^* and 2.9 follows from the definition of $\hat{f}_n$ and 2.8 by induction on y. 2.10 is also clear by induction on y. For 2.11 we have

$$\begin{aligned} f_n(f_n(y,x),x) &= \hat{f}_{n-1}(f_{n-1}^*(f_n(y,x)),x) \text{ (by 2.7)}, \\ &= \hat{f}_{n-1}(f_{n-1}^*(\hat{f}_{n-1}(f_{n-1}^*(y),x)),x) \text{ (by 2.7)}, \\ &= \hat{f}_{n-1}(f_{n-1}^*(y) + f_{n-1}^*(x),x) \text{ (by 2.9)}, \\ &= \hat{f}_{n-1}(f_{n-1}^*(y), \hat{f}_{n-1}(f_{n-1}^*(x),x)) \text{ (by 2.10)}, \\ &= f_n(y,f_n(x,x)) \text{ (by 2.7), as required.} \end{aligned}$$

For 2.12 (i) we have

$$\begin{aligned} f_{n+1}(x) &= \hat{f}_n(f_n^*(x),x) \text{ (by 2.7)}, \\ &\geqslant \hat{f}_n(1,x), \\ &= f_n(x), \text{ as required.} \end{aligned}$$

For 2.12(ii) we use induction on x. For $x = 0$ the result is clear since both sides are 1. Suppose $f_n(y) \leqslant x < f_n(y+1)$ and that the result is true for y. Then we have

$$\begin{aligned} f_{n+1}(x,0) &= \hat{f}_n(f_n^*(x),0) && \text{(by 2.7)}, \\ &= \hat{f}_n(1+f_n^*(y),0) && \text{(by definition of } f_n^*), \\ &= f_n(\hat{f}_n(f_n^*(y),0)) && \text{(by definition of } \hat{f}_n), \\ &= f_n(f_{n+1}(y,0)) && \text{(by 2.7)}, \\ &\leqslant f_n(f_n(y)) && \text{(by our inductive assumption)}, \\ &\leqslant f_n(x) && \text{(by the supposition on } y,x), \text{ as required.} \end{aligned}$$

We establish 2.13 by induction on y. Firstly,

$$\begin{aligned} \hat{f}_{n+1}(0,x) &= x && \text{(by definition of } \hat{f}_{n+1}), \\ &= \hat{f}_n(0,x) && \text{(by definition of } \hat{f}_n), \\ &= \hat{f}_n((2^0-1).f_n^*(x),x), && \text{as required.} \end{aligned}$$

Now suppose 2.13 is true for y. Then we have

$$\begin{aligned} \hat{f}_{n+1}(y+1,x) &= f_{n+1}(\hat{f}_{n+1}(y,x)) && \text{(by definition of } \hat{f}_{n+1}), \\ &= f_{n+1}(\hat{f}_{n+1}(y,x),\hat{f}_{n+1}(y,x)) && \text{(by convention)} \\ &= \hat{f}_n(f_n^*(\hat{f}_{n+1}(y,x)), \hat{f}_{n+1}(y,x)) && \text{(by 2.7)} \\ &= \hat{f}_n(f_n^*(\hat{f}_n((2^y-1)\cdot f_n^*(x),x)), \hat{f}_{n+1}(y,x)) && \text{(by inductive hyp.)}, \\ &= \hat{f}_n((2^y-1)\cdot f_n^*(x)+f_n^*(x), \hat{f}_{n+1}(y,x)) && \text{(by 2.9)}, \\ &= \hat{f}_n(2^y\cdot f_n^*(x), \hat{f}_n((2^y-1)\cdot f_n^*(x),x)) && \text{(by inductive hyp.)}, \\ &= \hat{f}_n((2^{y+1}-1)\cdot f_n^*(x),x) && \text{(by 2.10), as required.} \end{aligned}$$

□

We now define, for $n \geqslant 1$, the function F_n: IN → IN by

2.14 $F_n(x) = \hat{f}_n(f^*(x),0)$.

To relate the F_n's to f we introduce the functions $e_n, e_{\bar{n}}$: IN → IN defined by

2.15 (i) $e_0(x) = x$, (ii) $e_{n+1}(x) = 2^{e_n(x)}$.

2.16 (i) $e_{\bar{0}}(x) = x$ (ii) $e_{\overline{n+1}}(x) = 2^{e_{\bar{n}}(x)}-1$.

Lemma 4.

For all $n, x \in$ IN, $n \geqslant 1$, we have

2.17 $F_n(f(x)) = f_n(F_n(x))$.

2.18 $F_n(x) = \hat{f}(e_{\bar{n}}(f^*(x)),0)$.

2.19 $F_1^{(n)}(x) = \hat{f}(e_n(f^*(x))-1,0)$. (We prefer to write $F_1^{(n)}(x)$ for $\hat{F}_1(n,x)$ here.)

2.20 $F_1^{(n)}(x) \leqslant F_{n+1}(x)$ for $x \geqslant 2$.

Proof

For 2.17 we have

$$\begin{aligned} F_n(f(x)) &= \hat{f}_n(f^*(f(x)),0) && \text{(by 2.14)},\\ &= \hat{f}_n(1+f^*(x),0) && \text{(by 2.8)},\\ &= f_n(\hat{f}_n(f^*(x),0)) && \text{(by definition of } \hat{f}_n),\\ &= f_n(F_n(x)) && \text{(by 2.14)}, \text{ as required.} \end{aligned}$$

For 2.18 notice that by 2.13 with $x = 0$ (recall $f_n^*(0) = 1$) we have, for all $n,y \in \mathbb{N}$, $\hat{f}_{n+1}(y,0) = \hat{f}_n(2^y-1,0)$. Hence by induction on n, for all $n,y \in \mathbb{N}$, $\hat{f}_n(y,0) = \hat{f}(e_n^-(y),0)$, so 2.18 follows upon setting $y = f^*(x)$ here. We prove 2.19 also by induction on n. The case $n = 1$ is clear from 2.13 (with n=0) and 2.14. For $n \geqslant 1$ we have

$$\begin{aligned} F_1^{(n+1)}(x) &= F_1(F_1^{(n)}(x)),\\ &= \hat{f}_1(f^*(F_1^{(n)}(x)),0) && \text{(by 2.14)},\\ &= \hat{f}_1(f^*(\hat{f}(e_n(f^*(x))-1,0)),0) && \text{(by inductive hyp.)},\\ &= \hat{f}_1(e_n(f^*(x)),0) && \text{(by 2.9 since } f^*(0)=1),\\ &= \hat{f}(e_{n+1}(f^*(x))-1,0) && \text{(by 2.13, 2.15), as required.} \end{aligned}$$

For 2.20 first we note that for all $n,x \in \mathbb{N}$, with $x \geqslant 2$ we have $e_{n+1}^-(x) > e_n(x)$. For $e_1^-(x) = 2^x-1 > x = e_0(x)$ and if $e_{n+1}^-(x) > e_n(x)$ then $e_{n+2}^-(x) = e^{e_{n+1}^-(x)} - 1 \geqslant 2^{1+e_n(x)} - 1 = 2.e_{n+1}(x) - 1 > e_{n+1}(x)$.
2.20 now follows from 2.18 and 2.19since $x \geqslant 2$ implies $f^*(x) \geqslant 2$.

□

An example.

2.21 Suppose $f(x) = x + 1$.

Then it is easy to check that $f^*(x) = x + 1$, $\hat{f}(y,x) = y + x$ and hence that $f_1(y,x) = y + x + 1$ (so $f_1(x) = 2x + 1$). We therefore have, using 2.18 and 2.19, that for all $n \geqslant 1$:-

2.22 $F_n(x) = e_n^-(x+1)$, so $F_1(x) = 2^{x+1} - 1$, and

2.23 $F_1^{(n)}(x) = e_n(x+1)-1$.

It is rather tedious to compute f_n exactly for $n \geqslant 2$ (in terms of more familiar functions) but it can be easily estimated using the functions ω_n defined by:-

2.24 (i) $\omega_0(x) = x+1$, (ii) $\omega_{n+1}(x) = 2^{\omega_n(l(x))} - 1$, where $l(x) = [\log_2(x+1)]$.

(Thus $\omega_1(x) \approx 2x$, $\omega_2(x) \approx x^2$, $\omega_3(x) \approx x^{l(x)}$ $\omega_4(x) \approx x^{l(x)^{ll(x)}}$ etc.)

Then it is easy to see that $\omega_n(y) = e_n^-(x+1)$ for all $x,n,y \in \mathbb{N}$ such that $e_n^-(x) \leqslant y < e_n^-(x+1)$ and hence (by 2.17 and 2.22) we have for all $n,x \in \mathbb{N}$:-

2.25 $\omega_n(x) \leqslant f_n(x) < \omega_n(\omega_n(x))$.

3. THE MAP $\sigma \to \bar{\sigma}$ AND INTERPRETATIONS IN $T\sigma$

We now return to the situation of section 1 and let σ be a true Π_2 sentence, say $\sigma = \forall\vec{u}\exists\vec{v}\,\sigma'(\vec{u},\vec{v})$ where σ' is Δ_0. Define the Δ_0 formulas $f'(x,y)$,

$f(x,y)$ by:-

3.1 $f'(x,y) \Leftrightarrow \forall \vec{u} < x\ \exists \vec{v} < y\ \sigma'(\vec{u},\vec{v}) \wedge \forall w < y \neg \forall \vec{u} < x\ \exists \vec{v} < w\ \sigma'(\vec{u},\vec{v})$,

3.2 $f(x,y) \Leftrightarrow \exists z < y\ (f'(x,z) \wedge y = x^2 + 1 + z)$.

Clearly $T_\sigma \vdash \forall x\ \exists!\ y\ f(x,y)$ and it is harmless to write $f(x) = y$ for $f(x,y)$ and to regard f as a term of L when working in T_σ (since in T_σ one can prove induction for bounded formulas even if f is allowed to occur in the terms bounding quantifiers). Notice also that 2.1 and 2.2 (and 2.3) are provable from T_σ, indeed we have:-

3.3 $T_\sigma \vdash \forall x\ (f(x) \geqslant x^2 + 1)$.

We now want to represent $\hat{f}$ and f^* by Δ_0 formulas in T_σ. This can be done quite naturally using the fact that $I\Delta_0$ admits Δ_0-definable functions $(x)_y$ (= "the (y+1)st number of the sequence coded by x") and $|x|$ (= "the length of the sequence coded by x") satisfying for some suitable polynomial p, :-

3.4 $I\Delta_0 \vdash \forall u\ \exists x < p(u)\ (|x| = 1 \wedge (x)_0 = u)$, and

3.5 $I\Delta_0 \vdash \forall x,w,u\ (|x| = w \rightarrow \exists x' < x.p(u)\ (|x'| = w+1 \wedge$
$\wedge (x')_w = u \wedge \forall y < w\ ((x')_y = (x)_y))$.

Now because of the rapid growth of f given by 3.3 (which, together with 3.4 and 3.5, implies a polynomial bound, say $p'(x,y,z)$ on the code for the sequence $x, \hat{f}(1,x), \hat{f}(2,x), \ldots, \hat{f}(y,x)=z$) it is easy to show that if we define '$\hat{f}(x,y)=z$' by the Δ_0 formula:-

3.6 $\exists u < p'(x,y,z)(|u| = y + 1 \wedge (u)_0 = x \wedge (u)_y = z \wedge \forall w{<}y\ ((u)_{w+1} = f((u)_w)))$,

then, in any model M of T_σ $\hat{f}$ defines a partial function with the properties that for all $x \in M$ the set $I_x = \{y \in M: M \models \exists z\ z = \hat{f}(y,x)\}$ is an initial segment of M closed under $+1$ and the function $y \rightarrow \hat{f}(y,x)$ is total and increasing on I_x with range cofinal in M. Further, 2.4 is satisfied for all $x \in M$ and $y \in I_x$.

We may similarly define f^* (which will in fact be provably total in T_σ), the f_n's and the F_n's (which will be partial) by natural Δ_0 formulas and because only simple inductions were used, it is clear that all the identities and inequalities established in section 2 for IN will hold any model of T_σ whenever the functions under consideration are defined. In fact if one side of such an identity (or the "greater side" of an inequality) is defined then so is the other and the identity (or inequality) holds. (For example, for 2.11 one can prove, for each $n \in \mathbb{N}$ with $n \geqslant 1$, in T_σ the sentence

$\forall x,y,u,v\ ((f_n(y,x)=u \wedge f_n(u,x)=v) \rightarrow \exists z\ (z=f_n(x,x) \wedge v = f_n(y,z)))$.)

We can now introduce our main definitions:-

3.7 For each $n \in \mathbb{N}$, σ_n denotes the Π_2 sentence $\forall x \exists y\ y = f_n(x)$.
(Recall $f_0 = f$.)

3.8 $\bar{\sigma}$ denotes the Π_2 sentence $\forall x \exists y\ y = F_1(x)$.

Theorem 5.
For each $n \in \mathbb{N}$, σ_n is interpretable in T_σ by a Π_{n+1} formula.

Proof
We must construct, for each $n \in \mathbb{N}$, a Π_{n+1} formula $\phi_n(x)$ satisfying 1.1-1.3 (for $T = T_\sigma$, $\phi = \phi_n$) and 1.4:- $T_\sigma \vdash \forall x\ (\phi_n(x) \rightarrow \exists y((f_n(x)=y)^{\phi_n} \wedge \phi_n(y))$.
Clearly we may take $\phi_0(x)$ to be $x = x$. Suppose $\phi_n(x)$ has been constructed.
We define:-

3.9 $\phi_{n+1}(x) \Leftrightarrow \phi_n(x) \wedge \forall y\ (\phi_n(y) \rightarrow \exists z\ (z = f_{n+1}(y,x) \wedge \phi_n(z)))$.

As it stands ϕ_{n+1} is Π_{n+3}, but it is clearly equivalent in T_σ to $\phi_n(x) \wedge \forall y\,(\phi_n(y) \rightarrow (\exists z\,(z=f_{n+1}(y,x)) \wedge \forall z\,(z=f_{n+1}(y,x) \rightarrow \phi_n(z))))$, which can obviously be put into Π_{n+2} form.

To show 1.1 holds for ϕ_{n+1} note first that we have $\phi_n(0)$ (by 1.1 for ϕ_n). (We are working in T_σ throughout this proof.) Suppose $\phi_n(y)$. Then by 1.4 for ϕ_n we may choose z such that $\phi_n(z) \wedge z = f_n(y)$. But by 2.12(ii) there is $z' \leqslant z$ such that $z' = f_{n+1}(y,0)$ and, further, $\phi_n(z')$ holds (by 1.2 for ϕ_n) as required.

Now 1.2 for ϕ_{n+1} follows easily from the corresponding property of ϕ_n and the fact that f_{n+1} is increasing in its second argument, and 1.3 clearly follows from 1.2 and 1.4 (since $f_n(x) \geqslant f_0(x) \geqslant x^2+1$, by 2.12(i)). Hence it remains to show that if $\phi_{n+1}(x)$, then for some z, $z = f_{n+1}(x) \wedge \phi_{n+1}(z)$. (Note that the formula '$z = f_{n+1}(x)$' is Δ_0 so that we may drop the relativization.) Now $\phi_{n+1}(x)$ implies $\phi_n(x)$ and hence (using 3.9 with '$y=x$') we at least have a z such that $z = f_{n+1}(x,x) \wedge \phi_n(z)$. Now suppose $\phi_n(y)$. Then (by 3.9) for some z_0, $z_0 = f_{n+1}(y,x) \wedge \phi_n(z_0)$. Now using 3.9 again (with '$y=z_0$') there is a z_1 such that $z_1 = f_{n+1}(z_0,x) \wedge \phi_n(z_1)$. But by 2.11 $f_{n+1}(z_0,x) = f_{n+1}(f_{n+1}(y,x),x)$ $= f_{n+1}(y, f_{n+1}(x,x)) = f_{n+1}(y,z)$. Thus we have shown that $\phi_n(z) \wedge \forall y\,(\phi_n(y) \rightarrow \exists z_1\,(z_1 = f_{n+1}(y,z) \wedge \phi_n(z_1))$, i.e. $\phi_{n+1}(z)$, where $z = f_{n+1}(x,x)$, as required.

□

For $M \models T_\sigma$ and $n \in \mathbb{N}$, $n \geqslant 1$, we now define:-

3.10 $\overset{-1}{F_n}(M) = \{a \in M: M \models \exists y\; y = F_n(a)\}$.

The following lemma follows from 2.17 and we leave the easy details to the reader.

Lemma 6.

Suppose $M \models T_\sigma$. Then for all $n \in \mathbb{N}$, $n \geqslant 1$, $\overset{-1}{F_n}(M)$ is an initial segment of M (possibly with a greatest element). Further, $M \models \sigma_n$ (i.e. f_n is total in M) if and only if $F_n^{-1}(M) \models T_\sigma$ (i.e. f is total in $F_n^{-1}(M)$).

□

We are now ready to prove the main result of this section.

Theorem 7

Let A be a Π_1 sentence such that $T_\sigma \nvdash A$. Then $T_\sigma + A$ is interpretable in T_σ.

Proof

Suppose $A = \forall \vec{x}\, B(\vec{x})$, where $B(\vec{x})$ is Δ_0, and that $T_\sigma \vdash A$. Now by an easy compactness argument (and 3.8) it follows that:-

3.11 $I\Delta_0 \vdash \forall w,y\,(F_1^{(n)}(w) = y \rightarrow \forall \vec{x} < w\, B(\vec{x}))$, for some $n \in \mathbb{N}$.

Define $\psi(w) \Leftrightarrow \exists y\,(F_{n+1}(w) = y \wedge \phi_{n+1}(y))$, where ϕ_{n+1} is the formula (given by theorem 5) interpreting σ_{n+1} in T_σ. Let M be any model of T_σ and let $J = \{a \in M : M \models \psi(a)\}$ and $I = \{a \in M : M \models \phi_{n+1}(a)\}$. Then I is an initial segment of M satisfying σ_{n+1} and hence clearly $I \models T_\sigma$. Further $J = \overset{-1}{F_{n+1}}(I)$ and so by lemma 6 (with $M = I$), J is an initial segment of I (and hence of M) such that $J \models T_\sigma$. We are going to show that $J \models A$, so let $b \in J$. We must show $J \models \forall \vec{x} < b\, B(\vec{x})$ for which it suffices to show that $I \models \forall \vec{x} < b\, B(\vec{x})$. Now working in I we have for some $c \in I$, $c = F_{n+1}(b)$, and hence by 2.20, for some $c' \in I$, $c' = F_1^{(n)}(b)$. Since certainly $I \models I\Delta_0$, the required conclusion follows from 3.11. Thus we have shown that $T_\sigma + A$ is interpreted in T_σ by ψ.

□

The aim of the rest of this paper is to prove a converse of theorem 7 and to show that theorem 5 cannot be significantly improved, for which we require a theory of games played in models of $I\Delta_0 + \exp$.

4. CONSTRUCTING INITIAL SEGMENTS BY GAMES

We denote by L^r the relational language of arithmetic (i.e. L^r is the same as L except that $+, \cdot$ are ternary relation symbols) and by $L^r(J)$ the language obtained by adding to L^r the unary relation symbol J. We fix for the rest of this section a nonstandard countable model, M say, of $I\Delta_0 + \exp$ and a nonstandard $B \in M$, and we denote by $L^r(J,B)$ the language obtained by adding to $L^r(J)$ a constant symbol $\underline{d}$ for each $d \in M, d < B$. We assume that $L^r(J,B)$ is Gödel numbered in M in some standard way. The classes $\exists_n$, $\forall_n$ (for $n \in \mathbb{N}$) of $L^r(J)$ formulas are defined as follows:-

4.1 $\exists_0 = \forall_0 =$ the class of quantifier-free formulas of $L^r(J)$ that begin with at most one negation symbol. (This inessential restriction is for technical reasons explained below.)

4.2 $\exists_{n+1} = \{\exists\vec{x}\phi : \phi \in \forall_n\}$.

4.3 $\forall_{n+1} = \{\forall\vec{x}\phi : \phi \in \exists_n\}$.

For $s \in M$, $\exists_n(s)$ ($\forall_n(s)$) denotes the (M-coded) set of all $\exists_n$ ($\forall_n$) formulas having Gödel number $< s$. Thus if s is nonstandard $\exists_n(s)$ certainly contains all standard $\exists_n$ formulas. We also write $\exists_n(s,B)$ for the (M-coded) set of all formulas of $L^r(J,B)$ obtained by substituting constants of $L^r(J,B)$ for (some of) the variables of an $\exists_n(s)$ formula. Similarly for $\forall_n(s,B)$. If $\phi \in \exists_n(s,B) \cup \forall_n(s,B)$, ϕ^* denotes the prenex negation of ϕ obtained by interchanging $\forall$'s and $\exists$'s in ϕ and removing the initial negation symbol from the quantifier-free matrix of ϕ if there is one, or adding one if there is not. Clearly 4.1 ensures that $\phi^{**} = \phi$.

Let us suppose now that nonstandard $a,b,t \in M$ and an M-coded function g are given satisfying the following:-

4.4 $a < b$ and $e_n(b) < B$ for all $n \in \mathbb{N}$, and

4.5 g is strictly increasing on its domain, which includes all $x \in M$ with $x \leq b$, $\hat{g}(x,a)$ is defined for all $x \in M$ with $x \leq 2^t$, and $\hat{g}(2^t,a) = b$.

We are now ready to describe the game $\Gamma(n,s,\alpha)$, where $n \in \mathbb{N}$ and $s, \alpha \in M$ (and we shall always assume that $s \geq m$, where $m \in \mathbb{N}$ is chosen so that $\exists_0(m,B) \neq \emptyset$), which is "played in M" by two players, I and II. Let H be the set of sentences in $\exists_n(s,B)$ and suppose that for some $i < \alpha$ the players have constructed a (M-coded) sequence, $\phi_1,\phi_2,\ldots,\phi_{2i-1},\phi_{2i}$, of sentences from H, called a <u>play of length i</u>. The game is continued according to one of the following three rules:-

4.6 I chooses any $\phi_{2i+1} \in H$ and II must set either $\phi_{2i+2} = \phi_{2i+1}$ or $\phi_{2i+2} = \phi^*_{2i+1}$.

4.7 I chooses some $j \leq i$, where ϕ_{2j} is non-trivially of the form $\exists\vec{x}\,\psi(\vec{x})$ and ψ does not begin with an $\exists$, and sets $\phi_{2i+1} = \phi_{2j}$. Player II must then choose $\vec{c} < B$ and set $\phi_{2i+2} = \psi(\vec{\underline{c}})$.

4.8 I chooses some $j \leq i$, where ϕ_{2j} is non-trivially of the form $\forall\vec{x}\psi(\vec{x})$ and ψ does not begin with an $\forall$, and some $\vec{c} < B$. PLayer I then sets $\phi_{2i+1} = \psi(\vec{\underline{c}})$ and II must set $\phi_{2i+2} = \psi(\vec{\underline{c}})$.

The game is finished after α such moves when a play of length α, say $\phi_1,\phi_2,\ldots,\phi_{2\alpha-1},\phi_{2\alpha}$ will have been constructed. We then declare player I the winner if at least one of the following five conditions hold (II wins otherwise):-

4.9 there is no truth assignment, v, to the sentences $J(\underline{c})$ (for $c < B$) such that $v(\phi_{2i}) =$ True for all $i \leq \alpha$ with $\phi_{2i} \in \exists_0(s,B)$, where, for ψ an atomic sentence of H, $v(\psi) =$ True if and only if $(M,c)_{c<B} \models \psi$, or

4.10 Φ_{2i} is Φ^{*}_{2j} for some $i,j \leqslant \alpha$, or

4.11 Φ_{2i} is $\neg J(\underline{a})$ for some $i \leqslant \alpha$, or

4.12 Φ_{2i} is $J(\underline{b})$ for some $i \leqslant \alpha$, or

4.13 for some $c,d < B$ with $c \leqslant g(d)$, Φ_{2i} is $J(\underline{d})$ and Φ_{2j} is $\neg J(\underline{c})$ for some $i,j \leqslant \alpha$.

Now it is easy to show that there are Δ_0 formulas (involving some suitably large parameter, say $e_{20}(B)$) defining the sets of plays, winning plays for I and winning plays for II, and these sets will hence be M-coded. Further, strategies (for I or II) are simply M-coded functions (mapping plays $\dashrightarrow$ H) having the obvious (Δ_0) properties, and because the game is M-finite, an easy induction in M shows that either player I or player II has a winning strategy. We, of course, support player II so our next aim is to find a large α (as a function of n and s) for which II has a winning strategy for $\Gamma(n,s,\alpha)$.

<u>Lemma 8</u>.
Player II has a winning strategy for the game $\Gamma(0,s,\alpha)$ providing $\alpha < [{}^{t}/_{s}]$.

<u>Proof</u>.
Let H be the set of sentences in $\exists_0(s,B)$. Observe that if $\Phi \in H$ then clearly (assuming a reasonable Gödel numbering) there is an M-coded set $C(\Phi) = \{c_1,..,c_s\}$, where $c_i < c_{i+1} < B$ for $i = 1,..,s-1$, such that the truth value of Φ is naturally determined in M whenever truth values for $J(\underline{c}_1),..,J(\underline{c}_s)$ are given.

Now suppose that S is a strategy for I for $\Gamma(0,s,\alpha)$. It is sufficient to show how II can defeat this strategy, so suppose that $\Phi_1,\Phi_2,...,\Phi_{2i-1},\Phi_{2i}$ is a play of length i ($<\alpha$) of $\Gamma(0,s,\alpha)$ in which I has been using S. We suppose further that there are $a_i,b_i \in M$ such that the following three (Δ_0) conditions hold:-

4.14 $a \leqslant \hat{g}(2^{t-is}, a_i) \leqslant b_i \leqslant b$.

4.15 $\bigcup_{j=1}^{i} C(\Phi_{2j}) \cap (a_i,b_i) = \emptyset$.

4.16 If $J(\underline{c})$ is given the value true for $c \leqslant a_i$ and false for $c \geqslant b_i$, then Φ_{2j} gets the value true for each $j \leqslant i$.

Notice that by 4.5 these conditions are satisfied for $i=0$ if we define $a_0 = a$ and $b_0 = b$. Suppose that S now dictates that I plays the sentence $\Phi_{2i+1} \in H$. (Only rule 4.6 is relevant here.) Say $C(\Phi_{2i+1}) = \{c_1,...,c_s\}$ in increasing order and let us suppose that $a_i \leqslant c_1 < c_s \leqslant b_i$ (otherwise choose a convex subsequence of $c_1,...,c_s$ maximal with this property) and define $c_0 = a_i$ and $c_{s+1} = b_i$. I now claim that for some $j \leqslant s$ we have $\hat{g}(2^{t-(i+1)s}, c_j) \leqslant c_{j+1}$. For suppose not. Then for all $j \leqslant s$ we would have $\hat{g}(2^{t-(i+1)s+j}, c_0) > c_{j+1}$. For this is clear for $j = 0$, and if true for some $j < s$, then $c_{j+2} < \hat{g}(2^{t-(i+1)s}, c_{j+1}) < \hat{g}(2^{t-(i+1)s}, \hat{g}(2^{t-(i+1)s+j},c_0)) \leqslant \hat{g}(2^{t-(i+1)s+j+1}, c_0)$ (cf.2.10), so is true for $j+1$. But now (setting $j = s$) we have contradicted 4.14, and my claim is proved.

Now define j_0 to be the least $j \leqslant s$ such that $\hat{g}(2^{t-(i+1)s}, c_j) \leqslant c_{j+1}$ and set $a_{i+1} = c_{j_0}$ and $b_{i+1} = c_{j_0+1}$. Clearly 4.14 and 4.15 hold (for $i = i+1$), and and II may choose Φ_{2i+2} either Φ_{2i+1} or Φ^{*}_{2i+1} as dictated by 4.16 (for $i = i+1$).

Now notice that if $\alpha < [\frac{t}{s}]$, then $2^{t-\alpha s} \geqslant 2$ so that certainly $b_\alpha > g(a_\alpha)$ (by 4.14 with $i = \alpha$). Thus this construction can be continued for α steps and clearly the resulting play of length α satisfies none of 4.9–4.13, so II has defeated S as required.

□

<u>Lemma 9</u>
Suppose $n \in \mathbb{N}$, $s, \alpha \in M\setminus\{0\}$, and that player II has a wining strategy for

the game $\Gamma(n,s,\alpha)$. Then player II has a winning strategy for the game $\Gamma(n+1,s,\beta)$ for any $\beta \in M$ with $\beta < [\log_2\alpha]$.

Proof.
Let H be the set of sentences in $\exists_{n+1}(s,B)$, and let U be a winning strategy for II for $\Gamma(n,s,\alpha)$. We first introduce some notation concerning this game. If $i \leq \alpha$, then $P(i)$ denotes the (M-coded) set of plays of length i of $\Gamma(n,s,\alpha)$ in which II uses the strategy U and if $\rho \in \bigcup_{i\leq\alpha} P(i)$ then $E(\rho)$ denotes the set of moves made by II in ρ, i.e. $E(\rho)$ is the set of sentences occuring with even subscript in ρ. If $\rho \in P(i)$, $\rho' \in P(j)$ and $i \leq j$ then we write $\rho \leq \rho'$ if the sequence ρ is an initial segment of the sequence ρ'.

Now suppose that S is any strategy for I for $\Gamma(n+1,s,\beta)$. As in lemma 8 it is sufficient to show how II can defeat S, so suppose that $\Phi_1,\Phi_2,\ldots,\Phi_{2i-1}, \Phi_{2i}$ is a play of length i $(<\beta)$ of the game $\Gamma(n+1,s,\beta)$ in which I has been using S. We suppose further that there is some $k_i \in M$ and $\rho_i \in P(k_i)$ such that the following four (Δ_0) conditions hold:-

4.17 $k_i \leq \alpha \cdot (\frac{1}{2}+\frac{1}{4}+\ldots+\frac{1}{2^i})$.

4.18 For all $j \leq i$, if $\Phi_{2j} \in \exists_0(s,B)$ then $\Phi_{2j} \in E(\rho_i)$.

4.19 If $j \leq i$ and Φ_{2j} is non-trivially of the form $\exists\vec{x}\psi(\vec{x})$ (where $\psi(\vec{x})$ does not begin with an $\exists$) then for some $\vec{c} < B$, $\psi(\underset{\sim}{\vec{c}}) \in E(\rho_i)$.

4.20 If $j \leq i$ and Φ_{2j} is non-trivially of the form $\forall\vec{x}\psi(\vec{x})$ (where $\psi(\vec{x})$ does not begin with an $\forall$) and $\vec{c} < B$, the for no $h \in M$ with $h \leq \frac{\alpha}{2^i}$, and $\rho \in P(k_i+h)$ with $\rho_i \leq \rho$, do we have $\psi(\underset{\sim}{\vec{c}})^* \in E(\rho)$.

Now suppose that S dictates that I now plays Φ_{2i+1}. We must determine Φ_{2i+2} so that 4.17-4.20 are preserved.

Case 1 I is invoking rule 4.6.
Then $\Phi_{2i+1} \in H$ and II must set either $\Phi_{2i+2} = \Phi_{2i+1}$ or $\Phi_{2i+2} = \Phi_{2i+1}^*$. Now let us suppose for the moment that $\Phi_{2i+1} \notin \exists_0(s,B)$ so that we may clearly assume that Φ_{2i+1} is of the form $\exists\vec{x}\,\psi(\vec{x})$ where $\psi(\vec{x}) \in \exists_n(s,B)$ ($\psi(\vec{x})$ not beginning with an $\exists$) and where the quantification '$\exists\vec{x}$' is not vacuous. Now if there is some $h_i \leq \frac{\alpha}{2^{i+1}}$, $\vec{c} < B$ and $\rho \in P(k_i+h_i)$ with $\rho_i \leq \rho$ such that $\psi(\underset{\sim}{\vec{c}}) \in E(\rho)$, then we set $k_{i+1} = k_i + h_i$, $\rho_{i+1} = \rho$ and $\Phi_{2i+2} = \Phi_{2i+1}$. Clearly 4.17, 4.18 and 4.19 are preserved. To see that 4.20 is too, suppose $j \leq i+1$, Φ_{2j} is of the form $\forall\vec{x}\psi'(\vec{x})$ (non-trivially), $h \leq \frac{\alpha}{2^{i+1}}$, $\rho' \in P(k_{i+1}+h)$, $\rho_{i+1} \leq \rho'$ and $\psi'(\underset{\sim}{\vec{d}})^* \in E(\rho')$ for some $\vec{d} < B$. Then clearly $j \leq i$ and, setting $h' = h_i + h$, we have $h' \leq \frac{\alpha}{2^i}$, $\rho' \in P(k_i + h')$ and $\rho_i \leq \rho'$, which violates 4.20. If there are no h_i, $\vec{c}$, ρ with the above properties then we may clearly set $k_{i+1} = k_i$, $\rho_{i+1} = \rho_i$ and $\Phi_{2i+2} = \Phi_{2i+1}^*$.

Now for the case that $\Phi_{2i+1} \in \exists_0(s,B)$, let us first observe that since $i < \beta < [\log_2\alpha]$ we have:-

4.21 $4 \leq \frac{\alpha}{2^i}$ and hence $k_i \leq \alpha-4$.

Thus for some $\Phi_{2i+2} \in \{\Phi_{2i+1}, \Phi_{2i+1}^*\}$, there is $\rho_{i+1} \in E(1+k_i)$ such that $\rho_{i+1} = \langle \rho_i, \Phi_{2i+1}, \Phi_{2i+2} \rangle$ and we hence set $k_{i+1} = 1 + k_i$. The fact that 4.17-4.20 are preserved now follows easily from 4.21.

Case 2. I is invoking rule 4.7.
Then for some $j \leq i$, $\Phi_{2i+1} = \Phi_{2j} = \exists\vec{x}\psi(\vec{x})$ (ψ not beginning with an $\exists$), where '$\exists\vec{x}$' is non-vacuous. By 4.19, $\psi(\underset{\sim}{\vec{c}}) \in E(\rho_i)$ for some $\vec{c} < B$ and hence we may set $k_{i+1} = k_i$, $\rho_{i+1} = \rho_i$ and $\Phi_{2i+2} = \psi(\underset{\sim}{\vec{c}})$ so that 4.17-4.19 are trivially preserved and the only way 4.20 might not be is if $\psi(\vec{c}) = \forall\vec{y}\ \psi'(\underset{\sim}{\vec{c}},\vec{y})$ (ψ' not beginning with an $\forall$) and for some $\vec{d} < B$, $h \leq \frac{\alpha}{2^{i+1}}$, $\rho \in P(k_i + h)$ with $\rho_i \leq \rho$ we had $\psi'(\underset{\sim}{\vec{c}},\underset{\sim}{\vec{d}})^* \in E(\rho)$. However, $k_i + h < \alpha$ so, by 4.8, $\langle\rho,\psi'(\underset{\sim}{\vec{c}}\,\underset{\sim}{\vec{d}}),\ \psi'(\underset{\sim}{\vec{c}},\underset{\sim}{\vec{d}})\rangle$ is a play of length $k_i + h + 1$

$(\leqslant\alpha)$ of $\Gamma(n,s,\alpha)$ in which II has obviously used S but which is winning for I (by 4.10) - a contradiction

Case 3 I is invoking rule 4.8.

Then for some $j \leqslant i$, Φ_{2j} is $\forall\vec{x}\psi(\vec{x})$ (ψ not beginning with an $\forall$) and Φ_{2i+1} is $\psi(\vec{\underline{c}})$ for some $\vec{c} < B$. We are forced to set $\Phi_{2i+1} = \psi(\vec{\underline{c}})$. Now suppose I invokes rule 4.6 in the game $\Gamma(n,s,\alpha)$ following the play ρ_i to form $\langle\rho_i,\psi(\vec{\underline{c}})\rangle$. (This is allowed since $\psi(\vec{\underline{c}}) \in \exists_n(s,B)$.) If II now uses U then, by 4.20 and 4.21, the result will be $\langle\rho_i,\psi(\vec{\underline{c}}),\psi(\vec{\underline{c}})\rangle$. If $\psi(\vec{\underline{c}}) \in \exists_0(s,B)$ we let ρ_{i+1} be this play and $k_{i+1} = 1+k_i$. Otherwise, $\psi(\vec{\underline{c}}) = \exists\vec{y}\,\psi'(\vec{\underline{c}},\vec{y})$ (non-trivially, where $\psi'(\vec{\underline{c}},\vec{y})$ does not begin with an $\exists$) and we have, for some $\vec{d} < B$,

$\langle\rho_i,\psi(\vec{\underline{c}}),\psi(\vec{\underline{c}}),\psi(\vec{\underline{c}}),\psi'(\vec{\underline{c}},\vec{\underline{d}})\rangle \in P(2+k_i)$. (Player I has invoked rule 4.7 here and II has used U, which is permissible by 4.21.) We now let ρ_{i+1} be this play and set $k_{i+1} = 2 + k_i$. Now by 4.21, 4.17 is preserved. Also our construction guarantees that 4.18 and 4.19 are preserved and the same holds for 4.20 since if $j \leqslant i+1$ and Φ_{2j} is of the form stated in 4.20 then necessarily $j \leqslant i$.

Further, $h \leqslant \alpha/2^{i+1}$ and $\rho \in P(k_{i+1}+h)$ and $\rho_{i+1} \leqslant \rho$ clearly imply $\rho_i \leqslant \rho$ and

$\rho \in P(k_i+h')$ where $h' = 2 + h \leqslant 2 + \alpha/2^{i+1} \leqslant \alpha/2^i$ (by 4.21).

We have now shown how to construct a play $\vec{\Phi} = \Phi_1,\Phi_2,\ldots,\Phi_{2\beta-1},\Phi_{2\beta}$ of the game $\Gamma(n+1,s,\beta)$ in which I uses S, and a play ρ_β of $\Gamma(n,s,\alpha)$ in which II uses U, satisfying 4.17-4.20 (for $i=\beta$). Since ρ_β is a win for II it clearly follows that $\vec{\Phi}$ is a win for II, so II has defeated S as required.

□

Corollary 10
Suppose $n \in \mathbb{N}$, $s, \alpha \in M$, α nonstandard, and $e_n^-(\alpha) \leqslant [\frac{t}{s}]$. Then player II has a winning strategy for the game $\Gamma(n,s,\alpha-2)$.

Proof

By induction on n. For $n = 0$ the result follows from lemma 8. Suppose $e_{n+1}^-(\alpha) < [t/s]$. Since $e_{n+1}^-(\alpha) = e_n^-(2^\alpha-1)$, player II has a winning strategy for the game $\Gamma(n,s,2^\alpha-3)$ (by the inductive hypothesis). So by lemma 9, II has a winning strategy for $\Gamma(n+1,s,\beta)$ where $\beta = [\log_2(2^\alpha-3)]-1 = \alpha-2$ (since α is nonstandard), as required.

□

We now use corollary 10 to construct initial segments of M which are closed under g and in which definable initial segments are quite long. We first define (B,K) where K is an initial segment of M not containing B, to be the $L^r(J)$-structure with domain $\{c \in M: M \models c < B\}$ where L^r-relations are inhereted from M and J is interpreted as K.

Theorem 11
Suppose $n \in \mathbb{N} \setminus \{0\}$ and $\beta \in M \setminus \mathbb{N}$ satisfy $e_n^-(\beta) < t$. Then there is an initial segment K of M such that:-

4.22 K is closed under the function g (and g is total on K), and

4.23 $a \in K$, $b \notin K$, and

4.24 if $\Phi(x)$ is any (standard) $\exists_n$ or $\forall_n$ formula of $L^r(J)$ (possibly involving parameters from (B,K)) such that $(B,K)\models\Phi(0)$ and $(B,K)\models\forall x\,(\Phi(x) \longrightarrow (x+1))$, then $(B,K)\models\Phi\,(2^{[\beta/m]})$ for some $m \in \mathbb{N} \setminus \{0\}$.

Proof
Since $n \geqslant 1$ we can clearly find a nonstandard $s \in M$ such that $e_n^-([\beta/2] + 2) < [t/s]$. By corollary 10, player II has a winning strategy, U say, for the game $\Gamma(n,s,\alpha)$ where $\alpha = [\beta/2]$. Let $\Phi_1,\Phi_2,\ldots,\Phi_n\ldots$ $(n \in \mathbb{N})$

be an infinitely repetitive ennumeration of $\exists_n(s,B)$ (recall that M is countable). We construct plays $\rho_0 \leqslant \rho_1 \leqslant \ldots \leqslant \rho_r \leqslant \ldots$ of $\Gamma(n,s,\alpha)$ satisfying for each $r \in \mathbb{N}$:-

4.25 there is $n_r \in \mathbb{N} \setminus \{0\}$ such that $\rho_r \in P(i_r)$ for some $i_r \leqslant \alpha(1-\frac{1}{n_r})$,

where we are using here the same notation as in the proof of lemma 9 (in particular, P(i) denotes the set of plays of $\Gamma(n,s,\alpha)$ in which II uses U). Suppose ρ_r has been constructed for some $r \in \mathbb{N}$ (ρ_0 is the empty play).

Case 1 $r + 1 = 3q$.
If ϕ_q is a sentence (set $\rho_{r+1} = \rho_r$, $n_{r+1} = n_r$ if not) then for some $\phi_q' \in \{\phi_q, \phi_q^*\}$ we have $\langle \rho_r, \phi_q, \phi_q' \rangle \in P(1 + i_r)$ and we let ρ_{r+1} be this play and set $n_{r+1} = 1 + n_r$.
Case 2 $r + 1 = 3_q + 1$

If ϕ_q is a sentence (non-trivially) of the form $\exists\vec{x}\psi(\vec{x})$ (ψ not beginning with an $\exists$), and $\phi_q \in E(\rho_r)$ (set $\rho_{r+1} = \rho_r$, $n_{r+1} = n_r$ if not) then for some $\vec{c} < B$, $\langle \rho_r, \phi_q, \psi(\underline{\vec{c}}) \rangle \in P(1 + i_r)$ and we let ρ_{r+1} be this play and set $n_{r+1} = 1 + n_r$.

Case 3 $r + 1 = 3_q + 2$.

Suppose $\phi_q = \phi_q(x)$ contains just one free variable and that either (i) both $\phi_q(\underline{0})$ and $\phi_q(\underline{\delta})^*$ are in $E(\rho_r)$ or (ii) both $\phi_q(\underline{0})^*$ and $\phi_q(\underline{\delta})$ are in $E(\rho_r)$, where $\delta = 2^{[\alpha/2n_r]}$ (set $\rho_{r+1} = \rho_r$, $n_{r+1} = n_r$ if not). If (i) holds we consider the play $\rho_{r+1} = \langle \rho_r, X_1, X_1', \ldots, X_\gamma, X_\gamma' \rangle$ of $\Gamma(n,s,\alpha)$, where $\gamma = [\frac{\alpha}{2n_r}]$, defined as follows. Each X_i is of the form $\phi_q(\underline{c_i})$ for some $c_i < \delta$ and I is invoking rule 4.6, and X_i' is II's response as dictated by U. Further, c_1 is $2^{\gamma-1}$ and for $i = 2,\ldots,\gamma$, c_i is $c_{i-1} + 2^{\gamma-i}$ if $X_{i-1}' = X_{i-1}$ and c_i is $c_{i-1} - 2^{\gamma-i}$ if $X_{i-1}' = X_{i-1}^*$.

Then clearly $\rho_{r+1} \in P(i_{r+1})$ where $i_{r+1} = i_r + \gamma \leqslant \alpha(1-\frac{1}{2n_r})$ and it is not hard to see that:-

4.26 for some $i,j \leqslant \gamma$, X_i' is $\phi_q(\underline{c_i})$ and X_j' is $\phi_q(\underline{c_j})^*$ and $c_j = c_i + 1$ (and $c_i < \delta \leqslant 2^{[\beta/5n_r]}$).

Also, if (ii) above holds we could obtain ρ_{r+1} in a similar way and 4.26 would hold with the asterisk interchanged.

This completes the construction of ρ_{r+1}.

Let $T = \bigcup_{r \in \mathbb{N}} E(\rho_r)$ and define $K = \{c < B : J(\underline{c}) \in T\}$. Since no ρ_r satisfies either 4.9 or 4.10 it is easy to check (using cases 1 and 2 of the construction) that for any (standard) formula $\phi(\vec{x}) \in \exists_n \cup \forall_n$ and $\vec{c} < B$, we have $(B,K) \models \phi(\vec{c})$ if and only if $\phi(\underline{\vec{c}}) \in T$. Since no ρ_r satisfies any of 4.11-4.13, this implies 4.22 and 4.23. Further, 4.24 is clearly now implied by 4.26 (for a suitable q).

□

5 RESULTS ON NON-INTERPRETABILITY

Our aim in this final section is to use theorem 11 to show that theorems 5 and 7 are nearly best possible. Let us first fix a nonstandard model, M say, of $T_{\overline{\sigma}}$ + exp, so that clearly all the functions $f^*, f, f_1, \ldots F_1, F_2, \ldots$ are total in M and satisfy everywhere all the properties established in section 2. Let $n \in \mathbb{N} \setminus \{0\}$ and $a \in M \setminus \mathbb{N}$. We define t, b, g, and β as follows:-

5.1 $t = f^*(a)$.

5.2 $b = \hat{f}(2^t, a)$. (it is easy to show that the function $x \mapsto \hat{f}(2^{f^*(x)}, x)$ is total in M.)

5.3 $g = f \upharpoonright \{x \in M : M \models x \leqslant b\}$.

5.4 $e_n^-(\beta) < t \leqslant e_n^-(\beta+1)$.

We may further suppose that there is some $B \in M$ such that $e_m^-(b) < B$ for for all $m \in \mathbb{N}$ (by taking an elementary extension of M if necessary).

Now 4.4,4.5 and the hypotheses of theorem 11 are satisfied so there is an initial segment K of M satisfying 4.22-4.24 and in particular $K \models T_\sigma$. Now using a result of [1] (namely that Σ_0 formulas of L can be written (provably in $I\Delta_0$+exp) in the form $\exists \vec{y} \leq e_m^-(x)\ \psi$, where ψ is quantifier-free and $m \in \mathbb{N}$) it is easy to show that for any Σ_n (or Π_n) formula, $\chi(\vec{x})$, of L there is an $\exists_n$ (or $\forall_n$) formula, $\chi_*(\vec{x})$, of $L^r(J)$ possibly involving parameters from (B,K)) such that:-

5.5 for all $\vec{c} < B$, $(B,K) \models \chi_*(\vec{c})$ if and only if $\vec{c} \subseteq K$ and $K \models \chi(\vec{c})$.

We are now ready to prove:-

Theorem 12

For $n \geq 2$, σ_{n+1} is not interpretable in T_σ by a Σ_n or a Π_n formula.

Proof

Suppose $\phi(x)$ is a Σ_n or Π_n formula of L interpreting σ_{n+1} in T_σ. Define the formulas $\psi(x)$ and $\eta(x)$ by:-

5.6 $\psi(x) \Leftrightarrow \exists z\ (z = F_{n+1}(x) \wedge \phi(z))$.

5.7 $\eta(x) \Leftrightarrow \exists z\ (z = \hat{f}(x,0) \wedge \psi(z))$.

Notice that if $\phi(x)$ is Σ_n then so are $\psi(x)$ and $\eta(x)$. Also by an argument similar to the one used in the proof of theorem 5 (cf.3.9), if $\phi(x)$ is Π_n then so are $\psi(x)$ and $\eta(x)$ (since $n \geq 2$).

Consider the formulas $\phi_*(x)$, $\psi_*(x)$ and $\eta_*(x)$ (cf.5.5). By our supposition (we also use 5.5 without mention here) $\phi_*(x)$ defines in (B,K) an initial segment of K closed under the function f_{n+1}. Hence by 5.6,3.10 and lemma 6, $\psi_*(x)$ defines in (B,K) an initial segment of K closed under f. It now clearly follows from 5.7 that $\eta_*(x)$ defines in (B,K) an initial segment of K (containing 0 and) closed under + 1. Hence by 4.24 there is $m \in \mathbb{N} \setminus \{0\}$ such that:-

5.8 $(B,K) \models \eta_*(\delta)$, where

5.9 $\delta = 2^{[\beta/m]}$.

Now by 5.7 we obtain $(B,K) \models \psi_*(\hat{f}(\delta,0))$, and by 5.6 $(B,K) \models \phi_*(F_{n+1}(\hat{f}(\delta,0)))$. Clearly we have a contradiction (by 4.23) if we can show $b \leq F_{n+1}(\hat{f}(\delta,0))$.

Now by 2.9, $f^*(\hat{f}(\delta,0)) = \delta+1$ and hence by 2.18 $F_{n+1}(\hat{f}(\delta,0)) = \hat{f}(e_{n+1}^-(\delta+1),0)$. By 5.9 we certainly have $e_n^-(\delta+1) \geq 2 \cdot e_n^-(\beta+1)$ and so by 5.4, $F_{n+1}(\hat{f}(\delta,0)) \geq \hat{f}(2^{2t}-1,0)$ $= \hat{f}(2^t,\hat{f}(2^t-1,0))$ (by 2.10). Now it is easy to show that for all $x \in M$, $F_1(x) > x$ (by induction on x), so by 2.18 (for $n=1$), 5.1 and 5.2 we obtain $F_{n+1}(\hat{f}(\delta,0)) \geq \hat{f}(2^t,a) = b$, as required.

□

Of course there is still a gap between theorems 5 and 12 and we leave this as an open problem. However, they obviously imply:-

Theorem 13

The theory $\{\sigma_n : n \in \mathbb{N}\}$ is not interpretable in T_σ. In particular (cf. the example at the end of section 2) the theory $\{\forall x \exists y\ y = \omega_n(x) : n \in \mathbb{N}\}$ is not interpretable in $I\Delta_0$ (although every finite subset is).

□

(Strictly speaking, the function f of 2.21 does not satisfy 3.3, but this condition was only to guarantee that the functions $f_1, f_2, \ldots, F_1, F_2, \ldots$ have Δ_0 graphs and this is certainly the case for this particular f.)

We should remark here that J.B. Paris has modified the games used in section 4 to allow any parameters from M to occur in the sentences played (and not just those less than some fixed bound) provided $M \models PA$. In this case the games are M-infinite but the strategies for player II still exist as M-definable functions. Using a version of corollary 11, Paris then obtains the following considerable improvement of proposition 1:-

Proposition 14

Let $M \models P A$ be nonstandard and countable. Then there is an initial segment (closed under $\cdot$) I of M such that if J is any initial segment of I definable in (M,I) then J is not simultaneously closed under all the functions $\{\omega_n : n \in \mathbb{N}\}$.

Let us now turn to theorem 7. Unfortunately if the fuunction f associated with σ has very erratic growth it may be the case that no fixed number of iterations of F_1 dominates 2^x (even though for f the slowest growing of all uniform functions, namely f(x) = x + 1, we have $F_1(x) = 2^{x+1}-1$). Hence $T_{\bar{\sigma}}$ may not imply $I\Delta_0$ + exp so we have no hope of proving the converse of theorem 7 in general (see theorem 16 below). However, we do have the following result:-

Theorem 15

Suppose A is a Π_1 sentence of L such that $T_\sigma + A$ is interpretable in T_σ. Then $T_{\bar{\sigma}}$ + exp $\vdash$ A.

Proof

Let $\phi'(x)$ interpret $T_\sigma + A$ in T_σ. Let $\phi''(x)$ be the natural translation of $\phi'(x)$ into an L^r formula and choose $n \in \mathbb{N}$ and an $\exists_n$ formula of $L^r(J)$, $\phi(x)$ say, which is logically equivalent to $J(x) \wedge \phi''(x)^J$. Now suppose $A = \forall \vec{x}\, A'(\vec{x})$ ($A'(\vec{x})$ a Δ_0 formula) and, for contradiction, that M is a countable nonstandard model of $T_{\bar{\sigma}}$ + exp, $a \in M$, and $M \models \exists \vec{x} < a \neg A'(\vec{x})$. Now it is easy to show that $x \longmapsto \hat{f}(e_{n+1}(f^*(x)),x)$ is a total function in M (it is dominated by $F_1^{(n+2)}(x)$) so we may define:-

5.10 $t = e_n(f^*(a))$,

5.11 $b = \hat{f}(2^t,a)$, and

5.12 $g = f \restriction \{x \in M: M \models x \leqslant b\}$.

We may further suppose (by taking an elementary extension of M if necessary) that there is a $B \in M$ such that $B > e_m^-(b)$ for all $m \in \mathbb{N}$, so that by theorem 11 there is an initial segment K of M satisfying 4.22-4.24 with $\beta = f^*(a)-1$. Since $K \models T_\sigma$ it follows that ϕ defines in (B,K) an initial segment satisfying A and closed under f. In particular:-

5.13 $(B,K) \not\models \neg\phi(a)$.

As in the proof of theorem 12, we define $\eta(x) \Leftrightarrow \exists z\, (z = \hat{f}(x,0) \wedge \phi'(z))$ and conclude, by 4.24 and 5.5, that $(B,K) \models \eta_*(\delta)$, where $\delta = 2^{[\beta/m]}$, and $\beta = f^*(a)-1$. Clearly this implies $(B,K) \models \eta(f^*(a))$ and hence $(B,K) \models \phi(\hat{f}(f^*(a),0))$. However, by induction on x one can easily show that for all $x \in M$, $\hat{f}(f^*(x),0) > x$, and hence $(B,K) \models \phi(a)$ contradicting 5.13 as required.

□

We can now state our main result, which follows immediately from theorems 7 and 14 and 2.22.

Theorem 16

Suppose $T_{\bar{\sigma}} \not\vdash$ exp. Then for any Π_1 sentence A, $T_\sigma + A$ is interpretable in T_σ if and only if $T_{\bar{\sigma}} \vdash A$. In particular, $I\Delta_0 + A$ is interpretable in $I\Delta_0$ if and only if $I\Delta_0$ + exp $\vdash$ A.

REFERENCES

[1] H. Gaifman and C. Dimitracopoulos, Fragments of Peano's Arithmetic and the MRDP theorem, in Logic and Algorithmic Monographie No.30 de L'Enseignment Mathématique, Genève 1982, pp. 187-206.

[2] J.B. Paris and C. Dimitracopoulos, A note on the Undefinability of Cuts, J. of Symbolic Logic, Vol.48, No.3, Sept.1983, pp.564-569.

[3] J.B. Paris and A.J. Wilkie, On the Scheme of Induction for Bounded Arithmetic Formulas, submitted to the Annals of Pure and Applied Logic.

[4] Pavel Pudlak, Cuts, Consistency Statements and Interpretations, J. of Symbolic Logic, Vol.50, No.2, June 1985, pp.423-441.

LOGIC COLLOQUIUM '84
J.B. Paris, A.J. Wilkie, and G.M. Wilmers (Editors)

ON THE MODEL THEORY OF EXPONENTIAL FIELDS (SURVEY)

Helmut Wolter

Humboldt Universität zu Berlin
1086 Berlin
Unter den Linden 6-8

INTRODUCTION

The present paper is an extension of the lecture "Some results about exponential fields" which I gave at the Conference "Table Ronde de Logique" (Paris, 15/16 Octobre 1983). It provides a survey on some results concerning the theory of exponential fields.

The investigations of this theory are motivated by A. TARSKI's decidability problem concerning the field of real numbers with the additional exponential function e^x. In recent years a lot of people have been concerned with exponential fields and rings and have obtained many interesting results (see e.g. [R], [M], [Wi], [Dr1], [HR], [DW1], [DW2], [Da], [Wo1]), but TARSKI's problem is still open and a solution is not in sight at the moment. A complete solution seems to be very difficult.

In general, there are several methods to solve a decidability problem but in this special case attemps were made in two directions. On the one hand there are attemps to prove the elimination of quantifiers. This method has the advantage that one has to investigate only the standard model and hence one can use all the tools of analysis, but it has the disadvantage that one has to know almost all about this model. On the other hand it has been tried to approximate the theory of the standard model by suitable axiom systems in order to find perhaps a complete system by means of model theoretic tools as for instance: model completeness, prime model and others. In both cases great difficulties have to be overcome and as long as we do not know whether the theory is decidable we should not ignore that it could be undecidable. Even independent of TARSKI's decidability problem the class of exponential fields is a very interesting subject of investigation. Only the interplay of analytical and algebraic means yields fundamental results where the algebraic methods often have to be developed first.

Of course, it is impossible to give here a complete survey on all results concerning the theory of exponential fields. We want to restrict ourselves mainly to the results obtained by our research group in Berlin. We investigated different classes of exponential fields with the intention to obtain more information on such structures and classes and their theories in order to give perhaps a contribution to the solution of TARSKI's decidability problem.

<u>Definition.</u> If F is a field of characteristic o and E a unary function from F into F, then (F,E) is said to be an exponential field if $E(x+y) = E(x)E(y)$ for all $x,y \in F$ and if $E(o) = 1$, $E(1) \neq 1$. In this case E is said to be an exponential function on F.

We could also regard exponential fields of positive characteristic, but then $E(x) = 1$ for all x and this case is not interesting. In the following let L be a language for exponential fields, i.e. L contains symbols

$+, -, \cdot, {}^{-1}$ for the usual field operations and an additional unary function symbol E for an exponential function. Further let E_{ax} be the set of axioms $E(x+y) = E(x)E(y)$, $E(o) \neq 1$, $E(1) \neq 1$ and let EF be an $\forall$-axiom system for fields of characteristic o augmented by E_{ax}. Then EF determines the theory of exponential fields. The most important models of EF are (R,e) and (C,e), where R and C are the fields of real and complex numbers, respectively, and e is the usual exponential function in these fields.

In the following Q denotes the field of rational numbers, Z the set of integers, F an arbitrary field of characteristic o, $F = (F,E)$ an exponential field and, unless stated otherwise, m,n,k,l,i,j denote natural numbers. i can also be $\sqrt{-1}$, the actual meaning of i will be clear from the context. If F is an ordered field and $a \in F$, then $|a|$ is the absolute value of a. a is said to be infinitesimal if $|a|$ is smaller than every positive rational number, and a is finite if $|a| > q$ for some $q \in Q$. Notions and denotations not specially explained in this paper are used as usual.

Now our aim is to give a contribution to finding a recursive and complete axiom system of $Th(R,e)$ provided that such a system exists. So we try to approximate this theory by appropriate and natural axioms.

2. UNORDERED EXPONENTIAL FIELDS

First of all we want to provide some simple well-known facts.

1. In (F,E) E is not uniquely determined by F and EF. Indeed, if f is an additive function from F into F and $E(f(1)) \neq 1$, then $E^*(x) = E(f(x))$ is an exponential function on F, too.

2. (C,e) is strongly undecidable, i.e. every theory having (C,e) as a model is undecidable.

The field of rationals is definable in (C,e) by the formula

$$\varphi(x) := \exists y \; \exists z (E(y) = E(z) = 1 \wedge z \neq o \wedge x = y\,x).$$

In fact, $(C,e) \models e^y = 1$ iff $y = 2q\pi i$, where $q \in Z$ and $i = \sqrt{-1}$. Since Q is strongly undecidable (see e.g. [Sh]), we have the claim and, moreover, we obtain

3. EF is undecidable.

The next lemma implies that the range of the exponential function in every EF-existentially complete model is the whole field, excepting o.

<u>Lemma 1.</u> [DW1]

(i) If $a \in F$ and $a \neq 0$, then there exists an extension $F^* = (F^*, E^*)$ of F such that $F^* \models EF$ and $F^* \models \exists x(E^*(x) = a)$.

(ii) If F is EF-existentially complete, then $F \models \exists x(E(x) = a)$ for all $a \in F$, $a \neq o$.

Similar as for (C,e), there exists a formula in L defining the field of rationals in every EF-existentially complete model.

<u>Theorem 2.</u> [DW 1]

Let F be EF-existentially complete. Then, for all $a \in F$, $a \notin Q$ iff $F \models \exists x(E(x) = 1 \wedge E(ax) = 2) := \Psi(x)$.

Hence $\neg \Psi(x)$ defines Q in F. Since $\neg\Psi(x)$ does not define Q in (C,e), we obtain

Corollary 3. [DW1]

(C,e) is not EF-existentially complete.

By compactness arguments and the strong undecidability of Q we finally obtain from the above theorem:

Corollary 4. [DW1]

(i) EF is not companionable (and hence EF has no model completion).

(ii) Every existentially complete exponential field is strongly undecidable.

Theorem 5. [DW1]

(R,e) has no existential closure, i.e. there is no EF-existentially complete extension of (R,e) that can be embedded in every such extension.

Our results show that the theory of EF is rather complicated and since EF has models with quite different properties, EF is not a good approximation of Th(R,e). Therefore, in the next chapter we restrict ourselves to more special classes of such fields, namely to ordered exponential fields. Nevertheless, (C,e) is a very interesting but difficult subject of investigation Several attemps have been made to get a survey on the distribution of the zeros of such functions in (C ,e) defined by terms in L , but still without success. In[HR] C.W. HENSON and L.A. RUBEL proved, by means of very complicated analytical tools (NEVANLINNA-Theory), the following conjecture of S. SCHANUEL: If L' is the language of exponential fields with parameters from C but without division and f is a function from C^n into C defined by a term of L' and if f is nowhere equal to o , then f has the form e^g, where g is a definable function, too. Using analogous methods H. KATZBERG proved the following result for functions f,g on C defined by terms of L'.

Theorem 6. [K]

(i) Let $f\colon C \to C$. $f(x)$ has finitely many zeros iff it has the form $p(x)e^{g(x)}$, where p is a polynomial on C and g is definable.

(ii) Let $f\colon C^n \to C$, $n > 1$. $f(x)$ has no zeros in C (and hence $f(x) = e^{g(x)}$ for some definable g) or $f(x)$ has infinitely many zeros.

3. ORDERED EXPONENTIAL FIELDS

Next we are going to study some parts of the universal theory of the ordered field of real numbers with exponentiation. Let OF be an $\forall$-axiom system for ordered fields and

$$T = OF \cup E_{ax} \cup \{(1+1/n)^n \leq E(1) \leq (1+1/n)^{n+1}; n > 0\}.$$

Since the statement $\forall x > 0 \forall y(E(y) = 1 + 1/x \to E(xy) < E(1))$ is true in (R,e) but not in all non-Archimedean T-models, the $\forall$-theory of T is weaker than $Th_\forall(R,e)$. Hence we regard the better approximation

$OEF' = OF \cup E_{ax} \cup \{E(x) \geq 1 + x\}$.

The following theorem, which can be proved by standard arguments, shows that the theory of ordered exponential fields OEF' is sufficiently strong to characterize the exponential function uniquely in the standard model (R,e).

Theorem 7. [DW1]

If $(F,E) \models OEF'$, then it holds in (F,E) that

(i) $E(x) \geq 0,\ x \neq 0 \longrightarrow E(x) > 1 + x$, E is strictly monotonously increasing and takes arbitrarily small and large positive values (but not necessarily all positive values).

(ii) $x > 0 \wedge E(y) = 1 + 1/x \longrightarrow E(xy) < E(1) < E((x+1)y)$.

(iii) E is differentiable and $E'(x) = E(x)$.

Here, the derivative is defined by means of the ε - δ -technique. For proving the next results some special algebraic tools are necessary, in particular we need so-called partial exponential fields. These are ordered fields with a partial exponential function. Suitable extensions of the fields and the corresponding exponential functions finally yield

Theorem 8. [DW1]

(i) OEF'-existentially complete models are real closed fields.

(ii) In every OEF'-existentially complete model the statement $\forall x > 0\ \exists y(E(y) = x)$ is true, i.e. in such models E has the intermediate value property (for short: Int(E)).

OEF' is not sufficiently strong to prove the $\forall$-theory of (R.e)

Theorem 9. [DW1]

$OEF' \not\vdash \forall x > 0(E(x) \geq 1 + x + x^2/2)$.

On the other hand, $OEF' \vdash \forall x > 1/n(E(x) \geq 1 + x + x^2/2)$ for all $n > 0$. Now we regard a stronger $\forall$-axiom system OEF. For this let

$E_k(x) = \sum_{i \leq k} x^i/i!$ and $OEF = OEF' \cup \{E(x) \geq E_k(x)\text{: } k \text{ odd, } k \geq 3\}$.

Similar as above, OEF-existentially complete models are real closed fields. Furthermore, in such models the intermediate value property is true for all terms without iterated exponential function. It is an open question whether this property is true for all terms and it is also open whether OEF proves $Th_{\forall}(R,e)$.

Remark: One can prove that $Th(OEF) = Th(OEF' \cup$ $\cup \{\forall x(|x| < 1/n \longrightarrow E(x) \geq E_k(x)\text{: for arbitrary fixed } n > 0 \text{ and all odd } k \geq 3\}$, i.e. it suffices to approximate the exponential function only in some arbitrarily small standard neighbourhood in order to get the same theory.

4. ON THE STRUCTURE OF EXPONENTIAL FIELDS - A METHOD FOR CONSTRUCTING NEW EXPONENTIAL FUNCTIONS

In the following let OEFI = OEF $\cup$ $\{$Int(E)$\}$ and let OEF'I be similarly definded. Further let $F^0 = \{a \in F: a > 0\}$ be the positive part of the ordered field F and let $F^1 = \{a \in F: a > 1\}$.
We now want to split F^0 and F^1 into so-called additive and multiplicative Archimedean classes, respectively.

Definition.

(i) Let $a,b \in F^0$. $a \sim b$ iff $na \geq b$ and $nb \geq a$ for some n.

(ii) Let $a,b \in F^1$, $a \approx b$ iff $a^n \geq b$ and $b^n \geq a$ for some n.

Obviously, $\sim$ and $\approx$ are equivalence relations on F^0 and F^1, respectively. The corresponding sets of equivalence classes are denoted by $\tilde{F}^0$ and $\approx\!F^1$. If F is Archimedean ordered, then there exist exactly one additive and one multiplicative Archimedean class. In arbitrary ordered fields the Archimedean classes are segments of F, and hence the ordering in F induces orderings in $\tilde{F}^0$ and $\approx\!F^1$ by the following

Definition. Let $s,s' \in \tilde{F}^0$ and $S,S' \in \approx\!F^1$.

(i) $s < s'$ if there are elements $a \in s$ and $b \in s'$ such that $a < b$ (in F) and $\neg\, a \sim b$.

(ii) $S < S'$ if there are elements $a \in S$ and $b \in S'$ such that $a < b$ (in F) and $\neg\, a \approx b$.

Theorem 10. [Wo2]

Let (F,E) be a non-Archimedean exponential field such that for every infinite $a > 0$ there exists an infinite b with $b^2 \leq a$. Then we have

(i) $\tilde{F}^0$ is densely ordered without first and last element.

(ii) For all $a,b \in F^0$: $a \sim b$ iff $E(a) \approx E(b)$.

(iii) If $(F,E) \models$ OEF'I, then F^1 is densely ordered without first and last element.

(iv) E maps every class $s \in \tilde{F}^0$ into some class $S \in \approx\!F^1$. If moreover Int(E) is satisfied, then E maps s onto S (and F^0 onto F^1).

Hence, an exponential function is an order-preserving map of the additive Archimedean classes into the multiplicative ones and if Int(E) is satisfied, then the map is onto. Of course, we could extend the relations $\sim$ and $\approx$ to F and F^0, respectively, and modify the above results. Theorem 10 gives a little hint how to answer the question in which ordered fields exponential functions are definable.

Now we want to investigate the problem how well OEF describes the exponential function in a given model, or in other words, how great the difference between two such functions can be in the same field. First we are going to show that in Archimedean ordered OEF-models E is uniquely determined. For this purpose let L_2 be the language L augmented by a symbol E^* for a second exponential function and $OEF_2 = OEF(E) \cup OEF(E^*)$ be the union of the axiom system OEF formulated with E and E^*, respectively, OEF_2I is defined analogously.

Theorem 11. [DW 2]

Let (F,E,E^*) be a model of OEF_2.

(i) If $a \in F$ and a is finite, then $E(a), E^*(a)$ are finite and $E(a) - E^*(a)$ is infinitesimal.

(ii) If F is Archimedean, then $E = E^*$.

Now we regard an arbitrary model (F,E,E^*) of OEF_2 and investigate the relations between E and E^*. Theorem 7 implies that E, E^* are differentiable, strictly monotonously increasing (hence injective), and that E, E^* take only positive but arbitrarily small and large values. Moreover, let E take all positive values in F. Then, for every $a \in F$ there is exactly one $b \in F$ such that $E^*(a) = E(b)$. Defining $h(a) = b - a$ we obtain a function h from F into F such that $E^*(a) = E(a + h(a))$.

Proposition 12. [Wo2]

(i) h is additive, differentiable and the derivative of h is o everywhere.

(ii) $OEF \vdash \forall x(0 < |x| \leq 1 \longrightarrow E(x) < E_k(x) + x^{k+1})$ for all k odd.

Of course, if F is non-Archimedean, then h need not be constant. Now let h be an arbitrary additive map from F into F and E an exponential function on F. If $E^*(a) = E(a+h(a))$ and $E^*(a) \geq E_k(a)$ for all $a \in F$ and all k odd, then E^* is an exponential function on F in the sense of OEF, too. The next result is an extension of Theorem 6 from [DW2].

Theorem 13. [Wo2]

If $(F,E) \models OEFI$, then $(F,E^*) \models OEF$ iff there exists an additive function $h: F \longrightarrow F$ such that $E^*(a) = E(a + h(a))$ for all $a \in F$ and

(i) If a is infinitesimal, then $|h(a)| < |a|^n$ for all n.

(ii) If a is finite, then $h(a)$ is infinitesimal.

(iii) Let a be infinite and $a > 0$. If $h(a) \geq 0$, then $h(a)$ can be arbitrary. If $h(a) < 0$, then $a + h(a)$ has to be positive, infinite and $h(a) \geq -a + n \cdot Ln(a)$ for all n, where Ln is the inverse function of E.

(iv) Let a be infinite and $a < 0$. If $h(x) < 0$, then $h(x)$ can be arbitrary. If $h(a) \geq 0$, then $a + h(a)$ has to be negative, infinite and $h(a) \leq -a - n \cdot Ln(-a)$ for all n.

This theorem gives a good survey on the exponential functions E^* definable by E and certain additive functions h on F. The next results show that in all non-Archimedean models $(F,E) \models OEFI$ exponential functions E^* without and with the property $Int(E^*)$, respectively, are definable by means of E and h. For this, let $(F,E) \models$ OEFI and F be non-Archimedean. If we regard the additive group of F as a Q-vector space with a basis B, then we are able to define (by means of Theorem 13) at least card(F) different functions $h: B \longrightarrow B \cup \{0\}$ with the desired properties. Hence, these functions h yield card(F) different exponential functions on the same field F (with additional properties).

Theorem 14. [Wo2]

(i) For all infinite $a, b \in F$ with $0 < a < b$ there is an E^* of the form $E^*(x) = E(x + h(x))$ such that $(F,E^*) \models$ OEF and $\text{range}(E^*) \cap S = \emptyset$, where S is the segment $\{x \in F: x \approx a$ or $a < x < b$ or $x \approx b\}$.

(ii) If $E^*(x) = E(x + h(x))$ and $(F,E^*) \models$ OEF and for every $b \in B$ there exists an n such that $h(\ldots h(b)\ldots) = 0$, where h is iterated n-times, then $\text{Int}(E^*)$.

(iii) If $h: F \longrightarrow F$ is additive, $B_o = \{b_i: i < \omega\} \subseteq B$, $h(b_i) = b_{i+1}$ for all i and $h(b) = 0$ for $b \in B - B_o$, then $\neg\,\text{Int}(E^*)$.

The technique used here yields some further interesting results (see also [DW 2]).

Theorem 15. [Wo2]

(i). In every OEF_2I-existentially complete model the rationals can be defined by the formula

$$\varphi(x) := \forall y (E(y) = E^*(y) \longrightarrow E(xy) = E^*(xy)).$$

(ii) OEF_2I is not companionable.

(iii) OEF_2I is undecidable.

(iv) The theories of each OEF_2I-existentially complete model and of all such models are undecidable.

If we use stronger axiom systems than OEF_2I, it is unknown whether Theorem 15 remains true, because in the stronger case there is perhaps not enough freedom to define a suitable function E^*.

In connection with decidability investigations of a theory T the existence of a prime model of T (in the sense of A. ROBINSON) can be helpful. But with respect to OEFI the solution of this problem seems to be very difficult as we will see from the next theorem, which can be proved by means of the same technique as above.

Theorem 16. [Wo2]

Let e be the usual exponential function on R and $e_n = e(\ldots e(1)\ldots)$, where e is iterated n-times. If $1, e_1, \ldots, e_{n-1}$ are no rational multiples of e_n and if e_{n+1} is algebraic, then OEFI has no prime model.

Under the suppositions of Theorem 16 one obtains that e_{n+1} is algebraic and the corresponding constant term E^*_{n+1} can be transcendental in some suitable model (F,E^*), and this contradicts the existence of a prime model. In particular, the existence of such a model can depend on the unsolved question whether e^e is transcendental. By such results the investigations of exponential fields can become extremely difficult.

Further important questions are those concerning extensions of exponential fields. We know almost nothing about such problems (see also [Ge]). If for instance (F,E) is an exponential field and $\bar{F}$ the real closure of F, then it is unclear whether there is an exponential function $\bar{E} \supseteq E$ on $\bar{F}$. If F^* is the perfect closure of F, then, of course, there is a corresponding extension E^* of E with sufficiently good properties. Great difficulties also have to be overcome if we are given an exponential field (F,E) where F is real closed and we want to extend E to the algebraic closure of F such that E has appropriate properties. With respect to this problem H. KATZBERG obtained some partial results.

5. SOME ANALYTICAL PROPERTIES OF EXPONENTIAL FIELDS

Now we do not regard $\forall$-axiom systems any longer because we need stronger axioms if we want to investigate more interesting analytical properties of exponential fields. For this let $OEF^* = OEF \cup \{$Intermediate value property for terms$\} \cup \{$ROLLE's Theorem for terms$\}$. OEF^* is formulated for all terms t in L with respect to a fixed variable in such intervals where the terms are defined. By OEF^* the inequalities $E(x) \geq E_k(x)$ can be proved from the other axioms if k is odd and $k \geq 3$. It is an open problem whether the terms have the intermediate value property in OEFI-models.

By means of WILKIE's and RICHARDSON's results DAHN was able to prove

<u>Theorem 17.</u> [Da]

> If $F \models OEF$, $F \subseteq F^* \models OEF^*$ and if $t(x), s(x)$ are terms with parameters from F, then $F^* \models \exists y \forall x(x \geq y \longrightarrow t(x) \geq s(x)) := \varphi$ iff $\text{Diagram}(F) \cup OEF^* \vdash \varphi$.

This theorem finally implies

<u>Theorem 18.</u> [Da]

> If $F, F^* \models OEF^*$, $F \subseteq F^*$ and $\varphi(x)$ is a quantifier free formula with parameters from F, then $F \models \exists x \varphi(x)$ iff $F^* \models \exists x \varphi(x)$.

The last result is a little hint that OEF^* could be model complete but the problem is still open.

Now we come to the "Problem of the last root", which goes back to A. MACINTYRE (see [Dr1]). It is induced by the following question: Let $p(x)$ be a non-zero exponential polynomial over $\mathbb{R}$. Is there an intelligible function depending only on the real parameters of p which bounds the absolute values of the real roots of p ?

The next theorem answers this question positively not only for exponential polynomials in the standard model but also for all non-zero exponential terms with one variable in all OEF^*-models. Let $F \models OEF^*$ and $t(x,\bar{a})$ be a term with parameters $\bar{a}$ from F and the complexity k. Further let s_i be a constant term with the parameters $\bar{a}$ and the complexity $\leq k+1$ and let $s = \sum s_i^2$, where the sum is taken over the finite set of all these terms. (It would suffice to regard $s = \sum |s_i|$, but $|s_i|$ is no term).

Finally, let D be the corresponding finite part $\{s_i > 0: F \models s_i > 0\} \cup \{s_i = 0: F \models s_i = 0\}$ of Diagram(F) and $T = OEF^* \cup D$. Then we have

Theorem 19. [Wo1]

> There is a computable constant term t^* depending only on $\bar{a}$ such that
>
> (i) $F \models \exists y \ \forall x > y(t(x,\bar{a}) \gtrless c)$, then
> $T \vdash \forall x(x > t^* \longrightarrow t(x,\bar{a}) \gtrless c)$.
>
> (ii) $T \vdash \forall x(t(x,\bar{a}) = 0 \longrightarrow |x| \leq t^*)$.

Here "computable" means that there exists a computable natural number n such that $t^* = E(\ldots E(s)\ldots)$, where E is iterated n-times. This theorem was generalized by J. GEHNE and P. GÖRING (see [Ge], [Gö]). They extended the language L by the inverse of the exponential function and by a rather big class of algebraic functions and obtained the analogous result for the extended language L''.

In order to prove Theorem 19 we used some Lemmata and techniques from DAHN's paper [Da] where he approximated exponential terms in non-Archimedean OEF^*-models by appropriated series and obtained interesting results about the limit behaviour of exponential terms. Now we want to present the most important theorem of this paper (which is not necessary to prove Theorem 19).

Theorem 20. [Da]

> Let F, $t(x,\bar{a})$ be as in the previous theorem and $b \in F$. If $F \models \lim_{x \to \infty} t(x,\bar{a}) = b$, then there is a constant term t^* (with the same parameters and the same number of iteration steps of E as t) such that $F \models t^* = b$.

This implies that the limit of a term t belongs already to the exponential subfield generated by the paramters from t. The latter theorem (and the theorems 17, 18) as well as the techniques developed by DAHN in order to get this results were extended by DAHN, GEHNE and GÖRING to terms from the language L'' (see [DG], [Ge], [Gö]).

By means of Theorem 20 one finally obtains the converse of a theorem of RICHARDSON (see [R]). For this, let M be the smallest set of functions containing $1, x$ and let it be closed under $+, \cdot$ and t^s for functions $t, s \in M$ (obviously, the operation t^s can be expressed in L''). Further let D be the smallest set of constants such that $1 \in D$ and if $a, b \in D$, then

$a+b,\ a.b,\ e^a,\ a^{-1} \in D$.

Theorem 21.

(i) [R] If $a \in D$, then there are $f,g \in M$ such that $a = \lim_{x \to \infty} f/g$

(ii) [DG] If $f,g \in M$ and f/g is bounded (by a rational number), then $\lim_{x \to \infty} f/g \in D \cup \{0\}$.

(ii) generalizes a theorem of van den DRIES, who proved it for functions $< 2^{2^x}$ (from one point on). He obtained a lot of other interesting results which aim at the elimination of quantifiers for the theory of exponential fields (see [Dr2], [Dr 3]).

Finally we want to remark that for the investigation of our theory it was essential to have a suitable description of the exponential function, i.e. we needed an appropriate functional and differential equation and a sufficiently good approximation of E by rational terms. For other functions satisfying these conditions corresponding results could be possible but there are only few such nice functions. Arc tangens for instance has sufficiently good properties and hence, as P. PÄTZOLD proved in [P], many of the results presented here can be changed for the corresponding model class.

REFERENCES

[Da] Dahn, B.I., The limit behaviour of exponential terms, to appear in Fund. Math.

[DG] Dahn, B.I. and P. Göring, Notes on exponential-logarithmic terms, Preprint.

[Dr1] Van den Dries, L., Exponential rings, exponential polynomials and exponential functions, Preprint.

[Dr2] Van den Dries, L., Analytic Hardy fields and exponential curves in the real plane, Preprint.

[Dr3] Van den Dries, L., Bounding the rate of growth of solutions of algebraic differential equations and exponential equations in Hardy fields, Preprint.

[DW1] Dahn, B.I. and H. Wolter, On the theory of exponential fields, ZML 29 (1983), 465 - 480.

[DW2] Dahn, B.I. and H. Wolter, Ordered fields with several exponential functions, to appear in ZML.

[Ge] Gehne, J., Über Lösungsmengen gewisser exponentiell-algebraischer Gleichungen, Diss. A (Humboldt-Universität).

[Gö] Göring, P., Über Eigenschaften exponentiell-logarithmischer Terme, Diss. A (Humboldt-Universität).

[HR] Henson, C.W. and L.A. Rubel, Some applications of Nevanlinna theory to mathematical logic: Identities of exponential functions, Preprint.

[K] Katzberg, H., Complex exponential terms with only finitely many zeros, Seminarbericht Nr. 49, Humboldt-Universität zu Berlin, Sektion Mathematik.

[M] Macintyre, A., The laws of exponentiation, Preprint.

[P] Pätzold, P., Über Körper und Terme mit arcustangens-Funktionen, Preprint.

[R] Richardson, D., Solution of the identity problem for integral exponential functions, ZML 15 (1969), 333-340.

[Sh] Shoenfield, J.R., Mathematical logic, Addison-Wesley Publishing Company, 1967.

[Wi] Wilkie, A.J., On the exponential fields, Preprint.

[Wo1] Wolter, H., On the problem of the last root for exponential terms, to appear in ZML.

[Wo2] Wolter, H., Some remarks on exponential functions in ordered fields, Preprint.

LOGIC COLLOQUIUM '84
J.B. Paris, A.J. Wilkie, and G.M. Wilmers (Editors)

BOUNDED ARITHMETIC FORMULAS AND TURING MACHINES OF CONSTANT ALTERNATION

Alan Woods*
University of Malaya
Kuala Lumpur, Malaysia

A formula in the first order language for the natural numbers N with $=,\leqslant,+,\cdot,0,1$ is bounded arithmetic (or Δ_0) if all its quantifiers are of the forms $\exists y\leqslant x$, $\forall y\leqslant x$.

It is easily seen that every Δ_0 formula $\phi(x)$ in one free variable is equivalent to a formula in "prenex normal form":

$$\exists \vec{y}_1 \leqslant x \forall \vec{y}_2 \leqslant x \exists \vec{y}_3 \leqslant x \ldots Q\vec{y}_k \leqslant x \quad \psi(x,\vec{y}) \tag{0.1}$$

where $\psi(x,\vec{y})$ is quantifier free, $\exists \vec{y}_i \leqslant x$ denotes a block of quantifiers $\exists y_{i1} \leqslant x \exists y_{i2} \leqslant x \ldots \exists y_{im} \leqslant x$, Q is $\exists$ or $\forall$ according as k is odd or even, and $\vec{y} = \vec{y}_1\vec{y}_2 \ldots \vec{y}_k$, etc.. The classical hierarchy of arithmetic formulas with unbounded quantification suggests immediately the important question:

BOUNDED ARITHMETIC HIERARCHY PROBLEM: Can k in (0.1) be fixed independently of ϕ?

In other words, is there a proper Δ_0 hierarchy according to the number of alternations of bounded quantifiers needed to define a predicate? Although it has already been considered by many authors, among them Harrow [10], Lipton [16], Wilkie [26], Paris and Dimitracopoulos [23],[6], this question remains open. A solution, either way, would have many interesting consequences.

For example in Harrow [10] it is shown that if the Δ_0 hierarchy does not collapse, then DSPACE(n) (or equivalently the Grzegorczyk class E_*^2) contains a set of natural numbers which cannot be defined by any Δ_0 formula. This would answer a long-standing question of Parikh [24].

On the other hand suppose that M is a nonstandard model of Peano Arithmetic and identify $a \in M$ with $\{0,1,2, \ldots, a-1\}$. It is known (see,e.g.[23]) that treating $+,\cdot$ as relations on a, the theory of the structure $(a,+,\cdot,\leqslant)$ determines the theory of $(a^n,+,\cdot,\leqslant)$ for all $n \in N$. But is there a nondecreasing (slowly growing) Δ_0 definable function g(x) with $g(n)\to\infty$ as $n\to\infty$ through N,

---

*Current Address: Dept. of Mathematics, Yale University, Box 2155 Yale Station, New Haven, Connecticut 06520, U.S.A.

(e.g., $g(x) = [\log_2 x], [\log_2 \log_2 x]$, etc) such that for all $a \in M$, the theory of $(a, +, \cdot, \leq)$ determines the theory of $(a^{g(a)}, +, \cdot, \leq)$? By adapting methods of Lessan [15] it can be shown (see [23]) that if the Δ_0 hierarchy collapses there can be no such $g(x)$.

The special case k=1 of the Δ_0 hierarchy question is of particular interest. It can be reformulated in the following way. Call a set $X \subseteq N$ <u>bounded Diophantine</u> if X is defined by a formula $\phi(x)$ of the form

$$\exists \vec{y} \leq x \; (\; F(x,\vec{y}) = 0 \;)$$

where $F(x,\vec{y})$ is a polynomial with integer coefficients. Davis, Matijasevič, and Robinson [5] ask:

<u>BOUNDED DIOPHANTINE QUESTION</u>: Does X being bounded Diophantine imply that its complement X^c is also bounded Diophantine?

Closely related questions have been considered by Manders and Adleman [17], Hodgson and Kent [11], and Börger [3].

The main new result of the present paper is a weak hierarchy theorem for Δ_0 formulas.

<u>THEOREM 0.1</u> There is <u>no</u> fixed k such that every Δ_0 formula $\phi(x)$ is equivalent to a Δ_0 formula of the form

$$\exists y_1 \leq x \, \forall y_2 \leq x \, \exists y_3 \leq x \, \ldots \, Qy_k \leq x \quad \psi(x,\vec{y})$$

where $\psi(x,\vec{y})$ is quantifier free.

In other words, for every $k \in N$ there is a Δ_0 formula which is not equivalent to any Δ_0 formula containing only k quantifiers. This improves on earlier work of Alex Wilkie [26] which covered the case where $\psi(x,\vec{y})$ is allowed to contain at most a constant number (independent of ϕ) of occurrences of $\leq$. Note that an attempt to go from k <u>blocks</u> of quantifiers to k quantifiers via the classical coding argument runs into the difficulty that x^r code numbers are required in order to assign a different code to each ordered r-tuple $y_1, y_2, \ldots, y_r < x$.

There is a somewhat similar problem in proving theorem 0.1 (and Wilkie's theorem) in that for each $r \in N$, there are formulas $\psi(x,\vec{y})$ containing terms which can take values $\geq x^r$, and we wish to diagonalise over these using a formula with terms of fixed size. This is done by using a space-time-alternation trade-off for Turing machines of constant alternation. While it is certainly not <u>necessary</u> to introduce such machines, they do make the technique - and the important issues - easier to understand. Whether the method is suitable for attacking the full Δ_0 hierarchy problem is not yet clear, but it does seem to be a good way of separating the full hierarchy from its <u>low</u> levels.

Some preliminaries.

All logarithms $\log x = \log_2 x$ in this paper are to base 2. Subsets of N and relations on N will be called Δ_0 if they can be defined by Δ_0 formulas.

The class Δ_0 of all such sets can be characterised in several other ways. Let us adopt the convention, for our purposes harmless, that $\{0,1\}^*$ denotes the set of all nonempty finite strings over the alphabet $\{0,1\}$ which are either 0 or begin with 1, so that sets $S \subseteq N$ can be identified with languages (i.e., subsets) $L \subseteq \{0,1\}^*$ by identifying each $x \in S$ with its binary representation as a string of 0's and 1's. It turns out that this correspondence identifies the Δ_0 sets with the languages $L \subseteq \{0,1\}^*$ accepted in linear time by Turing machines of constant alternation.

Alternating Turing machines were introduced into computational complexity theory by Chandra, Kozen and Stockmeyer (see [4]). Since the idea of a Turing machine of constant alternation deserves to be more widely known amongst logicians, we give a detailed description of them and some of their properties.

1. TURING MACHINES OF CONSTANT ALTERNATION.

Σ_1 Turing machines.

A Σ_1- TM M is an offline nondeterministic Turing machine. M has a read only input tape, on which it is given $x \in \{0,1\}^*$ bounded on each side by a marker symbol #, and a finite number m of read/write work tapes. (Occasionally we will replace x by $x_1\# x_2\# \ldots \# x_k$ where $x_i \in \{0,1\}^*$.) M is characterised by its finite work tape alphabet A, finite set Q of state symbols, initial state q_1, accepting state q_a, and its program P, so we write $M = \langle P, A, q_1, q_a, Q \rangle$. P is a finite set of instructions:

$$a\, b_{old}\, q_{old}\, b_{new}\, q_{new}\, D_0 D_1 \ldots D_m \tag{1.1}$$

where $a \in \{0,1,\#\}$; $b_{old}, b_{new} \in A^m$; $q_{old}, q_{new} \in Q$; $D_i \in \{L_i, R_i, 0\}$ for $i=0,1,\ldots,m$, and not all $D_i=0$. The instruction (1.1) dictates that if a, b_{old} are the symbols currently scanned by the input and work tape heads, respectively, and q_{old} is M's current state, then M should replace b_{old} by b_{new}, q_{old} by q_{new}, and move the head on tape i one symbol left, right, or not at all, according as D_i is L_i, R_i, or 0. (Note that at least one head moves.)

An instantaneous description (ID) of M consists of the finite strings on the tapes, the head positions (relative to the tape contents) and the current state. An ID is a "snapshot" of M at an instant. The initial ID denoted by $ID_1(x)$ has blank work tapes, state q_1, and the input head on the left most symbol of x.

A <u>computation</u> C of a Σ_1- TM M on input x is a sequence (possibly infinite) of instantaneous descriptions $ID_1(x) = ID_1, ID_2, \ldots$ such that:

(i) ID_{i+1} is obtained from ID_i by performing some instruction in P.

(ii) If C ends in ID_i with a, b_{old}, q_{old} as scanned symbols and current state, then no instruction in P begins with $a\ b_{old}\ q_{old}$. (M must continue if it can.)

C is an <u>accepting computation</u> if

(i) C is finite, and

(ii) the final ID in C has current state q_a.

There may be many computations C on input x since more than one instruction may begin with $a\ b_{old}\ q_{old}$. (If there is only one, we say the instruction is <u>deterministic</u>.) The computations C form the branches of a rooted tree $\tau_M(x)$ with an ID at each node and $ID_1(x)$ as the root. (ID' is a <u>successor</u> of ID in $\tau_M(x)$ if ID' can be obtained from ID by performing a single instruction from P.) If all instructions in P are deterministic - so $\tau_M(x)$ has only one branch - then M is called a DTM.

<u>Σ_k and π_k Turing machines.</u>

$< P, A, q_1, q_a, Q_1, Q_2, \ldots, Q_k >$ is a description of a Σ_k- TM M if

(i) $Q = Q_1 \dot{\cup} Q_2 \dot{\cup} \ldots \dot{\cup} Q_k$ is a partition of the set of states Q of a Σ_1- TM $\bar{M} = <P, A, q_1, q_a, Q>$

(ii) For each instruction $a\ b_{old}\ q_{old}\ b_{new}\ q_{new}\ D_0\ D_1 \ldots D_m$ in P, if $q_{old} \in Q_i$ then $q_{new} \in Q_i \cup Q_{i+1}$.

We write $Q_i = \begin{cases} \exists_i & \text{if i is odd.} \\ \forall_i & \text{if i is even.} \end{cases}$

A description of a π_k- TM is exactly the same except that we take

$$Q_i = \begin{cases} \forall_i & \text{if i is odd.} \\ \exists_i & \text{if i is even.} \end{cases}$$

The states $q \in \exists = \bigcup_i \exists_i$ are called <u>existential states</u>. The states $q \in \forall = \bigcup_i \forall_i$ are called <u>universal states</u>.

A computation C of a Σ_k or π_k- TM M on input x is any subtree of $\tau_{\bar{M}}(x)$

such that

(I) $ID_1(x) \in C$.

(II) For all ID $\in$ C with at least one successor in $\tau_{\bar{M}}(x)$,

(a) if ID has an existential current state $q \in \exists$ then exactly one successor ID' of ID is in C,

(b) if ID has a universal current state $q \in \forall$ then all successors of ID are in C.

C is an accepting computation of M on input x if all branches in C are accepting computations of the Σ_1- TM $\bar{M}$.

M accepts x if there is some accepting computation C of M on input x.

Intuitively an alternating TM M works in the following way. Suppose M is scanning symbols a b_{old} on its tapes in state q_{old}. If q_{old} is existential, M chooses one of the instructions in P beginning with a b_{old} q_{old} to perform. (We say M "guesses" the move, or "branches existentially".) If q_{old} is universal, M "branches universally" by spawning enough copies of itself to try all instructions in P beginning with a b_{old} q_{old}. These clones have no way of communicating with each other - acceptance is achieved if an external observer finds that all the machines produced have halted in acceptance.

The language $L_M \subseteq \{0,1\}^*$ accepted by M is defined by

$$L_M = \left\{ x : x \text{ is accepted by } M \right\} .$$

Let T(n), S(n) denote functions of n. Consider a computation C of M on an input x of length n. (This means x has n symbols so n $\approx$ log x.)

C uses time T(n) if all branches in C have length $\leq$ max $\{T(n),n\} + 1$. (The length of a branch is the number of ID's on it.) Thus M performs at most max $\{T(n),n\}$ steps (i.e., instructions) along each branch of C.

C uses space S(n) if along any branch in C, the most distant cells visited by any work tape head are no more then S(n) cells apart.

M works in time T(n) and (or) space S(n) if for all sufficiently large n and all strings $x \in L_M$ of length n, there is an accepting computation C on input x which uses time T(n) and (or) space S(n) (respectively).

$$\Sigma_k\text{- TIME } (T(n)) = \{L_M : M \text{ is a } \Sigma_k\text{- TM which works in time } T(n)\}$$

$$\Sigma_k\text{- SPACE } (S(n)) = \{L_M : M \text{ is a } \Sigma_k\text{- TM which works in space } S(n)\}$$

Σ_k- TIME (T(n)) * SPACE (S(n))=

{ L_M: M is a Σ_k- TM which works in time T(n) and space S(n)}

Similar definitions of π_k- TIME (T(n)), DTIME (T(n)), etc. are assumed for π_k- TM's and DTM's.

A function T(n) is fully time constructible if T(n) $\geq$ n and there is some DTM M which, when given as input a string of n symbols 1, halts after exactly T(n) steps. All nonconstant polynomials with nonnegative integer coefficients are fully time constructible.

Remarks:

(i) We advocate the use of the * notation to denote simultaneous resource bounds in general. Strictly speaking we should perhaps write Σ_k*TIME(T(n)) etc. since alternation can be regarded as a resource-cf. remark (iii).

(ii) Provided T(n) is sufficiently honest (e.g., if T(n) is fully time constructible) a language L is accepted by some Σ_k- TM in time T(n) if and only if its complement $L^c = \{x \in \{0,1\}^* = x \notin L\}$ is accepted by some π_k- TM in time T(n).

(iii) We do not really have to introduce the partition $Q=Q_1 \dot{\cup} Q_2 \dot{\cup} \ldots \dot{\cup} Q_k$. An alternative approach is simply to define alternating Turing machines $M = \langle P, A, q_1, q_a, \exists, \forall \rangle$ with computations C given by (I) and (II) above. There is a natural partition of each branch in $\tau_{\bar{M}}(x)$ into segments according to whether the current state of the ID is in $\exists$ or $\forall$. We could say M is a Σ_k- TM if the initial state $q_1 \in \exists$ and for every $x \in L_M$ there is some accepting computation C on input x which has at most k segments. The classes Σ_k- TIME (T(n)) obtained using this alternative definition are the same as those defined above. This approach also allows bounding the number of alternations by a non-constant function A(n)- see [4] and [22].

Mainly for convenience in stating theorems, we observe that a linear "speed up" theorem of Book and Greibach [2] for Σ_1- TM's generalises:

PROPOSITION 1.1 Let T(n) $\geq$ n, k $\geq$ 1 and c $\geq$ 1. If L $\in$ Σ_k- TIME (c T(n)) then L = L_M for some Σ_k- TM M with only 2 work tapes (plus the input tape) which works in time T(n).

Note that the size of M's tape alphabet depends on c. If T(n) is fully time constructible the analogous proposition for π_k- TM's also holds.

In particular linear time on Σ_k and π_k- TM's (respectively) is characterised by

$$\bigcup_{c\geq 1} \Sigma_k\text{- TIME (cn)} = \Sigma_k\text{- TIME (n)}$$

$$\bigcup_{c\geq 1} \pi_k\text{- TIME (cn)} = \pi_k\text{- TIME (n)}$$

An obvious question is whether the classes Σ_k- TIME (T(n)), π_k- TIME (T(n)) form a proper hierarchy, and in particular:

LINEAR TIME HIERARCHY PROBLEM: Is there some k such that

$$\bigcup_{m\in N} \Sigma_m\text{- TIME}(n) = \Sigma_k\text{- TIME}(n) \ ?$$

Apparently all that is known in this direction is the theorem of Paul, Pippenger, Szemeredi and Trotter [21] that deterministic linear time $\bigcup_{c\geq 1}$ DTIME(cn)(= DTIME((1+ε)n) for any ε>0, see [13]) is properly contained in Σ_1- TIME(n). The linear time hierarchy problem is, of course, a linear analogue of the Stockmeyer polynomial time hierarchy problem (see e.g. [4]) which presumably generalises the P = NP question.

2. CONNECTIONS WITH THE Δ_0 HIERARCHY PROBLEM.

Although the connection between Δ_0 sets and languages in the linear time hierarchy has been folk law for a considerable time now, the literature on the subject takes a rather circuitous path. Bennett [1] proved that the Δ_0 sets are identical with the class RUD of rudimentary languages. We will not bother with the usual definition of these except to say that it is immediate that RUD $\subseteq \bigcup_{k\in N} \Sigma_k$- TIME(n). Wrathall [27] defined a linear time hierarchy by means of oracles, and showed that its union is identical to RUD. Modulo the equivalence between hierarchies defined using oracles and those defined using machines of constant alternation (see [4]), it follows that Δ_0 and $\bigcup_{k\in N} \Sigma_k$- TIME(n) are identical. This was noted explicitly by Lipton [16] and other authors.

The equivalence of the Δ_0 and linear time hierarchy problems can be obtained directly via a series of lemmas which we now sketch.

PROPOSITION 2.1 (Bennett [1]) The relation $z = x^y$ can be defined by a Δ_0 formula.

Besides Bennett several other writers have described Δ_0 definitions of the graph of exponentiation, among them Quincey, Paris, Dimitracopoulos, Gaifman, and Pudlak (see e.g. [6],[8],[25]) so we omit the proof.

LEMMA 2.2 There is a Δ_0 formula $\nu(x,y,i,k)$ such that for all $i< k$, $\nu(x,y,i,k) \leftrightarrow$ there are exactly y occurrences of the digit i in the k-ary expansion of x.

Proof: Suppose x,y,i satisfy the right hand side. Then $y \leq \left[\frac{\log x}{\log k}\right] + 1 \leq h^2$ where h is the least integer such that $h^2 \geq [\log x] + 1$. For $j \leq h^2$ let z_j denote the number of occurrences of i in the first j digits of the k-ary expansion of x (or in x if x has less than j digits).

The sequence $z_h, z_{2h}, \dots, z_{h^2}$ consists of h numbers each having at most $[\log h] + 1$ binary digits. Therefore it can be coded by a number z with $h([\log h] + 1)$ binary digits. Note that $z \leq 2^{h([\log h]+1)} \leq x$ (unless x is very small).

Similarly, for each $j < h$ there is a number $w \leq x$ such that w codes $z_{jh+1}, z_{jh+2}, \dots, z_{(j+1)h}$. Using proposition 2.1 it is easily seen there is a Δ_0 formula $\nu(x,y,i,k)$ which holds if and only if
$\exists h \leq x \exists z \leq x \forall j \leq h \exists w \leq x$ (h,i,j,k,w,x,y,z have the relationship just described).

<u>LEMMA 2.3</u> Consider the formula

$$\exists \vec{y}_1 \leq x^d \forall \vec{y}_2 \leq x^d \dots Q\vec{y}_k \leq x^d \; \psi(x,\vec{y})$$

where ψ is a quantifier free Δ_0 formula and $d \in N$ is a constant. This formula is equivalent to a Δ_0 formula of the form

$$\exists \vec{Y}_1 \leq x \forall \vec{Y}_2 \leq x \dots Q\vec{Y}_k \leq x \;\; \psi^+(x,\vec{Y})$$

where ψ^+ is quantifier free.

<u>Proof</u> : Let $\vec{Y}$ have d variables $V_1, V_2, \dots, V_d$ (say) for each variable v in $\vec{y}$. ψ^+ is obtained by substituting for each v in $\psi(x,\vec{y})$ the corresponding term $V_1x^{d-1} + V_2x^{d-2} + \dots + V_{d-1}x + V_d$ and appending the additional conjuncts $V_1x^{d-1} + V_2x^{d-2} + \dots + V_d \leq x^d$.

<u>THEOREM 2.4</u> There is a constant $c \in N$ such that if $L \in \Sigma_k$- TIME(n) then there is some quantifier free Δ_0 formula $\psi(x,\vec{y})$ for which

$$L = \{ x \in N: \exists \vec{y}_1 \leq x \forall \vec{y}_2 \leq x \dots Q\vec{y}_{k+c} \leq x \;\; \psi(x,\vec{y})\}.$$

<u>Proof</u>: Suppose $M = \langle P,A,q_1,q_a,Q_1, \dots, Q_k\rangle$ is a Σ_k- TM with m work tapes which works in time n, and that $L = L_M$. Each instruction

$$a\, b_{old}\, q_{old}\, b_{new}\, q_{new}\, D_0 D_1 \dots D_m$$

in P is a string of length 3m + 4 over the finite alphabet

$$A^+ = A \cup \{0,1,\#\} \cup Q_1 \cup \dots \cup Q_k \cup \{L_o,R_o,L_1,R_1, \dots, L_m,R_m\} ,$$

where we may assume that L_i and R_i do not occur in A or the Q_j's.

Now consider any $x \in L_M$ of length n and let C be an accepting computation which uses time n. Any branch of the tree C corresponds to sequence of at most n instructions. Placing these instructions end-to-end produces a word of length (3m + 4)n over the h element(say) alphabet A^+. By identifying A^+ with $\{0,1,\dots,h-1\}$ (and adding 1 at the left) we obtain the base h reperesentation of a number y which (for n large) satisfies

$$y \leq h^{(3m + 4)n + 1} \leq 2^{(n-1)d} \leq x^d$$

for some constant $d \in N$, since h and m are constants depending only on M (not on x).

Thus x is accepted by M if and only if

$$\exists y_1 \leq x^d \forall y_2 \leq x^d \exists y_3 \leq x^d \ldots Qy_k \leq x^d \ \theta(x,\vec{y})$$

where $\theta(x,\vec{y})$ says that

<u>if</u> (i) for i = 1,2, ..., k, each y_i codes a sequence of instructions from P with $q_{old} \in Q_i$

<u>and</u> (ii) the sequence of instructions coded by $y_1, y_2, \ldots, y_k$ determines a computation of $\bar{M}$ (the Σ_1- TM corresponding to M),

<u>then</u> (iii) this computation of $\bar{M}$ ends in acceptance.

By Lemma 2.3 it suffices to show $\theta(x,\vec{y})$ can be taken to be a Δ_o formula. With the help of proposition 2.1 parts (i) and (iii) are easily translated into Δ_o formulas. Observe that to verify (ii) (given that (i) holds) it is only necessary to check that:

(a) q_{old} for the first instruction is the initial state q_1, for all later instructions q_{old} is the same as q_{new} for the preceding instruction, and q_{new} of the last instruction is the accepting state q_a.

(b) For every tape cell the correct symbol appears in the a b_{old} portion of the instructions performed at the instants when that cell is visited by the head. In the case of an input tape cell the correct symbol is the appropriate digit of x (or #). For a work tape cell it appears in the b_{new} part of the instruction performed on the previous visit, it there was one, and is blank otherwise.

To check cell i on tape j it suffices to check the instructions at each pair of consecutive head visits. If w is the word formed by placing the instructions coded by $y_1, \ldots, y_k$ end-to-end, then the instructions at which cell i on tape j is visited can be located in w by means of the subwords w_o such that $w = w_o w_1$ (w_o is a prefix of w) and the number of occurrences of R_j in w_o exceeds the number of occurrences of L_j by exactly i.

By proposition 2.1 and lemma 2.2 (for "computing" the number of occurrences of R_j and L_j in w_o) the words w, w_o and their decoding can be described by means of Δ_o formulas. Thus $\theta(x,\vec{y})$ may be defined by a Δ_o formula and clearly the number of <u>blocks</u> of quantifiers in θ can be bounded independently of M.

REMARKS. (i) The construction of the formula $\theta(x,\vec{y})$ in the proof above is based on an argument for Σ_1- TM's shown to the author by Jeff Paris.
(ii) The definitions of Σ_k- TM's and Σ_k- TIME(n) can be generalised by using in place of x inputs of the form $x_1\# x_2\# \ldots \# x_m$ where $x_1,x_2,\ldots,x_m \in \{0,1\}^*$. Then if $L \in \Sigma_k$- TIME(n) there is some Δ_0 formula $\psi(x_1,\ldots,x_m)$ such that for all $\vec{x}$,

$$x_1\# \ x_2\# \ldots \# x_m \in L \leftrightarrow \psi(x_1,x_2,\ldots,x_m).$$

This can be proved by a straight-forward (but tedious) modification of the arguments above.

In order to prove a "converse" of theorem 2.4 we need to show there is some fixed k such that for each quantifier free Δ_0 formula $\psi(\vec{x})$ the language

$$\{x_1\# \ x_2\# \ldots \# x_m: \ \psi(x_1,x_2,\ldots,x_m)\} \text{ is in } \Sigma_k\text{- TIME(n)}.$$

This is a special case of results in section 5 and could be postponed until then. Different proofs have also been given by Lipton and Pippenger (see [16]) and (essentially) by Bennett [1]. However it may be of some interest that we can take k = 2.

LEMMA 2.5 Let $L = \{x_1\# \ x_2\# \ldots \# x_m: \ \psi(x_1,x_2,\ldots,x_m)\}$ where $\psi(\vec{x})$ is a quantifier free Δ_0 formula. Then

$$L \in \Sigma_2\text{- TIME(n)} \cap \pi_2\text{- TIME(n)}.$$

Discussion : Clearly + and testing of $\leq$ can be performed deterministically by Turing machines in linear time using essentially the primary school algorithms. How quickly multiplication of two n digit numbers can be done by multitape deterministic (or apparently even π_1 or Σ_1) TM's is an open question.

Proof of lemma 2.5: Throughout lower case letters c_k denote easy to compute constants. Consider a Σ_2- TM M which works as follows. Given $\vec{x}$ of length n, M first guesses the values of all the terms (including subterms) which appear in ψ. Operating deterministically M reduces the problem to one of checking a finite number of multiplications. For each of these M must check that values w,y,z (say) stored as binary strings of length $\leq c_1 n$ satisfy $w = y.z$.

Let B be the power of 2 satisfying $n \leq B < 2n$, and let $h = \left[\frac{c_1 n}{\log B}\right] + 1$.

Then

$$y = \sum_{i=0}^{h} y_i \ B^i, \quad z = \sum_{i=0}^{h} z_i \ B^i \text{ and } w = \sum_{i=0}^{h} w_i \ B^i$$

where $0 \leq y_i, z_i, w_i < B$. Since B is a power of 2, the binary representations of y_i, z_i, w_i can be read off from those of y,z,w.

If $y.z = w$ then $w = \sum_{i=0}^{h} \left(\sum_{j=0}^{i} z_{i-j} \ y_j \right) B^i$ so there are "carries" C_i such that for $i = 0,1, \ldots,h$, $w_i \equiv \sum_{j=0}^{i} z_{i-j} y_j + C_{i-1} \pmod B$

$$C_i = \left\lfloor \frac{\sum_{j=0}^{i} z_{i-j} y_j + C_{i-1}}{B} \right\rfloor \quad \text{and } C_h = 0 = C_{-1} .$$

(w_i may be found by adding the carry C_{i-1} from the calculation of w_{i-1} to $\sum_{j=0}^{i} z_{i-j} y_j$ and then dividing by B. The remainder is w_i and the quotient C_i is carried over to the calculation of w_{i+1}.)

M guesses the value of the carries $C_0, C_1, \ldots, C_{h-1}$. By induction on i, $C_i \lesssim 2i\,B$ (provided $B \geq 2$). Therefore the space (and time) required to store the C_i's is $\lesssim h\,[\log (2h\,B) + 1] \lesssim \left(\frac{c_1 n}{\log n} + 1\right) \log n^2 \lesssim c_2 n$.

M also guesses the residues of each y_i and z_j for all primes $p \lesssim c_3 \log n$. By the prime number theorem (or its weak versions-see proposition 4.3 below) these residues take total space $\lesssim c_4 n$ to store. (This is reasonable - they contain essentially the same information as y and z.) The small primes p can, of course, be generated in negligible time.

M does this for all the multiplications it has to check. Then in order to check $w = y.z$ (and similarly all the other multiplications in ψ) M branches universally over:

(i) all y_i, z_j and then, for each of these, over all p to check that the guessed residues (mod p) are correct, and

(ii) all i and then, for each of these, over all p to check

$$\sum_{j=0}^{i} y_{i-j} z_j + C_{i-1} \equiv C_i B + w_i \pmod p \tag{2.1}$$

using the guessed residues of y_i, z_j (mod p).

To check (i), the residue (mod p) of

$y_i = \sum_{j=0}^{s} Y_j 2^j = Y_0 + 2\,(Y_1 + 2\,(Y_2 + \ldots))$ may be computed by $c_5 s$ additions (mod p). These take time $\lesssim c_5 s \log p \lesssim c_6 \log n \log\log n \lesssim n$. Similarly the right hand side of (2.1) can be found (mod p) in time $<< n$. To complete the checking of a branch of type (ii), M uses a trivial multiplication algorithm to compute the left hand side (mod p) of (2.1) in time

$$\lesssim c_7\, i(\log p)^2 \lesssim c_8 \frac{n}{\log n} (\log\log n)^2 << n.$$

If M accepts then

$$\sum_{j=0}^{i} y_{i-j} z_j + C_{i-1} \equiv C_i B + w_i \pmod P \tag{2.2}$$

where P is the product of all primes $p < c_3 \log n$. But if c_3 was chosen large enough, then by the prime number theorem (or proposition 4.3) P is larger than both sides of (2.2) so

$$\sum_{j=0}^{i} y_{i-j} z_j + C_{i-1} = C_i B + w_i \ .$$

Since this holds for all i, it follows that $w = y.z$. Thus (by proposition 1.1) $L \in \Sigma_2$- TIME(n). Similarly, since $\neg\psi(\vec{x})$ is quantifier free, $L^c \in \Sigma_2$- TIME(n). Hence $L \in \pi_2$- TIME (n).

<u>THEOREM 2.6</u> Let $\phi(x)$ be a Δ_o formula of the form

$$\exists \vec{y}_1 \leq x \forall \vec{y}_2 \leq x \ldots Q\vec{y}_k \leq x \quad \psi(x,\vec{y}) \tag{2.3}$$

where $k \geq 1$ and ψ is quantifier free. Then

$$\{ x : \phi(x)\} \in \Sigma_{k+1} - \text{TIME}(n).$$

<u>Proof:</u> To test, given x, whether (2.3) holds, a linear time Σ_{k+1}-TM M first branches existentially to guess the binary digits of $\vec{y}_1$, then universally over all possibilities for $\vec{y}_2$, etc.. Finally M tests that $\psi(x,\vec{y})$ holds by simulating a Σ_2- TM if k is odd, and a π_2- TM if k is even.

Theorems 2.4 and 2.6 together show $\bigcup_{k\in N} \Sigma_k$- TIME(n) = Δ_o and the full Δ_o and linear time hierarchies are equivalent.

3. <u>SPACE-TIME-ALTERNATION TRADE-OFF.</u>

Theorem 3.2 below gives a way of reducing the <u>product</u> of the time and space used by a computation at the expense of increasing the space bound and the number of alternations.

Consider a Σ_k- TM $M = < P,A,q_1,q_a, \exists_1,\forall_2,\exists_3, \ldots,Q_k >$ which will be fixed throughout this section. Given two ID's of M (and therefore of the corresponding Σ_1- TM $\bar{M}$) we say <u>there is a path from ID_1 to ID_2</u> if ID_2 can be obtained from ID_1 by $\bar{M}$ preforming a finite number of instructions from P. The time and space used by paths is defined analogously to that used by computations of $\bar{M}$. The idea for the following lemma goes back to Nepomnyaschii [19].

<u>LEMMA 3.1</u> Suppose T(n) is fully time constructible and $m \in N$. There is a Σ_{2m+1}- TM M'_m such that given an integer f satisfying $1 \leq f \leq \frac{T(n)}{\log n}$ and ID_1, ID_2 (suitably coded and including x) :

(i) M'_m accepts if and only if there is a path from ID_1 to ID_2 which uses space $T(n)/f$ and time $T(n)f^m$,

(ii) M'_m works in time $T(n)$,

where n is the length of x.

Proof: By proposition 1.1 it suffices to construct a Σ_{2m+1}- TM M'_m which works in time $c_m T(n)$. Since $T(n)$ is constructible, a clock counting to $T(n)$ and marking off space $T(n)$ can be included by using extra tapes.

m = 0. M'_0 guesses $s = [T(n)/f]$. (s can be checked by marking off s cells, f times, in $> T(n)-f$ but $\leq T(n)$ steps.) M'_0 then simulates $\bar{M}$ for (at most) $T(n)$ steps starting with ID_1. If ID_2 is reached without exceeding work space s , M'_0 accepts.

Induction step. Suppose M'_m exists. We construct M'_{m+1}. This $\Sigma_{2(m+1)+1}$- TM first guesses (and checks) $s = [T(n)/f]$. Then to check there is a suitable path from ID_1 to ID_2, M'_{m+1} guesses a sequence of f +1 key ID's of $\bar{M}$:

$$ID_1 = ID'_0, ID'_1, \ldots, ID'_f = ID_2 \tag{3.1}$$

Each of these involves work tape space $\leq s$ in $\bar{M}$. The additional information in ID'_i other than x (which M'_{m+1} already has) can therefore be stored in space $\leq s$ in M'_{m+1}.($\bar{M}$'s input head position requires space roughly $\log n \leq s$ since $f \leq T(n)/\log n$). Thus writing these ID's takes time $f\,s \leq T(n)$. M'_{m+1} then branches universally over $i = 0,1, \ldots, f-1$ and simulates the Σ_{2m+1}- TM M'_m on ID'_i, ID'_{i+1} to test whether there is a path from ID'_i to ID'_{i+1} using time $T(n)f^m$ and space $T(n)/f$. Clearly it is possible to choose the sequence (3.1) so that this will be the case for all i, if and only if there is a path from ID_1 to ID_2 which uses time $T(n)f^{m+1}$ and space $T(n)/f$.

REMARK: Since $T(n)$ is fully time constructible, it follows that there is a π_{2m+1}- TM M''_m which works in time $T(n)$ and accepts f, ID_1, ID_2 if and only if there is no path from ID_1 to ID_2 which uses space $T(n)/f$ and time $T(n)f^m$.

THEOREM 3.2 Let $T(n)$ be fully time constructible, $1 \leq f(n) \leq \frac{T(n)}{\log n}$, and $k,m \in N$. Then there exists K such that

$$\Sigma_k\text{- TIME } (T(n)\ f(n)^m) * \text{ SPACE } (T(n)/f(n)) \subseteq \Sigma_K\text{- TIME } (T(n)).$$

Proof: We take $K = 2k + 2m + 1$. (A more careful argument gives $K = 2(k+m)$.) We will assume $f(n)$ is integer valued. (If not replace $f(n)$ by $[f(n)] + 1$, $T(n)$ by $2\,T(n)$ and use speedup.) Suppose C is an accepting computation of the Σ_k- TM M on an input x of length n, and that all branches in C use space $T(n)/f(n)$

and time $T(n)\ f(n)^m$. All ID's of M can be classified as being $\exists_1, \forall_2, \exists_3, \ldots$, or Q_k according to whether the current state q satisfies $q \in \exists_1, q \in \forall_2, q \in \exists_3, \ldots$, or $q \in Q_k$. This divides the branches of C into segments which can be classified similarly. An ID (or more precisely a node) in the tree C is Q_i minimal if it is the first ID in Q_i on the path to the ID from $ID_1(x)$ or (for convenience) if it is a halting ID in Q_{i-1}.

Consider a Σ_k- TM M_1 which, working in time $c\,T(n)$, first quesses f and s, checks $s = [T(n)/f]$, and then simulates the $\exists_1, \forall_2, \exists_3, \ldots$ parts of M's computation C as follows.

An $\exists_{2i+1}$ segment. Unless $i = 0$, M_1 already has in storage an $\exists_{2i+1}$ minimal ID of C. Call this ID_{2i+1}. (For $i = 0$, take $ID_1 = ID_1(x)$, the initial ID of M.) M_1 simulates M by guessing ID_{2i+2}- the $\forall_{2i+2}$ minimal ID following ID_{2i+1} in C. This is verified by branching universally (except if ID_{2i+2} is a halting ID). Along one branch, M_1 guesses an immediate predecessor ID' of ID_{2i+2} in $\exists_{2i+1}$ and checks there is a path in $\tau_M(x)$ between ID_{2i+1} and ID' which uses space $T(n)/f$ and time $T(n)f^m$ by simulating the Σ_{2m+1}- TM M'_m from lemma 3.1. Along the other branch M_1 continues to branch universally to simulate segments of C in $\forall_{2i+2}$. (If ID_{2i+2} is a halting ID this is omitted.)

$\forall_{2i+2}$ segments. M_1 branches universally over

(a) all possibilities for $\exists_{2i+3}$ minimal ID's ID_{2i+3} which use work space s (at most) ,

(b) all possibilities for an ID in $\forall_{2i+2}$ which uses work space s, and which by the performance of some instruction from P yields an ID which uses space exactly $s + 1$, and

(c) all possiblities for an ID in $\forall_{2i+2}$ which uses work space at most s and is not halting.

Branches of type (a). For each ID_{2i+3} M_1 guesses whether or not there is a path in $\forall_{2i+2}$ from ID_{2i+2} to ID_{2i+3} which uses time $T(n)f^m$ and space $T(n)/f$. If it guesses "no" then M_1 verifies this by simulating the π_{2m+1}- TM M''_m. If it guesses "yes" then M_1 branches universally (except if ID_{2i+3} is a halting ID). Along one branch it checks the existence of the path by simulating the Σ_{2m+1}- TM M'_m. Along the other branch (which is not used if ID_{2i+3} is a halting ID) M_1 simulates M on a segment in $\exists_{2i+3}$.

Branches of type (b). M_1 checks there is no path to the relevant ID from ID_{2i+2}

which uses only time $T(n)f^m$ and space $T(n)/f$ by simulating the π_{2m+1}- TM M''_m.
Branches of type (c). M_1 checks there is no path to the non-halting ID in $\forall_{2i+2}$ which uses time exactly $T(n)f^m$ and space $T(n)/f$ by simulating a π_{2m+1}- TM which is a minor variant of M''_m.
Correctness. Clearly if M accepts x then M_1 can also accept by guessing f = f(n). We must also check that if M_1 accepts x than M will also accept. The only difficulty that could arise is in the simulation of $\forall_{2i+2}$ segments, where a path to some ID_{2i+3} on a non-accepting branch might be overlooked by M_1 if f is guessed too large (because them the simulator used in the type (a) branches may allow too little space) or if f is guessed to small (because the simulator may allow too little time). However if the whole computation of M_1 is accepting, then the branches of type (b) ensure space s is sufficient, and those of type (c) ensure that enough time is available.

Ignoring the simulation of M'_m and M''_m, we see that each pair $\exists, \forall$ of segments of M's computation contributes 4 segments $\exists,\forall, \exists,\forall$ to the branches of M_1's computation. Taking into account M'_m and M''_m contributes at most 2m + 1 additional segments. Thus M_1 is indeed a $\Sigma_{2k+2m+1}$- TM.

REMARK The condition $f(n) \leq T(n)/\log n$ can be removed at the expense of increasing K. To see this note that without loss of generality we may suppose $f(n) \leq T(n)$. Let $f_1(n) = f(n)^{m/(m+1)} \leq T(n)/\log n$. Then

$$\Sigma_k\text{- TIME } (T(n)f(n)^m)\text{* SPACE } (T(n)/f(n)) \subseteq \Sigma_k\text{- TIME } (T(n)f_1(n)^{m+1})\text{*SPACE } (T(n)/f_1(n)).$$

Some special cases of theorem 3.2 can be found in the literature. Kannan [14] states (modulo $\bigcup_{k\geq 1} \Sigma_k\text{- TIME}(n) = \Delta_0$):

COROLLARY 3.3 Let c and ε be constants with $0 < \varepsilon < 1$. Then

$$\Sigma_k\text{- TIME } (n^c)\text{* SPACE } (n^{1-\varepsilon}) \subseteq \Delta_0.$$

For Σ_1- TM's this was proved earlier by Nepomnyashcii [19]. Since there is no point in a Σ_1- TM repeating an ID, any Σ_1- TM which works in space log n also works (simultaneously) in space log n and time n^c for some c, since there are only that many different ID's.

COROLLARY 3.4 (Nepomnyashcii [20]) $\Sigma_1\text{-SPACE } (\log n) \subseteq \Delta_0$

The space-time-alternation trade-off approach to hierarchy problems.

COROLLARY 3.5 Let k,m be positive integers, T(n) be fully time constructible, and $f(n) \to \infty$ as $n \to \infty$. Then

$$\Sigma_k\text{- TIME}(T(n)f(n)^m)\text{*SPACE}(T(n)/f(n)) \subsetneq \bigcup_{K\geq 1} \Sigma_K\text{-TIME}(T(n)).$$

Proof: Let $f_1(n) = f(n)^{m/(m+\frac{1}{2})}$. There is a Σ_k- TM U which works in time $T(n)f_1(n)^{m+1}$ and space $T(n)/f_1(n)$, and is a universal machine for the class

$$Y = \Sigma_k\text{- TIME } (T(n)f(n)^m)* \text{ SPACE } (T(n)/f(n)).$$

By theorem 3.2, $L_U \in \Sigma_K$- TIME(T(n)) and can thus be used to diagonalise over Y.

In particular if T(n) is fully time constructible and for each k there is some K,m and some function $f(n) \to \infty$ such that

$$\Sigma_k\text{- TIME } (T(n)) \subseteq \Sigma_K\text{- TIME } (T(n)f(n)^m)* \text{ SPACE } (T(n)/f(n))$$

then the Σ_k- TIME (T(n)) hierarchy is strict. Similarly if

$$X \subseteq \Sigma_k\text{- TIME } (n\ f(n)^m)* \text{ SPACE } (n/f(n))$$

then there is a Δ_o formula which is universal for (a superset) of X and consequently $X \neq \Delta_o$.

Let Σ_1- TIME_1(T(n)) denote the class of all languages L which are accepted by Σ_1- TM's with only 1 work tape. Nepomnyashcii [20] has observed that, done with care, the method of Hopcroft and Ullman [12] shows that for $0 < \varepsilon < 1$, and some c,

$$\Sigma_1\text{- TIME}_1\ (n^{2-\varepsilon}) \subseteq \Sigma_1\text{- TIME } (n^c)* \text{ SPACE } (n^{1-\varepsilon'})$$

where $0 < \varepsilon' < \frac{\varepsilon}{2}$, and that consequently for some k,

$$\Sigma_1\text{- TIME}_1(n^{2-\varepsilon}) \subseteq \Sigma_k\text{- TIME } (n) \subseteq \Delta_o .$$

This has the curious consequence:

CORALLARY 3.6 If $\Delta_o = \Sigma_1$- TIME (n) then

$$\forall \varepsilon > 0\ (\ \Sigma_1\text{- TIME}_1\ (n^{2-\varepsilon}) \subseteq \Sigma_1\text{- TIME } (n)).$$

It is of course known that any nondeterministic multitape TM which works in linear time can be simulated by a 1 tape Σ_1- TM which works in time n^2 (see e.g. [13]) and that this quadratic increase is close to optimal for certain languages. (See Mass [18].) However a converse in the form of the nonlinear speed up suggested by corollary 3.6 seems somewhat implausible, so this may represent a possible approach towards proving Σ_1- TIME (n) $\neq \pi_1$- TIME (n).

4. A WEAK Δ_o HIERARCHY THEOREM

For $p \geq 1$ odd, and x an integer, let $|x|_p$ denote the integer satisfying:

$$|x|_p \equiv x \pmod p,\ -\frac{p}{2} < |x|_p < \frac{p}{2} .$$

LEMMA 4.1 Let $k \in N$. There is a language $L_k \in$ DSPACE (log n) with the property that for any quantifier free Δ_o formula $\psi(x_o,x_1,\ldots,x_k)$ there exists $e_\psi \in N$ such that

$$\forall \vec{x}\ (x_0\#\ x_1\#\ \ldots\ \#x_k\ \#e_\psi \in L_k \leftrightarrow \psi(x_0,x_1,\ldots,x_k))$$

Discussion. Wilkie's hierarchy theorem [26] is based on the special case in which $\psi(x_o,x_1,\ldots,x_k)$ is of the form

$$F(x_o,x_1,\ldots,x_k) = 0 \tag{4.1}$$

where $F(\vec{x})$ is a polynomial with integer coefficients. One considers a DTM M which, given $x_o,x_1,\ldots,x_k$ and a Gödel number for F, computes $|F(\vec{x})|_p$ for a sufficient number of small primes p. If the product $P = \prod_p p$ is so large that $|F(\vec{x})| < P$ then

$$F(\vec{x}) = 0 \leftrightarrow F(\vec{x}) \equiv 0 \pmod{P} \leftrightarrow \forall p(F(\vec{x}) \equiv 0 \pmod{p}).$$

In our case $\psi(\vec{x})$ can be assumed to be a boolean combination of a finite number of polynomial inequalities:

$$F(x_o,x_1,\ldots,x_k) \geq 0 \tag{4.2}$$

We can still calculate the numbers $|F(\vec{x})|_p$. The problem is how to use the representation of a number X (= $F(\vec{x})$, say) by its residue digits $|X|_p$ to determine whether or not $X \geq 0$ while still keeping the space requirement sublinear so that the trade-off theorem applies.

The difficulty of determining the sign of X from its residue digits is one of two fundamental problems which have limited the use of modular arithmetic in practical computation. (The other is the related problem of detecting overflows.) These problems were studied in the 1960's in connection with the design of arithmetic units for computers. The algorithm described below is based on a sign detection scheme given by Eastman [7] who details its practical implementation, complete with instructions as to which wires to feed through which cores to build the arithmetic unit.

If Y is an integer relatively prime to p, let $\left|\frac{X}{Y}\right|_p$ denote the integer Z satisfying

$$YZ \equiv X \pmod{p}, \quad -\frac{p}{2} < Z \leq \frac{p}{2} .$$

If Y_1,Y_2,Y are relatively prime to p then

$$\frac{X}{Y} = \frac{X_1}{Y_1} \pm \frac{X_2}{Y_2} \rightarrow \left|\frac{X}{Y}\right|_p \equiv \left|\frac{X_1}{Y_1}\right|_p \pm \left|\frac{X_2}{Y_2}\right|_p \pmod{p}$$

so it makes sense to write $\left|\frac{X_1}{Y_1} \pm \frac{X_2}{Y_2}\right|_p$ etc..

Consider a system of positive, odd, pairwise relatively prime, moduli $p_o,p_1,\ldots,p_r$. Let

$$P_o = 1 , \quad P_i = \prod_{j=1}^{i} p_j , \quad P_{ij} = P_i/p_j .$$

Then $X \equiv \sum_{j=1}^{i} P_{ij} \left| \frac{X}{P_{ij}} \right|_{p_j} \pmod{P_i}$ (4.3)

This is just the familiar formula for solving "Chinese remainder" problems - it can be verified by evaluating both sides (mod p_j) for all j.

(4.3) says there is some integer t such that

$$X - \sum_{j=1}^{i} P_{ij} \left| \frac{X}{P_{ij}} \right|_{p_j} = t\, P_i \ .$$

We denote this integer t by $\left[\!\left[\frac{X}{P_i} \right]\!\right]$.

<u>Remark</u>. It would not matter for our purposes, but it is easy to check that cancelling a common factor of X and P_i does not effect the value of $\left[\!\left[\frac{X}{P_i} \right]\!\right]$. If we were only interested in prime power moduli we could define $\left[\!\left[\frac{X}{Y} \right]\!\right]$ for any rational number $\frac{X}{Y}$, $Y > 0$, by taking $p_1, \ldots, p_r$ to be the prime powers in the decomposition of Y.

$$\left| \sum_{j=1}^{i} P_{ij} \left| \frac{X}{P_{ij}} \right|_{p_j} \right| < \sum_{j=1}^{i} \frac{P_i}{p_j} \cdot \frac{p_j}{2} = i \frac{P_i}{2}$$

That is, $\left| X - \left[\!\left[\frac{X}{P_i} \right]\!\right] P_i \right| < i \frac{P_i}{2}$ so $\left| \frac{X}{P_i} - \left[\!\left[\frac{X}{P_i} \right]\!\right] \right| < \frac{i}{2}$.

If i is small, then $\left[\!\left[\frac{X}{P_i} \right]\!\right]$ is a good integer approximation to $\frac{X}{P_i}$. The modulus p_0 was deliberately left out of the products P_i so we can use $|X|_{p_0}$ to calculate $\left[\!\left[\frac{X}{P_i} \right]\!\right]$ by means of

<u>Definition</u>. $$\left[\!\left[\frac{X}{P_i} \right]\!\right]_{p_0} = \left| \frac{|X|_{p_0}}{P_i} - \sum_{j=1}^{i} \frac{1}{p_j} \left| \frac{X}{P_{ij}} \right|_{p_j} \right|_{p_0}$$

<u>LEMMA 4.2</u> Suppose $p_0 > i(2p_i + 1)$. If $|X| < \frac{p_0 - i}{2} P_i$ then

(i) $\left[\!\left[\frac{X}{P_i} \right]\!\right]_{p_0} \geq \frac{i}{2} \rightarrow X > 0$

(ii) $\left[\!\left[\frac{X}{P_i} \right]\!\right]_{p_0} \leq -\frac{i}{2} \rightarrow X < 0$

(iii) $-\frac{i}{2} < \left[\!\left[\frac{X}{P_i} \right]\!\right]_{p_0} < \frac{i}{2} \rightarrow |X| < \frac{p_0 - (i-1)}{2} P_{i-1}$

Proof: $\left[\!\left[\frac{X}{P_i}\right]\!\right] = \frac{X}{P_i} - \sum_{j=1}^{i} \frac{1}{p_j} \left|\frac{X}{P_{ij}}\right|_{p_j}$ so

$$\left|\left[\!\left[\frac{X}{P_i}\right]\!\right]\right| \leq \left|\frac{X}{P_i}\right| + \sum_{j=1}^{i} \frac{1}{p_j} \cdot \frac{p_j}{2} < \frac{p_o - i}{2} + \frac{i}{2} = \frac{p_o}{2}$$

Therefore, $\left[\!\left[\frac{X}{P_i}\right]\!\right] = \left|\left[\!\left[\frac{X}{P_i}\right]\!\right]\right|_{p_o} = \left[\!\left[\frac{X}{P_i}\right]\!\right]_{p_o}$.

Since $\left[\!\left[\frac{X}{P_i}\right]\!\right]$ approximates $\frac{X}{P_i}$ to within $\frac{i}{2}$, the sign of $\left[\!\left[\frac{X}{P_i}\right]\!\right]_{p_o}$ is the same as the sign of $\frac{X}{P_i}$, and therefore of X, except if $\left|\left[\!\left[\frac{X}{P_i}\right]\!\right]_{p_o}\right| < \frac{i}{2}$. This proves (i) and (ii) .

In the exceptional case $\left|\left[\!\left[\frac{X}{P_i}\right]\!\right]\right| < \frac{i}{2}$, so

$$\left|\frac{X}{P_i}\right| = \left|\sum_{j=1}^{i} \frac{1}{p_j} \left|\frac{X}{P_{ij}}\right|_{p_j} + \left[\!\left[\frac{X}{P_i}\right]\!\right]\right| < \frac{i}{2} + \frac{i}{2} = i .$$

Therefore $|X| < i\,P_i = i\,p_i\,P_{i-1}$ and since, by assumption, $i p_i < \frac{p_o - i}{2}$, it follows that

$$|X| < \left(\frac{p_o - i}{2}\right) P_{i-1}$$

which establishes (iii).

The sign test.

Suppose we are given $|X|_{p_o}, |X|_{p_1}, \ldots, |X|_{p_r}$ and $p_o, p_1, \ldots, p_r$ such that $p_o > i\,(2p_i + 1)$ for all i, and that it is known initially that $|X| < \frac{p_o - r}{2} P_r$. Apply the lemma with i = r to test the sign of X. Either the test succeeds or (iii) ensures that $|X| < \frac{p_o - (r-1)}{2} P_{r-1}$ so we can apply the lemma again with i=r-1. Continuing in this way through i = r,r-1,r-2,...,1 (as far as necessary) we either discover the sign of X or deduce (when i = 1) that $|X| < \frac{p_o}{2}$, in which case the sign of X is given by $|X|_{p_o}$. The important point is that by using arithmetic modulo the p_i's, computation of the numbers $\left[\!\left[\frac{X}{P_i}\right]\!\right]_{p_o}$ from $|X|_{p_o}, |X|_{p_1}, \ldots, |X|_{p_r}$ involves storing (apart from the input) only a constant number of integers whose magnitude is less than p_o^2 . Therefore the space used is less than c log p_o .

The existence of suitable moduli for proving lemma 4.1 is guaranteed by weak versions of the prime number theorem (see e.g. [9]). We write f(n) $\approx$ h(n) if there are constants $c_1 > 0, c_2$ such that $c_1 h(n) < f(n) < c_2 h(n)$ for all sufficiently large n.

PROPOSITION 4.3 (Chebychev) The number $\pi(n)$ of prime number $\leq n$ satisfies $\pi(n) \approx \frac{n}{\log n}$. Also, $\sum_{\substack{p \leq n \\ p \text{ prime}}} \log p \approx n$, that is $\prod_{\substack{p \leq n \\ p \text{ prime}}} p = 2^{h(n)}$ where $h(n) \approx n$.

Proof of lemma 4.1:

Consider a DTM M which is given input $x_0 \# x_1 \# \ldots \# x_k \# \ulcorner\psi\urcorner$ where $\ulcorner\psi\urcorner$ is Gödel number for some quantifier free formula ψ in some sufficiently easy to decode Gödel numbering, with each ψ having infinitely many code numbers. M will be designed so that for inputs of length $n > c_\psi$, M accepts $x_0 \# x_1 \# \ldots \# x_k \# \ulcorner\psi\urcorner$ if and only if $\psi(x_0, x_1, \ldots, x_k)$. Since the constant c_ψ depends only on ψ (not on $\ulcorner\psi\urcorner$) choosing $e_\psi = \ulcorner\psi\urcorner$ large enough will ensure $n > c_\psi$ and therefore

$$\forall \vec{x} \; (M \text{ accepts } x_0 \# x_1 \# \ldots \# x_k \# e_\psi \leftrightarrow \psi(\vec{x}))$$

M first breaks ψ up into a boolean combination of a finite number of polynomial inequalities of the form $F(\vec{x}) > 0$. These require negligible storage space and are tested sequentially, so only the space required for a single test need be considered. M generates an odd prime $p_0 \approx n^2$ (the existence of p_0 is guaranteed by proposition 4.3) and uses the distinct odd primes $p_1 < p_2 < \ldots < p_r$ less than (some easy to compute approximation of) $c_1 n \log \log n$. Only a fixed number of the p_i's are stored at any one time, but they can of course be generated in sequence using space $c_2 \log n$ whenever necessary. Similarly, using arithmetic (mod p_i) M can generate $|F(x_0, x_1, \ldots, x_k)|_{p_i}$ in space $c_3 \log n$ whenever required.

By proposition 5.3, $r \approx \frac{n \log \log n}{\log n}$ so $r(2p_r + 1) \approx (\frac{n \log \log n}{\log n})(n \log \log n)$
As $p_0 \approx n^2$ it follows that for n sufficiently large, $p_0 > r(2p_r + 1)$
so the first condition of lemma 4.2 is satisfied. Since $F(\vec{x})$ is a polynomial ,

$$|F(\vec{x})| \leq (\max(\vec{x}))^{c_F} \leq 2^{c_F n} \quad \text{for some constant } c_F \text{ which depends on } F.$$

But by proposition 4.3, $\frac{p_0 - r}{2} \; p_r \geq 2^{cn \log \log n}$. Therefore for $n > c_\psi$,

$$|F(\vec{x})| \leq 2^{c_F n} \leq 2^{cn \log \log n} \leq \frac{p_0 - r}{2} \; p_r$$

so the initial condition for applying the sign test is satisfied. Thus M can compute the truth value of $\psi(x)$ using only space c log n, which can be reduced to log n by the standard tape compression argument (see e.g. [13]).

THEOREM 4.4 Let $k \in N$. There is a Δ_0 formula $\theta_k(x,e)$ with the property that, for every Δ_0 formula $\phi(x)$ with at most k quantifiers, there is some $e_\phi \in N$ such that

$$\forall x \; (\theta_k(x, e_\phi) \leftrightarrow \phi(x)) .$$

Proof: By increasing k (if necessary) it suffices to consider the case where $\phi(x)$ is of the form

$$\exists y_1 \leq x \forall y_2 \leq x \ldots Qy_k \leq x \ \psi(x,\vec{y})$$

with ψ quantifier free.

Let L_k be the language in lemma 4.1 and let $\beta(x_0,x_1,\ldots,x_k,e)$ be the Δ_0 formula corresponding to L_k which exists by Nepomnyashcii's theorem (corollary 3.4 plus remark (ii) after theorem 2.4). Then for every quantifier free Δ_0 formula $\psi(x,y_1,y_2,\ldots,y_k)$ there is some $e_\psi \in N$ such that

$$\forall x \forall \vec{y} \ (\ \psi(x,\vec{y}) \ \leftrightarrow \ \beta(x,\vec{y},e_\psi \ ))$$

Take $\theta_k(x,e)$ to be $\exists y_1 \leq x \forall y_2 \leq x \ldots Qy_k \leq x \ \beta(x,\vec{y},e)$.

Of course theorem 4.4 generalises to the case where x is replaced by a sequence of variables $x_1,x_2,\ldots,x_m$.

We are now ready to prove the existence of a Δ_0 formula $\phi(x)$ which cannot be defined using only k quantifiers.

Proof of theorem 0.1:

Let $\theta_k(x,e)$ be as in theorem 4.4 and let $\phi(x)$ be the Δ_0 formula $\neg\theta_k(x,x)$. Then $\phi(x)$ is not equivalent to any Δ_0 formula $\psi(x)$ with a most k quantifiers. For by theorem 4.4 there exists $e_\psi \in N$ such that $\forall x(\ \psi(x) \leftrightarrow \theta_k(x,e_\psi \ ))$. Therefore

$$\psi(e_\psi) \leftrightarrow \theta_k(e_\psi,e_\psi) \leftrightarrow \neg\phi(e_\psi) \ ,$$

so the truth values of $\phi(x)$ and $\psi(x)$ differ at $x = e_\psi$.

5. A CONJECTURE.

The author believes that by an extension of the methods in this paper it should be possible to prove:

CONJECTURE: There is a Δ_0 formula $\theta(x,e)$ with the property that for every quantifier free Δ_0 formula $\psi(x,\vec{y})$, there is some $e_\psi \in N$ such that

$$\forall x \ (\ \theta(x,e_\psi) \ \ \leftrightarrow \ \exists \vec{y} \leq x \ \ \psi(x,\vec{y})).$$

From this it would follow that for each k there is a Δ_0 formula $\phi(x)$ which is not equivalent to any formula of the form

$$\exists y_1 \leq x \forall y_2 \leq x \ldots \forall y_{2k} \leq x \ \exists \vec{y}_{2k+1} \leq x \ \ \psi(x,\vec{y})$$

with ψ quantifier free. Since an arbitrarily large final block of existential quantifiers is allowed, a proof would show the existence of a bounded Diophantine set whose complement is not bounded Diophantine.

Acknowledgements: The author would like to thank Professors Paris and Wilkie for pointing out that theorem 0.1 was open, and Professor John Crossley and the logic group at Monash for making available the facilities which made initial work on this research possible.

REFERENCES.

[1] Bennett,J.H., On Spectra, Ph.D. Dissertation, Princeton(1962).

[2] Book, R.V., and Greibach, S.A., Quasi-realtime languages, Math. Systems Theory, 4(1970) 97-111.

[3] Börger, E., Note on bounded Diophantine representation of subrecursive sets, in Recursion Theory (Ithaca, New York 1982), Proc. Thirteenth Summer Research Institute Amer. Math Soc..

[4] Chandra, A.K., Kozen, D.C., and Stockmeyer, L.J., Alternation, J.Assoc. Comp. Mach. 28(1981) 114-133.

[5] Davis, M.,Matijasevič,Y., and Robinson, J., Hilbert's tenth problem. Diophantine equations: positive aspects of a negative solution. Amer. Math. Soc. Proc. of Symposia in Pure Math. 28(1976) 323-378.

[6] Dimitracopoulos, C., Matijasevič's Theorem and Fragments of Arithmetic, Ph.D thesis, University of Manchester (1980).

[7] Eastman, W.L., Sign determination in a modular number system, in Proceedings of Harvard Symposium on Digital Computers and their Application, 3-6 April 1961. Harvard University Press (1962) 136-162.

[8] Gaifman, H., and Dimitracopoulos, C., Fragments of Peano's arithmetic and the MRDP theorem, in Logic and Algorithmic, Monographie No.30, L'Enseignement Mathematique.

[9] Hardy, G.H., and Wright, E.M., An Introduction to the Theory of Numbers, Oxford University Press.

[10] Harrow, K., The bounded arithmetic hierarcy, Information and Control 36(1978) 102-117.

[11] Hodgson, B.R., and Kent, C.F., A normal form for arithmetical representation of NP-sets, J.Computer System Sci 27(1983) 378-388.

[12] Hopcroft, J.E., and Ullman, J.D., Relations between time and tape complexities. J.Assoc. Comput. Mach. 15(1968) 414-427.

[13] ________, Introduction to Automata Theory, Languages, and Computation. Addison-Wesley (1979).

[14] Kannan, R., Towards separating nondeterministic time from deterministic time, IEEE 22nd Annual Symposium on Foundations of Computer Science (1981) 235-243.

[15] Lessan, H., Models of Arithmetic, Ph.D. thesis, University of Manchester (1978).

[16] Lipton, R.J., Model theoretic aspects of computational complexity, IEEE 19th Annual Symposium on Foundations of Computer Science (1978).

[17] Manders, K.L., and Adleman, L., NP-complete decision problems for binary quadratics, J.Computer Systems Sci. 15(1978) 168-184.

[18] Mass, W., Quadratic lower bounds for deterministic and nondeterministic one-tape Turing machines, Proc. of 16th Annual ACM Symp. on Theory of Computing (1984) 401-408.

[19] Nepomnyashcii, V.A., Rudimentary interpretation of two-tape Turing Computation, Kibernetika (1970) No.2, 29-35. Translated in Cybernetics (1972) 43-50.

[20] Nepomnjascii, V.A., Rudimentary predicates and Turing calculations, Dokl. Akad. Nauk SSSR 195(1970). Translated in Soviet Math. Dokl. 11(1970) 1462-1465.

[21] Paul, W.J., Pippenger, N.,Szemeredi, E., and Trotter, W., On determinism versus nondeterminism and related problems,IEEE 24th Annual Symposium on Foundations of Computer Science (1983), 429-438.

[22] Paul, W.J., Prauss, E.J., and Reischuk, R., On alternation, Acta Inform. 14 (1980) 243-255.

[23] Paris, J.B., and Dimitracopoulos, C., Truth definitions for Δ_0 formulae, in Logic and Algorithmic (Zuric 1980), Enseign. Math. 30(1982) 317-329.

[24] Parikh, R., Existence and feasibility in arithmetic, J.Symbolic Logic 36 (1971) 494-508.

[25] Pudlak, P., A definition of exponentiation by a bounded arithmetical formula, Commentationes Mathematicae Universitatis Caroline, 24(1983) 667-671.

[26] Wilkie, A.J., Applications of complexity theory to Σ_0- definability problems in arithmetic, in Model Theory of Algebra and Arithmetic (Proceedings, Karpacz, Poland 1979), Lecture Notes in Math. 834(1980) 363-369.

[27] Wrathall,C., Rudimentary predicates and relative computation, SIAM J.Comput. 7(1978) 194-209.